T0231774

3D Engine Design for Virtual Globes

Patrick Cozzi
Kevin Ring

Cover design by Francis X. Kelly.
3D cover art by Jason K. Martin.

CRC Press
Taylor & Francis Group
6000 Broken Sound Parkway NW, Suite 300
Boca Raton, FL 33487-2742

© 2011 by Taylor & Francis Group, LLC
CRC Press is an imprint of Taylor & Francis Group, an Informa business

No claim to original U.S. Government works

Printed in the United States of America on acid-free paper
Version Date: 20110505

International Standard Book Number: 978-1-56881-711-8 (Hardback)

This book contains information obtained from authentic and highly regarded sources. Reasonable efforts have been made to publish reliable data and information, but the author and publisher cannot assume responsibility for the validity of all materials or the consequences of their use. The authors and publishers have attempted to trace the copyright holders of all material reproduced in this publication and apologize to copyright holders if permission to publish in this form has not been obtained. If any copyright material has not been acknowledged please write and let us know so we may rectify in any future reprint.

Except as permitted under U.S. Copyright Law, no part of this book may be reprinted, reproduced, transmitted, or utilized in any form by any electronic, mechanical, or other means, now known or hereafter invented, including photocopying, microfilming, and recording, or in any information storage or retrieval system, without written permission from the publishers.

For permission to photocopy or use material electronically from this work, please access www.copyright.com (http://www.copyright.com/) or contact the Copyright Clearance Center, Inc. (CCC), 222 Rosewood Drive, Danvers, MA 01923, 978-750-8400. CCC is a not-for-profit organization that provides licenses and registration for a variety of users. For organizations that have been granted a photocopy license by the CCC, a separate system of payment has been arranged.

Trademark Notice: Product or corporate names may be trademarks or registered trademarks, and are used only for identification and explanation without intent to infringe.

Library of Congress Cataloging-in-Publication Data

Cozzi, Patrick.
    3D Engine design for virtual globes / Patrick Cozzi, Kevin Ring.
        p. cm. -- (An A K Peters book)
    Includes bibliographical references and index.
    ISBN 978-1-56881-711-8 (hardback)
    1. Globes--Computer-assisted instruction. 2. Digital mapping. 3. Texture mapping. 4. Computer graphics. I. Ring, Kevin. II. Title.

    G3170.C69 2011
    912.0285'6693--dc22   2011010460

Visit the Taylor & Francis Web site at
http://www.taylorandfrancis.com

and the CRC Press Web site at
http://www.crcpress.com

To my parents, who bought me my first computer in 1994. Honestly, I just wanted to play games; I didn't think anything productive would come of it.

○○○○ Patrick Says

When I was seven years old, I declared that I wanted to make my own computer games. This book is dedicated to my mom and dad, who thought that was a neat idea.

○○○○ Kevin Says

Contents

Foreword

Do not let the title of this book fool you. What the title tells you is that if you have an interest in learning about high-performance and robust terrain rendering for games, this book is for you. If you are impressed by the features and performance of mapping programs such as NASA World Wind or Google Earth and you want to know how to write software of this type, this book is for you.

Some authors write computer books that promise to tell you everything you need to know about a topic, yet all that is delivered is a smattering of high-level descriptions but no low-level details that are essential to help you bridge the gap between theoretical understanding and practical source code. This is not one of those books. You are given a quality tutorial about globe and terrain rendering; the details about real-time 3D rendering of high-precision data, including actual source code to work with; and the mathematical foundations needed to be an expert in this field. Moreover, you will read about state-of-the-art topics such as geometry clipmapping and other level-of-detail algorithms that deal efficiently with massive terrain datasets. The book's bibliography is extensive, allowing you to investigate the large body of research on which globe rendering is built.

What the title of the book does not tell you is that there are many more chapters and sections about computing with modern hardware in order to exploit parallelism. Included are discussions about multithreaded engine design, out-of-core rendering, task-level parallelism, and the basics necessary to deal with concurrency, synchronization, and shared resources. Although necessary and useful for globe rendering, this material is invaluable for any application that involves scientific computing or visualization and processing of a large amount of data. Effectively, the authors are providing you with two books for the price of one. I prefer to keep only a small number of technical books at my office, opting for books with large information-per-page density. This book is now one of those.

—Dave Eberly

Preface

Planet rendering has a long history in computer graphics. Some of the earliest work was done by Jim Blinn at NASA's Jet Propulsion Laboratory (JPL) in the late 1970s and 80s to create animations of space missions. Perhaps the most famous animations are the flybys of Jupiter, Saturn, Uranus, and Neptune from the Voyager mission.

Today, planet rendering is not just in the hands of NASA. It is at the center of a number of games, such as *Spore* and *EVE Online*. Even non-planet-centric games use globes in creative ways; for example, *Mario Kart Wii* uses a globe to show player locations in online play.

The popularity of virtual globes such as Google Earth, NASA World Wind, Microsoft Bing Maps 3D, and Esri ArcGIS Explorer has also brought significant attention to globe rendering. These applications enable viewing massive real-world datasets for terrain, imagery, vector data, and more.

Given the widespread use of globe rendering, it is surprising that no single book covers the topic. We hope this book fills the gap by providing an in-depth treatment of rendering algorithms utilized by virtual globes. Our focus is on accurately rendering real-world datasets by presenting the core rendering algorithms for globes, terrain, imagery, and vector data.

Our knowledge in this area comes from our experience developing Analytical Graphics, Inc.'s (AGI) Satellite Tool Kit (STK) and Insight3D. STK is a modeling and analysis application for space, defense, and intelligence systems that has incorporated a virtual globe since 1993 (admittedly, we were not working on it back then). Insight3D is a 3D visualization component for aerospace and geographic information systems (GIS) applications. We hope our real-world experience has resulted in a pragmatic discussion of virtual globe rendering.

Intended Audience

This book is written for graphics developers interested in rendering algorithms and engine design for virtual globes, GIS, planets, terrain, and massive worlds. The content is diverse enough that it will appeal to a wide audience: practitioners, researchers, students, and hobbyists. We hope that our survey-style explanations satisfy those looking for an overview or a more theoretical treatment, and our tutorial-style code examples suit those seeking hands-on "in the trenches" coverage.

No background in virtual globes or terrain is required. Our treatment includes both fundamental topics, like rendering ellipsoids and terrain representations, and more advanced topics, such as depth buffer precision and multithreading.

You should have a basic knowledge of computer graphics, including vectors and matrices; experience with a graphics API, such as OpenGL or Direct3D; and some exposure to a shading language. If you understand how to implement a basic shader for per-fragment lighting, you are well equipped. If you are new to graphics—welcome! Our website contains links to resources to get you up to speed: http://www.virtualglobebook.com/.

This is also the place to go for the example code and latest book-related news.

Finally, you should have working knowledge of an object-oriented programming language like C++, C#, or Java.

Acknowledgments

The time and energy of many people went into the making of this book. Without the help of others, the manuscript would not have the same content and quality.

We knew writing a book of this scope would not be an easy task. We owe much of our success to our incredibly understanding and supportive employer, Analytical Graphics, Inc. We thank Paul Graziani, Frank Linsalata, Jimmy Tucholski, Shashank Narayan, and Dave Vallado for their initial support of the project. We also thank Deron Ohlarik, Mike Bartholomew, Tom Fili, Brett Gilbert, Frank Stoner, and Jim Woodburn for their involvement, including reviewing chapters and tirelessly answering questions. In particular, Deron played an instrumental role in the initial phases of our project, and the derivations in Chapter 2 are largely thanks to Jim. We thank Francis Kelly, Jason Martin, and Glenn Warrington for their fantastic work on the cover.

This book may have never been proposed if it were not for the encouragement of our friends at the University of Pennsylvania, namely Norm Badler, Steve Lane, and Joe Kider. Norm initially encouraged the idea and suggested A K Peters as a publisher, who we also owe a great deal of thanks to. In particular, Sarah Cutler, Kara Ebrahim, and Alice and Klaus Peters helped us through the entire process. Eric Haines (Autodesk) also provided a great deal of input to get us started in the right direction.

We're fortunate to have worked with a great group of chapter reviewers, whose feedback helped us make countless improvements. In alphabetical order, they are Quarup Barreirinhas (Google), Eric Bruneton (Laboratoire Jean Kuntzmann), Christian Dick (Technische Universität München), Hugues Hoppe (Microsoft Research), Jukka Jylänki (University of Oulu), Dave Kasik (Boeing), Brano Kemen (Outerra), Anton Frühstück Malischew (Greentube Internet Entertainment Solutions), Emil Persson (Avalanche Studios), Aras Pranckevičius (Unity Technologies), Christophe Riccio (Imagination Technologies), Ian Romanick (Intel), Chris Thorne (VRshed), Jan Paul Van Waveren (id Software), and Mattias Widmark (DICE).

Two reviewers deserve special thanks. Dave Eberly (Geometric Tools, LLC), who has been with us since the start, reviewed several chapters multiple times and always provided encouraging and constructive feedback. Aleksandar Dimitrijević (University of Niš) promptly reviewed many chapters; his enthusiasm for the field is energizing.

Last but not least, we wish to thank our family and friends who have missed us during many nights, weekends, and holidays. (Case-in-point: we are writing this section on Christmas Eve.) In particular, we thank Kristen Ring, Peg Cozzi, Margie Cozzi, Anthony Cozzi, Judy MacIver, David Ring, Christy Rowe, and Kota Chrome.

Dataset Acknowledgments

Virtual globes are fascinating because they provide access to a seemingly limitless amount of GIS data, including terrain, imagery, and vector data. Thankfully, many of these datasets are freely available. We graciously acknowledge the providers of datasets used in this book.

Natural Earth

Natural Earth (http://www.naturalearthdata.com/) provides public domain raster and vector datasets at 1 : 10, 1 : 50, and 1 : 110 million scales. We use the image in Figure 1 and Natural Earth's vector data throughout this book.

Figure 1. Satellite-derived land imagery with shaded relief and water from Natural Earth.

NASA Visible Earth

NASA Visible Earth (http://visibleearth.nasa.gov/) provides a wide array of satellite images. We use the images shown in Figure 2 throughout this book. The images in Figure 2(a) and 2(b) are part of NASA's Blue Marble collection and are credited to Reto Stockli, NASA Earth Observatory. The city lights image in Figure 2(c) is by Craig Mayhew and Robert Simmon, NASA GSFC. The data for this image are courtesy of Marc Imhoff, NASA GSFC, and Christopher Elvidge, NOAA NGDC.

NASA World Wind

We use NASA World Wind's `mergedElevations` terrain dataset (http://worldwindcentral.com/wiki/World_Wind_Data_Sources) in our terrain implementation. This dataset has 10 m resolution terrain for most of the United States, and 90 m resolution data for other parts of the world. It is derived from three sources: the Shuttle Radar Topography Mission (SRTM) from NASA's Jet Propulsion Laboratory;[1] the National Elevation

---

[1] http://www2.jpl.nasa.gov/srtm/

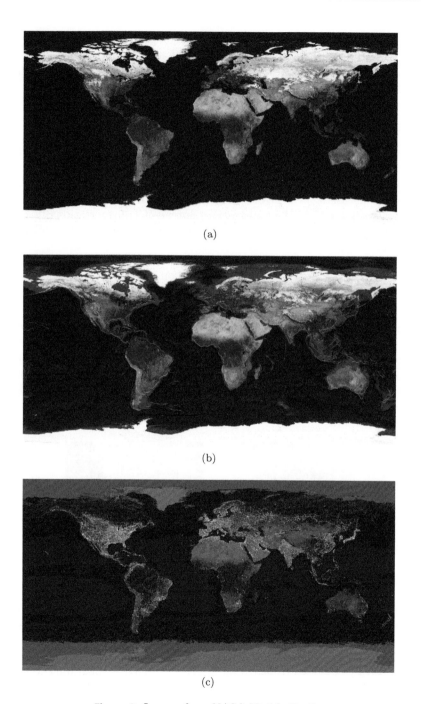

(a)

(b)

(c)

Figure 2. Images from NASA Visible Earth.

Dataset (NED) from the United States Geological Survey (USGS);[2] and
SRTM30_PLUS: SRTM30, coastal and ridge multibeam, estimated topography, from the Institute of Geophysics and Planetary Physics, Scripps
Institution of Oceanography, University of California San Diego.[3]

National Atlas of the United States of America

The National Atlas of the United States of America (http://www.national
atlas.gov/atlasftp.html) provides a plethora of map data at no cost. In
our discussion of vector data rendering, we use their airport and Amtrak
terminal datasets. We acknowledge the Administration's Research and
Innovative Technology Administration/Bureau of Transportation Statistics
(RITA/BTS) National Transportation Atlas Databases (NTAD) 2005 for
the latter dataset.

Georgia Institute of Technology

Like many developers working on terrain algorithms, we've used the terrain dataset for Puget Sound in Washington state, shown in Figure 3.
These data are part of the Large Geometric Models Archive at the Georgia
Institute of Technology (http://www.cc.gatech.edu/projects/large_models/
ps.html). The original dataset[4] was obtained from the USGS and made
available by the University of Washington. This subset was extracted by
Peter Lindstrom and Valerio Pascucci.

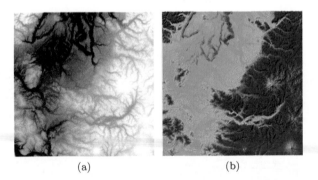

(a) (b)

Figure 3. (a) A height map and (b) color map (texture) of Puget Sound from the
Large Geometric Models Archive at the Georgia Institute of Technology.

---

[2]http://ned.usgs.gov/
[3]http://topex.ucsd.edu/WWW_html/srtm30_plus.html
[4]http://rocky.ess.washington.edu/data/raster/tenmeter/onebytwo10/

Yusuke Kamiyamane

The icons used in our discussion of vector data rendering were created by Yusuke Kamiyamane, who provides a large icon collection under the Creative Commons Attribution 3.0 license (http://p.yusukekamiyamane. com/).

Feedback

You are encouraged to email us with feedback, suggestions, or corrections at authors@virtualglobebook.com.

○○○○ **1**

Introduction

Virtual globes are known for their ability to render massive real-world terrain, imagery, and vector datasets. The servers providing data to virtual globes such as Google Earth and NASA World Wind host datasets measuring in the terabytes. In fact, in 2006, approximately 70 terabytes of compressed imagery were stored in Bigtable to serve Google Earth and Google Maps [24]. No doubt, that number is significantly higher today.

Obviously, implementing a 3D engine for virtual globes requires careful management of these datasets. Storing the entire world in memory and brute force rendering are certainly out of the question. Virtual globes, though, face additional rendering challenges beyond massive data management. This chapter presents these unique challenges and paves the way forward.

1.1   Rendering Challenges in Virtual Globes

In a virtual globe, one moment the viewer may be viewing Earth from a distance (see Figure 1.1(a)); the next moment, the viewer may zoom in to a hilly valley (see Figure 1.1(b)) or to street level in a city (see Figure 1.1(c)). All the while, real-world data appropriate for the given view are paged in and precisely rendered.

The freedom of exploration and the ability to visualize incredible amounts of data give virtual globes their appeal. These factors also lead to a number of interesting and unique rendering challenges:

- *Precision.* Given the sheer size of Earth and the ability for users to view the globe as a whole or zoom in to street level, virtual globes require a large view distance and large world coordinates. Trying to render a massive scene by naïvely using a very close near plane; very

1

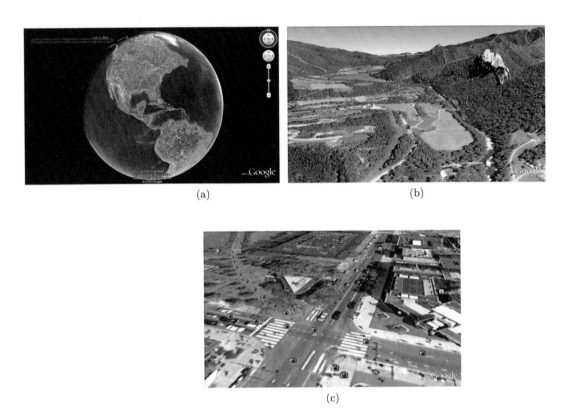

(a)   (b)

(c)

Figure 1.1. Virtual globes allow viewing at varying scales: from (a) the entire globe to (b) and (c) street level. (a) © 2010 Tele Atlas; (b) © 2010 Europa Technologies, US Dept of State Geographer; (c) © 2010 Google, US Census Bureau, Image USDA Farm Service Agency. (Images taken using Google Earth.)

distant far plane; and large, single-precision, floating-point coordinates leads to z-fighting artifacts and jittering, as shown in Figures 1.2 and 1.3. Both artifacts are even more noticeable as the viewer moves. Strategies for eliminating these artifacts are presented in Part II.

- *Accuracy.* In addition to eliminating rendering artifacts caused by precision errors, virtual globes should also model Earth accurately. Assuming Earth is a perfect sphere allows for many simplifications, but Earth is actually about 21 km longer at the equator than at the poles. Failing to take this into account introduces errors when positioning air and space assets. Chapter 2 describes the related mathematics.

(a)   (b)

Figure 1.2. (a) Jitter artifacts caused by precision errors in large worlds. Insufficient precision in 32-bit floating-point numbers creates incorrect vertex positions. (b) Without jittering. (Images courtesy of Brano Kemen, Outerra.)

- *Curvature.* The curvature of Earth, whether modeled with a sphere or a more accurate representation, presents additional challenges compared to many graphics applications where the world is extruded from a plane (see Figure 1.4): lines in a planar world are curves on Earth, oversampling can occur as latitude approaches 90° and −90°, a singularity exists at the poles, and special care is often needed to handle the International Date Line. These concerns are addressed throughout this book, including in our discussion of globe rendering in Chapter 4, polygons in Chapter 8, and mapping geometry clipmapping to a globe in Chapter 13.

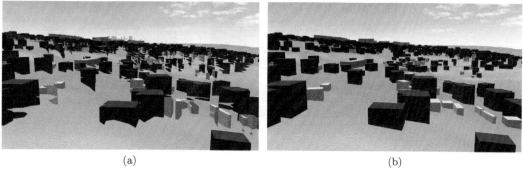

(a)   (b)

Figure 1.3. (a) Z-fighting and jittering artifacts caused by precision errors in large worlds. In z-fighting, fragments from different objects map to the same depth value, causing tearing artifacts. (b) Without z-fighting and jittering. (Images courtesy of Aleksandar Dimitrijević, University of Niš.)

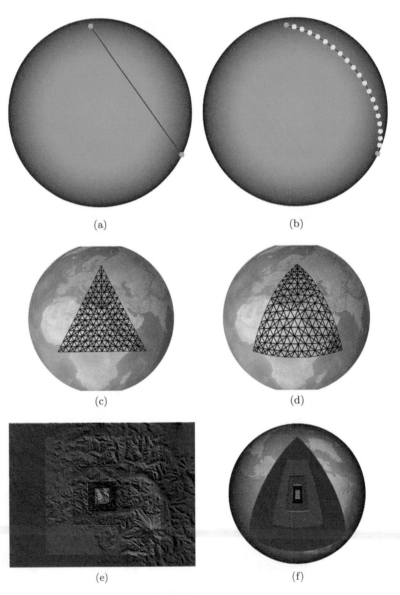

Figure 1.4. (a) Lines connecting surface points cut underneath a globe; instead, (b) points should be connected with a curve. Likewise, (c) polygons composed of triangles cut under a globe unless (d) curvature is taken into account. Mapping flat-world algorithms, (e) like geometry clipmapping terrain, to a globe can lead to (f) oversampling near the poles. (a) and (c) are shown without depth testing. (b) and (d) use the depth-testing technique presented in Chapter 7 to avoid z-fighting with the globe.

- *Massive datasets.* Real-world data have significant storage requirements. Typical datasets will not fit into GPU memory, system memory, or a local hard disk. Instead, virtual globes rely on server-side data that are paged in based on view parameters using a technique called *out-of-core rendering*, which is discussed in the context of terrain in Chapter 12 and throughout Part IV.

- *Multithreading.* In many applications, multithreading is considered to be only a performance enhancement. In virtual globes, it is an essential part of the 3D engine. As the viewer moves, virtual globes are constantly paging in data and processing it for rendering. Doing so in the rendering thread causes severe stalls, making the application unusable. Instead, virtual globe resources are loaded and processed in one or more separate threads, as discussed in Chapter 10.

- *Few simplifying assumptions.* Given their unrestrictive nature, virtual globes cannot take advantage of many of the simplifying assumptions that other graphics applications can.

  A viewer may zoom from a global view to a local view or vice versa in an instant. This challenges techniques that rely on controlling the viewer's speed or viewable area. For example, flight simulators know the plane's top speed and first-person shooters know the player's maximum running speed. This knowledge can be used to prefetch data from secondary storage. With the freedom of virtual globes, these techniques become more difficult.

  Using real-world data also makes procedural techniques less applicable. The realism in virtual globes comes from higher-resolution data, which generally cannot be synthesized at runtime. For example, procedurally generating terrains or clouds can still be done, but virtual globe users are most often interested in *real* terrains and clouds.

This book address these rendering challenges and more.

1.2 Contents Overview

The remaining chapters are divided into four parts: fundamentals, precision, vector data, and terrain.

1.2.1 Fundamentals

The fundamentals part contains chapters on low-level virtual globe components and basic globe rendering algorithms.

- *Chapter 2: Math Foundations.* This chapter introduces useful math for virtual globes, including ellipsoids, common virtual globe coordinate systems, and conversions between coordinate systems.

- *Chapter 3: Renderer Design.* Many 3D engines, including virtual globes, do not call rendering APIs such as OpenGL directly, and instead use an abstraction layer. This chapter details the design rationale behind the renderer in our example code.

- *Chapter 4: Globe Rendering.* This chapter presents several fundamental algorithms for tessellating and shading an ellipsoidal globe.

1.2.2   Precision

Given the massive scale of Earth, virtual globes are susceptible to rendering artifacts caused by precision errors that many other 3D applications are not. This part details the causes and solutions to these precision problems.

- *Chapter 5: Vertex Transform Precision.* The 32-bit precision on most of today's GPUs can cause objects in massive worlds to jitter, that is, literally bounce around in a jerky manner as the viewer moves. This chapter surveys several solutions to this problem.

- *Chapter 6: Depth Buffer Precision.* Since virtual globes call for a close near plane and a distant far plane, extra care needs to be taken to avoid z-fighting due to the nonlinear nature of the depth buffer. This chapter presents a wide range of techniques for eliminating this artifact.

1.2.3   Vector Data

Vector data, such as political boundaries and city locations, give virtual globes much of their richness. This part presents algorithms for rendering vector data and multithreading techniques to relieve the rendering thread of preparing vector data, or resources in general.

- *Chapter 7: Vector Data and Polylines.* This chapter includes a brief introduction to vector data and geometry-shader-based algorithms for rendering polylines.

- *Chapter 8: Polygons.* This chapter presents algorithms for rendering filled polygons on an ellipsoid using a traditional tessellation and subdivision approach and rendering filled polygons on terrain using shadow volumes.

- *Chapter 9: Billboards.* Billboards are used in virtual globes to display text and highlight places of interest. This chapter covers geometry-shader-based billboards and texture atlas creation and usage.

- *Chapter 10: Exploiting Parallelism in Resource Preparation.* Given the large datasets used by virtual globes, multithreading is a must. This chapter reviews parallelism in computer architecture, presents software architectures for multithreading in virtual globes, and demystifies multithreading in OpenGL.

1.2.4 Terrain

At the heart of a virtual globe is a terrain engine capable of rendering massive terrains. This final part starts with terrain fundamentals, then moves on to rendering real-world terrain datasets using level of detail (LOD) and out-of-core techniques.

- *Chapter 11: Terrain Basics.* This chapter introduces height-map-based terrain with a discussion of rendering algorithms, normal computations, and shading, both texture-based and procedural.

- *Chapter 12: Massive-Terrain Rendering.* Rendering real-world terrain accurately mapped to an ellipsoid requires the techniques discussed in this chapter, including LOD, culling, and out-of-core rendering. The next two chapters build on this material with specific LOD algorithms.

- *Chapter 13: Geometry Clipmapping.* Geometry clipmapping is an LOD technique based on nested, regular grids. This chapter details its implementation, as well as out-of-core and ellipsoid extensions.

- *Chapter 14: Chunked LOD.* Chunked LOD is a popular terrain LOD technique that uses hierarchical levels of detail. This chapter discusses its implementation and extensions.

There is also an appendix on implementing a message queue for communicating between threads.

We've ordered the parts and chapters such that the book flows from start to finish. You don't have to read the chapters in order though; we certainly didn't write them in order. Just ensure you are familiar with the terms and high level-concepts in Chapters 2 and 3, then jump to the chapter that interests you most. The text contains cross-references so you know where to go for more information.

There are *Patrick Says* and *Kevin Says* boxes throughout the text. These are the voices of the individual authors and are used to tell a story, usually an implementation war story, or to inject an opinion without clouding the main text. We hope these lighten up the text and provide deeper insight into our experiences.

The text also includes *Question* and *Try This* boxes that provide questions to think about and modifications or enhancements to make to the example code.

1.3 OpenGlobe Architecture

A large amount of example code accompanies this book. These examples were written from scratch, specifically for this book. In fact, just as much effort went into the example code as went into the book you hold in your hands. As such, treat the examples as an essential part of your learning— take the time to run them and experiment. Tweaking code and observing the result is time well spent.

Together, the examples form a solid foundation for a 3D engine designed for virtual globes. As such, we've named the example code *OpenGlobe* and provide it under the liberal MIT License. Use it as is in your commercial products or select bits and pieces for your personal projects. Download it from our website: http://www.virtualglobebook.com/.

The code is written in C# using OpenGL[1] and GLSL. C#'s clean syntax and semantics allow us to focus on the graphics algorithms without getting bogged down in language minutiae. We've avoided lesser-known C# language features, so if your background is in another object-oriented language, you will have no problem following the examples. Likewise, we've favored clean, concise, readable code over micro-optimizations.

Given that the OpenGL 3.3 core profile is used, we are taking a modern, fully shader-based approach. In Chapter 3, we build an abstract renderer implemented with OpenGL. Later chapters use this renderer, nicely tucking away the OpenGL API details so we can focus on virtual globe and terrain specifics.

OpenGlobe includes implementations for many of the presented algorithms, making the codebase reasonably large. Using the conservative metric of counting only the number of semicolons, it contains over 16,000 lines of C# code in over 400 files, and over 1,800 lines of GLSL code in over 80 files. We strongly encourage you to build, run, and experiment with the code. As such, we provide a brief overview of the engine's organization to help guide you.

OpenGlobe is organized into three assemblies:[2] OpenGlobe.Core.dll, OpenGlobe.Renderer.dll, and OpenGlobe.Scene.dll. As shown in Figure 1.5,

---

[1] OpenGL is accessed from C# using OpenTK: http://www.opentk.com/.

[2] *Assembly* is the .NET term for a compiled code library (i.e., an .exe or .dll file).

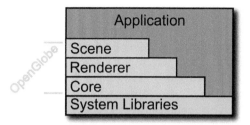

Figure 1.5. The stack of OpenGlobe assemblies.

these assemblies are layered such that Renderer depends on Core, and Scene depends on Renderer and Core. All three assemblies depend on the .NET system libraries, similar to how an application written in C depends on the C standard library.

Each OpenGlobe assembly has types that build on its dependent assemblies:

- *Core.* The Core assembly exposes fundamental types such as vectors, matrices, geographic positions, and the Ellipsoid class discussed in Chapter 2. This assembly also contains geometric algorithms, including the tessellation algorithms presented in Chapters 4 and 8, and engine infrastructure, such as the message queue discussed in Appendix A.

- *Renderer.* The Renderer assembly contains types that present an abstraction for managing GPU resources and issuing draw calls. Its design is discussed in depth in Chapter 3. Instead of calling OpenGL directly, an application built using OpenGlobe uses types in this assembly.

- *Scene.* The Scene assembly contains types that implement rendering algorithms using the Renderer assembly. This includes algorithms for globes (see Chapter 4), vector data (see Chapters 7–9), terrain shading (see Chapter 11), and geometry clipmapping (see Chapter 13).

Each assembly exposes types in a namespace corresponding to the assembly's filename. Therefore, there are three public namespaces: `OpenGlobe.Core`, `OpenGlobe.Renderer`, and `OpenGlobe.Scene`.

An application may depend on one, two, or all three assemblies. For example, a command line tool for geometric processing may depend just on Core, an application that implements its own rendering algorithms may depend on Core and Renderer, and an application that uses high-level objects like globes and terrain would depend on all three assemblies.

The example applications generally fall into the last category and usually consist of one main .cs file with a simple `OnRenderFrame` implementation that clears the framebuffer and issues `Render` for a few objects created from the Scene assembly.

OpenGlobe requires a video card supporting OpenGL 3.3, or equivalently, Shader Model 4. These cards came out in 2006 and are now very reasonably priced. This includes the NVIDIA GeForce 8 series or later and ATI Radeon 2000 series or later GPUs. Make sure to upgrade to the most recent drivers.

All examples compile and run on Windows and Linux. On Windows, we recommend building with any version of Visual C# 2010, including the free Express Edition.[3] On Linux, we recommend MonoDevelop.[4] We have tested on Windows XP, Vista, and 7, as well as Ubuntu 10.04 and 10.10 with Mono 2.4.4 and 2.6.7, respectively. At the time of this writing, OpenGL 3.3 drivers were not available on OS X. Please check our website for the most up-to-date list of supported platforms and integrated development environments (IDEs).

To build and run, simply open `Source\OpenGlobe.sln` in your .NET development environment, build the entire solution, then select an example to run.

We are committed to filling these pages with descriptive text, figures, and tables, not verbose code listing upon listing. Therefore, we've tried to provide relevant, concise code listings that supplement the core content. To keep listings concise, some error checking may be omitted, and #version 330 is always omitted in GLSL code. The code on our website includes full error checking and #version directives.

1.4   Conventions

This book uses a few conventions. Scalars and points are lowercase and italicized (e.g., s and p), vectors are bold (e.g., $\mathbf{v}$), normalized vectors also have a hat over them (e.g., $\hat{\mathbf{n}}$), and matrices are uppercase and bold (e.g., $\mathbf{M}$).

Unless otherwise noted, units in Cartesian coordinates are in meters (m). In text, angles, such as longitude and latitude, are in degrees (°). In code examples, angles are in radians because C# and GLSL functions expect radians.

---

[3] http://www.microsoft.com/express/Windows/
[4] http://monodevelop.com/

Part I

○○○○

Fundamentals

2

Math Foundations

At the heart of an accurately rendered virtual globe is an ellipsoidal representation of Earth. This chapter introduces the motivation and mathematics for such a representation, with a focus on building a reusable Ellipsoid class containing functions for computing surface normals, converting between coordinate systems, computing curves on an ellipsoid surface, and more.

This chapter is unique among the others in that it contains a significant amount of math and derivations, whereas the rest of the book covers more pragmatic engine design and rendering algorithms. You don't need to memorize the derivations in this chapter to implement a virtual globe; rather, aim to come away with a high-level understanding and appreciation of the math and knowledge of how to use the presented Ellipsoid methods.

Let's begin by looking at the most common coordinate systems used in virtual globes.

2.1 Virtual Globe Coordinate Systems

All graphics engines work with one or more coordinate systems, and virtual globes are no exception. Virtual globes focus on two coordinate systems: geographic coordinates for specifying positions on or relative to a globe and Cartesian coordinates for rendering.

2.1.1 Geographic Coordinates

A geographic coordinate system defines each position on the globe by a $(longitude, latitude, height)$-tuple, much like a spherical coordinate system defines each position by an $(azimuth, inclination, radius)$-tuple. Intuitively, longitude is an angular measure west to east, latitude is an angular measure south to north, and height is a linear distance above or below the surface. In Section 2.2.3, we more precisely define latitude and height.

Geographic coordinates are widely used; most vector data are defined in geographic coordinates (see Part III). Even outside virtual globes, geographic coordinates are used for things such as the global positioning systems (GPS).

We adopt the commonly used convention of defining longitude in the range $[-180°, 180°]$. As shown in Figure 2.1(a), longitude is zero at the prime meridian, where the western hemisphere meets the eastern. Increasing longitude moves to the east, and decreasing longitude moves to the west; longitude is positive in the eastern hemisphere and negative in the western. Longitude increases or decreases until the antimeridian, $\pm 180°$ longitude, which forms the basis for the International Date Line (IDL) in the Pacific Ocean. Although the IDL turns in places to avoid land, for our purposes, it is approximated as $\pm 180°$. Many algorithms need special consideration for the IDL.

Longitude is sometimes defined in the range $[0°, 360°]$, where it is zero at the prime meridian and increases to the east through the IDL. To convert longitude from $[0°, 360°]$ to $[-180°, 180°]$, simply subtract $360°$ if longitude is greater than $180°$.

Latitude, the angular measure south to north, is in the range $[-90°, 90°]$. As shown in Figure 2.1(b), latitude is zero at the equator and increases from south to north. It is positive in the northern hemisphere and negative in the southern.

Longitude and latitude should not be treated as 2D x and y Cartesian coordinates. As latitude approaches the poles, lines of constant longitude converge. For example, the extent with southwest corner $(0°, 0°)$ and northwest corner $(10°, 10°)$ has much more surface area than the extent from

(a)   (b)

Figure 2.1. Longitude and latitude in geographic coordinates. (a) Longitude spanning west to east. (b) Latitude spanning south to north.

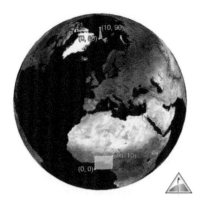

Figure 2.2. Extents with the same square number of degrees do not necessarily have the same surface area. (Image taken using STK. The Blue Marble imagery is from NASA Visible Earth.)

$(0°, 80°)$ to $(10°, 90°)$, even though they are both one square degree (see Figure 2.2). Therefore, algorithms based on a uniform longitude/latitude grid oversample near the poles, such as in the geographic grid tessellation in Section 4.1.4.

As mentioned in the Introduction, we use degrees for longitude and latitude, except in code examples, where they are in radians because C# and GLSL functions expect radians. Conversion between the two is straightforward: there are 2π rad or $360°$ in a circle, so one radian is $\frac{\pi}{180}°$, and one degree is $\frac{180}{\pi}$ rad. Although not used in this book, longitude and latitude are sometimes measured in *arc minutes* and *arc seconds*. There are 60 arc minutes in a degree and 60 arc seconds in an arc minute.

In OpenGlobe, geographic coordinates are represented using Geodetic 2D and Geodetic3D, the difference being the former does not include height, implying the position is on the surface. A static class, Trig, provides **ToRadians** and **ToDegrees** conversion functions. Simple examples for these types are shown in Listing 2.1.

```
Geodetic3D p = Trig.ToRadians(new Geodetic3D(180.0, 0.0, 5.0));

Console.WriteLine(p.Longitude); // 3.14159...
Console.WriteLine(p.Latitude);  // 0.0
Console.WriteLine(p.Height);    // 5.0

Geodetic2D g = Trig.ToRadians(new Geodetic2D(180.0, 0.0));
Geodetic3D p2 = new Geodetic3D(g, 5.0);

Console.WriteLine(p == p2);     // True
```

Listing 2.1. Geodetic2D and Geodetic3D examples.

2.1.2   WGS84 Coordinate System

Geographic coordinates are useful because they are intuitive—intuitive to humans at least. OpenGL doesn't know what to make of them; OpenGL uses Cartesian coordinates for 3D rendering. We handle this by converting geographic coordinates to Cartesian coordinates for rendering.

The Cartesian system used in this book is called the World Geodetic System 1984 (WGS84) coordinate system [118]. This coordinate system is fixed to Earth; as Earth rotates, the system also rotates, and objects defined in WGS84 remain fixed relative to Earth. As shown in Figure 2.3, the origin is at Earth's center of mass; the x-axis points towards geographic $(0°, 0°)$, the y-axis points towards $(90°, 0°)$, and the z-axis points towards the north pole. The equator lies in the xy-plane. This is a right-handed coordinate system, hence $x \times y = z$, where x, y, and z are unit vectors along their respective axis.

In OpenGlobe, Cartesian coordinates are most commonly represented using Vector3D, whose interface surely looks similar to other vector types you've seen. Example code for common operations like normalize, dot product, and cross product is shown in Listing 2.2.

The only thing that may be unfamiliar is that a Vector3D's X, Y, and Z components are doubles, indicated by the D suffix, instead of floats, which are standard in most graphics applications. The large values used in virtual globes, especially those of WGS84 coordinates, are best represented by

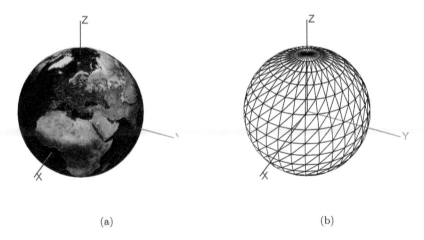

(a)   (b)

Figure 2.3. WGS84 coordinate system. (a) WGS84 coordinate system shown with a textured globe. (b) A wireframe globe shows the WGS84 coordinate system origin.

```
Vector3D  x  =  new  Vector3D (1.0 ,   0.0 ,   0.0);
//  (Same  as  Vector3D . UnitX )
Vector3D  y  =  new  Vector3D (0.0 ,   1.0 ,   0.0);
//  (Same  as  Vector3D . UnitY )

double  s  =  x . X  +  x . Y  +  x . Z ;        //  1.0
Vector3D  n  =  (y  −  x). Normalize ();   //  (1.0  /  Sqrt (2.0) ,
                                          //   −1.0  /  Sqrt (2.0) ,   0.0)
double  p  =  n . Dot (y);                //  1.0  /  Sqrt (2.0)
Vector3D  z  =  x . Cross (y);            //  (0.0 ,   0.0 ,   1.0)
```

Listing 2.2. Fundamental Vector3D operations.

double precision as explained in Chapter 5. OpenGlobe also contains vector types for 2D and 4D vectors and float, Half (16-bit floating point), int, and bool data types.[1]

We use meters for units in Cartesian coordinates and for height in geodetic coordinates, which is common in virtual globes.

Let's turn our attention to ellipsoids, which will allow us to more precisely define geographic coordinates, and ultimately discuss one of the most common operations in virtual globes: conversion between geographic and WGS84 coordinates.

2.2   Ellipsoid Basics

A sphere is defined in 3-space by a center, c, and a radius, r. The set of points r units away from c define the sphere's surface. For convenience, the sphere is commonly centered at the origin, making its implicit equation:

$$x_s^2 + y_s^2 + z_s^2 = r^2. \tag{2.1}$$

A point (x_s, y_s, z_s) that satisfies Equation (2.1) is on the sphere's surface. We use the subscript s to denote that the point is on the surface, as opposed to an arbitrary point (x, y, z), which may or may not be on the surface.

In some cases, it is reasonable to model a globe as a sphere, but as we shall see in the next section, an ellipsoid provides more precision and flexibility. An ellipsoid centered at $(0, 0, 0)$ is defined by three radii (a, b, c) along the x-, y-, and z-axes, respectively. A point (x_s, y_s, z_s) lies on the surface of an ellipsoid if it satisfies Equation (2.2):

$$\frac{x_s^2}{a^2} + \frac{y_s^2}{b^2} + \frac{z_s^2}{c^2} = 1. \tag{2.2}$$

---

[1] In C++, templates eliminate the need for different vector classes for each data type. Unfortunately, C# generics do not allow math operations on generic types.

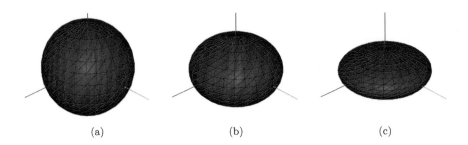

(a)   (b)   (c)

Figure 2.4. Oblate spheroids with different semiminor axes. All of these oblate spheroids have a semimajor axis $= 1$ and a semiminor axis along the z-direction (blue). (a) Semiminor axis $= 1$. The oblate spheroid is a sphere. (b) Semiminor axis $= 0.7$. (c) Semiminor axis $= 0.4$.

When $a = b = c$, Equation (2.2) simplifies to Equation (2.1), hence the ellipsoid is a sphere. An *oblate spheroid* is a type of ellipsoid that is particularly useful for modeling Earth. An oblate spheroid has two radii of equal length (e.g., $a = b$) and a smaller third radius (e.g., $c < a, c < b$). The larger radii of equal length are called the *semimajor axes* and the smaller radius is called the *semiminor axis*. Figure 2.4 shows oblate spheroids with varying semiminor axes. The smaller the semiminor axis compared to the semimajor axis, the more *oblate* the spheroid.

2.2.1   WGS84 Ellipsoid

For many applications, particularly games, it is acceptable to represent Earth or a planet using a sphere. In fact some celestial bodies, such as the Moon, with a semimajor axis of 1,738.1 km at its equator and a semiminor axis of 1,736 km at its poles, are almost spherical [180]. Other celestial bodies are not even close to spherical, such as Phobos, one of Mars's moons, with radii of $27 \times 22 \times 18$ km [117].

Although not as oddly shaped as Phobos, Earth is not a perfect sphere. It is best represented as an oblate spheroid with an equatorial radius of 6,378,137 m, defining its semimajor axis, and a polar radius of 6,356,752.3142 m, defining its semiminor axis, making Earth about 21,384 m longer at the equator than at the poles.

This ellipsoid representation of Earth is called the WGS84 ellipsoid [118]. It is the National Geospatial-Intelligence Agency's (NGA) latest model of Earth as of this writing (it originated in 1984 and was last updated in 2004).

```
public class Ellipsoid
{
  public static readonly Ellipsoid Wgs84 =
    new Ellipsoid(6378137.0, 6378137.0, 6356752.314245);
  public static readonly Ellipsoid UnitSphere =
    new Ellipsoid(1.0, 1.0, 1.0);

  public Ellipsoid(double x, double y, double z) { /* ... */ }
  public Ellipsoid(Vector3D radii)  { /* ... */ }

  public Vector3D Radii
  {
    get { return _radii; }
  }

  private readonly Vector3D _radii;
}
```

Listing 2.3. Partial Ellipsoid implementation.

The WGS84 ellipsoid is widely used; we use it in STK and Insight3D, as do many virtual globes. Even some games use it, such as Microsoft's Flight Simulator [163].

The most flexible approach for handling globe shapes in code is to use a generic ellipsoid class constructed with user-defined radii. This allows code that supports the WGS84 ellipsoid and also supports other ellipsoids, such as those for the Moon, Mars, etc. In OpenGlobe, Ellipsoid is such a class (see Listing 2.3).

2.2.2 Ellipsoid Surface Normals

Computing the outward-pointing surface normal for a point on the surface of an ellipsoid has many uses, including shading calculations and precisely defining height in geographic coordinates. For a point on a sphere, the surface normal is found by simply treating the point as a vector and normalizing it. Doing the same for a point on an ellipsoid yields a *geocentric surface normal.* It is called geocentric because it is the normalized vector from the center of the ellipsoid through the point. If the ellipsoid is not a perfect sphere, this vector is not actually normal to the surface for most points.

On the other hand, a *geodetic surface normal* is the actual surface normal to a point on an ellipsoid. Imagine a plane tangent to the ellipsoid at the point. The geodetic surface normal is normal to this plane, as shown in Figure 2.5. For a sphere, the geocentric and geodetic surface normals are equivalent. For more oblate ellipsoids, like the ones shown in Figures 2.5(b) and 2.5(c), the geocentric normal significantly diverges from the geodetic normal for most surface points.

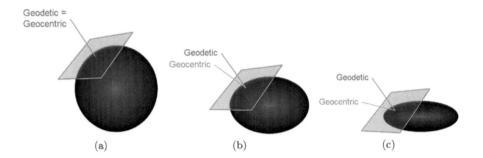

Figure 2.5. Geodetic versus geocentric surface normals. The geocentric normal diverges from the geodetic normal as the ellipsoid becomes more oblate. All figures have a semimajor axis $= 1$. (a) Semiminor axis $= 1$. (b) Semiminor axis $= 0.7$. (c) Semiminor axis $= 0.4$.

The geodetic surface normal is only slightly more expensive to compute than its geocentric counterpart

$$\mathbf{m} = \left(\frac{x_s}{a^2}, \frac{y_s}{b^2}, \frac{z_s}{c^2} \right),$$

$$\hat{\mathbf{n}}_s = \frac{\mathbf{m}}{\|\mathbf{m}\|},$$

where (a, b, c) are the ellipsoid's radii, (x_s, y_s, z_s) is the surface point, and $\hat{\mathbf{n}}_s$ is the resulting surface normal.

In practice, $\left(\frac{1}{a^2}, \frac{1}{b^2}, \frac{1}{c^2} \right)$ is precomputed and stored with the ellipsoid. Computing the geodetic surface normal simply becomes a component-wise multiplication of this precomputed value and the surface point, followed by normalization, as shown in Ellipsoid.`GeodeticSurfaceNormal` in Listing 2.4.

Listing 2.5 shows a very similar GLSL function. The value passed to `oneOverEllipsoidRadiiSquared` is provided to the shader by a uniform, so it is precomputed on the CPU once and used for many computations on the GPU. In general, we look for ways to precompute values to improve performance, especially when there is little memory overhead like here.

```
public Ellipsoid(Vector3D radii)
{
  // ...
  _oneOverRadiiSquared = new Vector3D(
      1.0 / (radii.X * radii.X),
      1.0 / (radii.Y * radii.Y),
      1.0 / (radii.Z * radii.Z));
}
```

```
public Vector3D GeodeticSurfaceNormal(Vector3D p)
{
  Vector3D normal = p.MultiplyComponents(_oneOverRadiiSquared);
  return normal.Normalize();
}

// ...
private readonly Vector3D _oneOverRadiiSquared;
```

Listing 2.4. Computing an ellipsoid's geodetic surface normal.

```
vec3 GeodeticSurfaceNormal(vec3 p,
                           vec3 oneOverEllipsoidRadiiSquared)
{
  return normalize(p * oneOverEllipsoidRadiiSquared);
}
```

Listing 2.5. Computing an ellipsoid's geodetic surface normal in GLSL.

> Run Chapter02EllipsoidSurfaceNormals and increase and decrease the oblateness of the ellipsoid. The more oblate the ellipsoid, the larger the difference between the geodetic and geocentric normals.

○○○○ **Try This**

2.2.3 Geodetic Latitude and Height

Given our understanding of geodetic surface normals, latitude and height in geographic coordinates can be precisely defined. *Geodetic latitude* is the

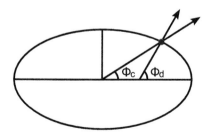

Figure 2.6. Comparison of geodetic latitude, ϕ_d, and geocentric latitude, ϕ_c.

angle between the equatorial plane (e.g., the xy-plane in WGS84 coordinates) and a point's geodetic surface normal. On the other hand, geocentric latitude is the angle between the equatorial plane and a vector from the origin to the point. At most points on Earth, geodetic latitude is different from geocentric latitude, as shown in Figure 2.6. Unless stated otherwise in this book, latitude always refers to geodetic latitude.

Height should be measured along a point's geodetic surface normal. Measuring along the geocentric normal introduces error, especially at higher heights, like those of space assets [62]. The larger the angular difference between geocentric and geodetic normals, the higher the error. The angular difference is dependent on latitude; on the WGS84 ellipsoid, the maximum angular difference between geodetic and geocentric normals is at $\approx 45°$ latitude.

2.3   Coordinate Transformations

Given that so much virtual globe data are stored in geographic coordinates but are rendered in WGS84 coordinates, the ability to convert from geographic to WGS84 coordinates is essential. Likewise, the ability to convert in the opposite direction, from WGS84 to geographic coordinates, is also useful.

Although we are most interested in the Earth's oblate spheroid, the conversions presented here work for all ellipsoid types, including a *triaxial ellipsoid*, that is, an ellipsoid where each radius is a different length ($a \neq b \neq c$).

In the following discussion, longitude is denoted by λ, geodetic latitude by ϕ, and height by h, so a (*longitude, latitude, height*)-tuple is denoted by (λ, ϕ, h). As before, an arbitrary point in Cartesian coordinates is denoted by (x, y, z), and a point on the ellipsoid surface is denoted by (x_s, y_s, z_s). All surface normals are assumed to be geodetic surface normals.

2.3.1   Geographic to WGS84

Fortunately, converting from geographic to WGS84 coordinates is a straightforward and closed form. The conversion is the same regardless of whether the point is above, below, or on the surface, but a small optimization can be made for surface points by omitting the final step.

Given a geographic point (λ, ϕ, h) and an ellipsoid (a, b, c) centered at the origin, determine the point's WGS84 coordinate, $r = (x, y, z)$.

The conversion takes advantage of a convenient property of the surface normal, $\hat{\mathbf{n}}_s$, to compute the location of the point on the surface, r_s; then, the height vector, $\mathbf{h}$, is computed directly and added to the surface point to produce the final position, r, as shown in Figure 2.7.

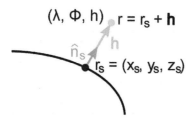

Figure 2.7. The geographic point (λ, ϕ, h) is converted to WGS84 coordinates by using the surface normal, $\hat{\mathbf{n}}_s$, to compute the surface point, r_s, which is offset by the height vector, $\mathbf{h}$, to produce the final point, r.

Given the surface position (λ, ϕ), the surface normal, $\hat{\mathbf{n}}_s$, is defined as

$$\hat{\mathbf{n}}_s = \cos\phi\cos\lambda\hat{\mathbf{i}} + \cos\phi\sin\lambda\hat{\mathbf{j}} + \sin\phi\hat{\mathbf{k}}. \tag{2.3}$$

Given the surface point $r_s = (x_s, y_s, z_s)$, the unnormalized surface normal, $\mathbf{n}_s$, is

$$\mathbf{n}_s = \frac{x_s}{a^2}\hat{\mathbf{i}} + \frac{y_s}{b^2}\hat{\mathbf{j}} + \frac{z_s}{c^2}\hat{\mathbf{k}}. \tag{2.4}$$

We are not given r_s but can determine it by relating $\hat{\mathbf{n}}_s$ and $\mathbf{n}_s$, which have the same direction but likely different magnitudes:

$$\hat{\mathbf{n}}_s = \gamma\mathbf{n}_s. \tag{2.5}$$

By substituting Equation (2.4) into $\mathbf{n}_s$ in Equation (2.5), we can rewrite $\hat{\mathbf{n}}_s$ as

$$\hat{\mathbf{n}}_s = \gamma\left(\frac{x_s}{a^2}\hat{\mathbf{i}} + \frac{y_s}{b^2}\hat{\mathbf{j}} + \frac{z_s}{c^2}\hat{\mathbf{k}}\right). \tag{2.6}$$

We know $\hat{\mathbf{n}}_s$, a^2, b^2, and c^2 but do not know γ, x_s, y_s, and z_s. Let's rewrite Equation (2.6) as three scalar equations:

$$\begin{aligned}
\hat{n}_x &= \frac{\gamma x_s}{a^2}, \\
\hat{n}_y &= \frac{\gamma y_s}{b^2}, \\
\hat{n}_z &= \frac{\gamma z_s}{c^2}.
\end{aligned} \tag{2.7}$$

Ultimately, we are interested in determining (x_s, y_s, z_s), so let's rearrange

Equation (2.7) to solve for x_s, y_s, and z_s:

$$x_s = \frac{a^2 \hat{n}_x}{\gamma},$$

$$y_s = \frac{b^2 \hat{n}_y}{\gamma}, \tag{2.8}$$

$$z_s = \frac{c^2 \hat{n}_z}{\gamma}.$$

The only unknown on the right-hand side is γ; if we compute γ, we can solve for x_s, y_s, and z_s. Recall from the implicit equation of an ellipsoid in Equation (2.2) in Section 2.2 that a point is on the surface if it satisfies

$$\frac{x_s^2}{a^2} + \frac{y_s^2}{b^2} + \frac{z_s^2}{c^2} = 1.$$

We can use this to solve for γ by substituting Equation (2.8) into this equation, then isolating γ:

$$\frac{(\frac{a^2 \hat{n}_x}{\gamma})^2}{a^2} + \frac{(\frac{b^2 \hat{n}_y}{\gamma})^2}{b^2} + \frac{(\frac{c^2 \hat{n}_z}{\gamma})^2}{c^2} = 1$$

$$a^2 \hat{n}_x^2 + b^2 \hat{n}_y^2 + c^2 \hat{n}_z^2 = \gamma^2 \tag{2.9}$$

$$\gamma = \sqrt{a^2 \hat{n}_x^2 + b^2 \hat{n}_y^2 + c^2 \hat{n}_z^2}.$$

Since γ is now written in terms of values we know, we can solve for x_s, y_s, and z_s using Equation (2.8). If the original geographic point is on the surface (i.e., $h = 0$) then the conversion is complete. For the more general case when the point may be above or below the surface, we compute a height vector, $\mathbf{h}$, with the direction of the surface normal and the magnitude of the point's height:

$$\mathbf{h} = h\hat{\mathbf{n}}_s.$$

```
public class Ellipsoid
{
  public Ellipsoid(Vector3D radii)
  {
    // ...
    _radiiSquared = new Vector3D(
        radii.X * radii.X,
        radii.Y * radii.Y,
        radii.Z * radii.Z);
  }

  public Vector3D GeodeticSurfaceNormal(Geodetic3D geodetic)
  {
    double cosLatitude = Math.Cos(geodetic.Latitude);
```

```
      return new Vector3D(
          cosLatitude * Math.Cos(geodetic.Longitude),
          cosLatitude * Math.Sin(geodetic.Longitude),
          Math.Sin(geodetic.Latitude));
  }

  public Vector3D ToVector3D(Geodetic3D geodetic)
  {
      Vector3D n = GeodeticSurfaceNormal(geodetic);
      Vector3D k = _radiiSquared.MultiplyComponents(n);
      double gamma = Math.Sqrt(
          k.X * n.X +
          k.Y * n.Y +
          k.Z * n.Z);

      Vector3D rSurface = k / gamma;
      return rSurface + (geodetic.Height * n);
  }

  // ...
  private readonly Vector3D _radiiSquared;
}
```

Listing 2.6. Converting from geographic to WGS84 coordinates.

The final WGS84 point is computed by offsetting the surface point, $r_s = (x_s, y_s, z_s)$, by $\mathbf{h}$:

$$r = r_s + \mathbf{h}. \tag{2.10}$$

The geographic to WGS84 conversion is implemented in Ellipsoid.ToVector3D, shown in Listing 2.6. First, the surface normal is computed using Equation (2.3), then γ is computed using Equation (2.9). The converted WGS84 point is finally computed using Equations (2.8) and (2.10).

2.3.2   WGS84 to Geographic

Converting from WGS84 to geographic coordinates in the general case is more involved than conversion in the opposite direction, so we break it into multiple steps, each of which is also a useful function on its own.

First, we present the simple, closed form conversion for points on the ellipsoid surface. Then, we consider scaling an arbitrary WGS84 point to the surface using both a geocentric and geodetic surface normal. Finally, we combine the conversion for surface points with scaling along the geodetic surface normal to create a conversion for arbitrary WGS84 points.

The algorithm presented here uses only two inverse trigonometric functions and converges quickly, especially for Earth's oblate spheroid.

WGS84 surface points to geographic. Given a WGS84 point (x_s, y_s, z_s) on the surface of an ellipsoid (a, b, c) centered at the origin, the geographic point (λ, ϕ) is straightforward to compute.

```
public class Ellipsoid
{
  public Vector3D GeodeticSurfaceNormal(Vector3D p)
  {
    Vector3D normal = p.MultiplyComponents(_oneOverRadiiSquared);
    return normal.Normalize();
  }

  public Geodetic2D ToGeodetic2D(Vector3D p)
  {
    Vector3D n = GeodeticSurfaceNormal(p);
    return new Geodetic2D(
        Math.Atan2(n.Y, n.X),
        Math.Asin(n.Z / n.Magnitude));
  }

  // ...
}
```

Listing 2.7. Converting surface points from WGS84 to geographic coordinates.

Recall from Equation (2.4) that we can determine the unnormalized surface normal, $\mathbf{n}_s$, given the surface point:

$$\mathbf{n}_s = \frac{x_s}{a^2}\hat{\mathbf{i}} + \frac{y_s}{b^2}\hat{\mathbf{j}} + \frac{z_s}{c^2}\hat{\mathbf{k}}$$

The normalized surface normal, $\hat{\mathbf{n}}_s$, is simply computed by normalizing $\mathbf{n}_s$:

$$\hat{\mathbf{n}}_s = \frac{\mathbf{n}_s}{\|\mathbf{n_s}\|}.$$

Given $\hat{\mathbf{n}}_s$, longitude and latitude are computed using inverse trigonometric functions:

$$\lambda = \arctan \frac{\hat{\mathbf{n}}_y}{\hat{\mathbf{n}}_x},$$

$$\phi = \arcsin \frac{\hat{\mathbf{n}}_z}{\|\mathbf{n_s}\|}.$$

This is implemented in Ellipsoid.ToGeodetic2D, shown in Listing 2.7.

Scaling WGS84 points to the geocentric surface. Given an arbitrary WGS84 point, $r = (x, y, z)$, and an ellipsoid, (a, b, c), centered at the origin, we wish to determine the surface point, $r_s = (x_s, y_s, z_s)$, along the point's geocentric surface normal, as shown in Figure 2.8(a).

This is useful for computing curves on an ellipsoid (see Section 2.4) and is a building block for determining the surface point using the geodetic

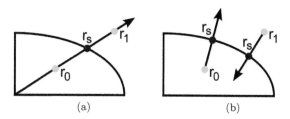

(a)   (b)

Figure 2.8. Scaling two points, r_0 and r_1, to the surface. (a) When scaling along the geocentric normal, a vector from the center of the ellipsoid is intersected with the ellipsoid to determine the surface point. (b) When scaling along the geodetic normal, an iterative process is used.

normal, as shown in Figure 2.8(b). Ultimately, we want to convert arbitrary WGS84 points to geographic coordinates by first scaling the arbitrary point to the geodetic surface and then converting the surface point to geographic coordinates and adjusting the height.

Let the position vector, $\mathbf{r}$, equal $r - 0$. The geocentric surface point, r_s, will be along this vector; that is

$$r_s = \beta \mathbf{r},$$

where r_s represents the intersection of the vector $\mathbf{r}$ and the ellipsoid. The variable β determines the position along the vector and is computed as

$$\beta = \frac{1}{\sqrt{\frac{x^2}{a^2} + \frac{y^2}{b^2} + \frac{z^2}{c^2}}}. \tag{2.11}$$

```
public class Ellipsoid
{
  public Vector3D ScaleToGeocentricSurface(Vector3D p)
  {
    double beta = 1.0 / Math.Sqrt(
        (p.X * p.X) * _oneOverRadiiSquared.X +
        (p.Y * p.Y) * _oneOverRadiiSquared.Y +
        (p.Z * p.Z) * _oneOverRadiiSquared.Z);
    return beta * position;
  }

  // ...
}
```

Listing 2.8. Scaling a point to the surface along the geocentric surface normal.

Therefore, r_s is determined with

$$x_s = \beta x,$$
$$y_s = \beta y, \qquad (2.12)$$
$$z_s = \beta z.$$

Equations (2.11) and (2.12) are used to implement Ellipsoid.Sc aleToGeocentricSurface shown in Listing 2.8.

Scaling to the geodetic surface. Using the geocentric normal to determine a surface point doesn't have the accuracy required for WGS84 to geographic conversion. Instead, we seek the surface point whose *geodetic normal* points towards the arbitrary point, or in the opposite direction for points below the surface.

More precisely, given an arbitrary WGS84 point, $r = (x, y, z)$, and an ellipsoid, (a, b, c), centered at the origin, we wish to determine the surface point, $r_s = (x_s, y_s, z_s)$, whose geodetic surface normal points towards r, or in the opposite direction.

We form r_s in terms of a single unknown and use the Newton-Raphson method to iteratively approach the solution. This method converges quickly for Earth's oblate spheroid and doesn't require any trigonometric functions, making it efficient. It will not work for points very close to the center of the ellipsoid, where multiple solutions are possible, but these cases are rare in practice.

Let's begin by considering the three vectors in Figure 2.9: the arbitrary point vector, $\mathbf{r} = r - 0$; the surface point vector, $\mathbf{r}_s = r_s - 0$; and the height vector, $\mathbf{h}$. From the figure,

$$\mathbf{r} = \mathbf{r}_s + \mathbf{h}. \qquad (2.13)$$

Recall that we can compute the unnormalized normal, $\mathbf{n}_s$, for a surface point

$$\mathbf{n}_s = \frac{x_s}{a^2}\hat{\mathbf{i}} + \frac{y_s}{b^2}\hat{\mathbf{j}} + \frac{z_s}{c^2}\hat{\mathbf{k}}. \qquad (2.14)$$

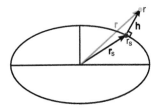

Figure 2.9. Computing r_s given r using the geodetic normal.

Observe that $\mathbf{h}$ has the same direction as $\mathbf{n}_s$, but likely a different magnitude. Let's relate them:

$$\mathbf{h} = \alpha \mathbf{n}_s.$$

We can substitute $\alpha \mathbf{n}_s$ for $\mathbf{h}$ in Equation (2.13):

$$\mathbf{r} = \mathbf{r}_s + \alpha \mathbf{n}_s.$$

Let's rewrite this as three scalar equations and substitute Equation (2.14) in for $\mathbf{n}_s$:

$$x = x_s + \alpha \frac{x_s}{a^2},$$
$$y = y_s + \alpha \frac{y_s}{b^2},$$
$$z = z_s + \alpha \frac{z_s}{c^2}.$$

Next, factor out x_s, y_s, and z_s:

$$x = x_s\left(1 + \frac{\alpha}{a^2}\right),$$
$$y = y_s\left(1 + \frac{\alpha}{b^2}\right),$$
$$z = z_s\left(1 + \frac{\alpha}{c^2}\right).$$

Finally, rearrange to solve for x_s, y_s, and z_s:

$$
\begin{aligned}
x_s &= \frac{x}{1 + \frac{\alpha}{a^2}}, \\
y_s &= \frac{y}{1 + \frac{\alpha}{b^2}}, \\
z_s &= \frac{z}{1 + \frac{\alpha}{c^2}}.
\end{aligned}
\tag{2.15}
$$

We now have $r_s = (x_s, y_s, z_s)$ written in terms of the known point, $r = (x, y, z)$; the ellipsoid radii, (a, b, c); and a single unknown, α. In order to determine α, recall the implicit equation of an ellipsoid, which we can write in the form $F(x) = 0$:

$$S = \frac{x_s^2}{a^2} + \frac{y_s^2}{b^2} + \frac{z_s^2}{c^2} - 1 = 0. \tag{2.16}$$

Substitute the expressions for x_s, y_s, and z_s in Equation (2.15) into Equation (2.16):

$$S = \frac{x^2}{a^2\left(1 + \frac{\alpha}{a^2}\right)^2} + \frac{y^2}{b^2\left(1 + \frac{\alpha}{b^2}\right)^2} + \frac{z^2}{c^2\left(1 + \frac{\alpha}{c^2}\right)^2} - 1 = 0. \tag{2.17}$$

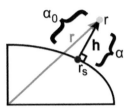

Figure 2.10. The Newton-Raphson method is used to find α from the initial guess, α_0.

Since this equation is no longer written in terms of the unknowns x_s, y_s, and z_s, we only have one unknown, α. Solving for α will allow us to use Equation (2.15) to find r_s.

We solve for α using the Newton-Raphson method for root finding; we are trying to find the root for S because when $S = 0$, the point lies on the ellipsoid surface. Initially, we guess a value, α_0, for α, then iterate until we are sufficiently close to the solution.

We initially guess r_s is the geocentric r_s computed in the previous section. Recall β from Equation (2.11):

$$\beta = \frac{1}{\sqrt{\frac{x^2}{a^2} + \frac{y^2}{b^2} + \frac{z^2}{c^2}}}.$$

For a geocentric r_s, $r_s = \beta \mathbf{r}$, so our initial guess is

$$x_s = \beta x,$$
$$y_s = \beta y,$$
$$z_s = \beta z.$$

The surface normal for this point is

$$\mathbf{m} = \left(\frac{x_s}{a^2}, \frac{y_s}{b^2}, \frac{z_s}{c^2} \right),$$

$$\hat{\mathbf{n}}_s = \frac{\mathbf{m}}{\|\mathbf{m}\|}.$$

Given our guess for r_s and $\hat{\mathbf{n}}_s$, we can now determine α_0. The unknown α scales $\hat{\mathbf{n}}_s$ to produce the height vector, $\mathbf{h}$. Our initial guess, α_0, simply scales $\hat{\mathbf{n}}_s$ to represent the distance between the ellipsoid surface and r as measured along the arbitrary point vector, $\mathbf{r}$, as shown in Figure 2.10. Therefore,

$$\alpha_0 = (1 - \beta) \frac{\|\mathbf{r}\|}{\|\mathbf{n_s}\|}.$$

We can now set $\alpha = \alpha_0$ and begin to iterate using the Newton-Raphson method. To do so, we need the function S from Equation (2.17) and its derivative with respect to α:

$$S = \frac{x^2}{a^2 \left(1 + \frac{\alpha}{a^2}\right)^2} + \frac{y^2}{b^2 \left(1 + \frac{\alpha}{b^2}\right)^2} + \frac{z^2}{c^2 \left(1 + \frac{\alpha}{c^2}\right)^2} - 1 = 0,$$

$$\frac{\partial S}{\partial \alpha} = -2 \left[\frac{x^2}{a^4 (1 + \frac{\alpha}{a^2})^3} + \frac{y^2}{b^4 (1 + \frac{\alpha}{b^2})^3} + \frac{z^2}{c^4 (1 + \frac{\alpha}{c^2})^3} \right].$$

We iterate to find α by evaluating S and $\frac{\partial S}{\partial \alpha}$. If S is sufficiently close to zero (i.e., within a given epsilon) iteration stops and α is found. Otherwise, a new α is computed:

$$\alpha = \alpha - \frac{S}{\frac{\partial S}{\partial \alpha}}.$$

Iteration continues until S is sufficiently close to zero. Given α, r_s is computed using Equation (2.15).

The whole process for scaling an arbitrary point to the geodetic surface is implemented using Ellipsoid.ScaleToGeodeticSurface, shown in Listing 2.9.

```
public class Ellipsoid
{
  public Ellipsoid(Vector3D radii)
  {
    // ...
    _radiiToTheFourth = new Vector3D(
        _radiiSquared.X * _radiiSquared.X,
        _radiiSquared.Y * _radiiSquared.Y,
        _radiiSquared.Z * _radiiSquared.Z);
  }

  public Vector3D ScaleToGeodeticSurface(Vector3D p)
  {
    double beta = 1.0 / Math.Sqrt(
        (p.X * p.X) * _oneOverRadiiSquared.X +
        (p.Y * p.Y) * _oneOverRadiiSquared.Y +
        (p.Z * p.Z) * _oneOverRadiiSquared.Z);
    double n = new Vector3D(
        beta * p.X * _oneOverRadiiSquared.X,
        beta * p.Y * _oneOverRadiiSquared.Y,
        beta * p.Z * _oneOverRadiiSquared.Z).Magnitude;
    double alpha = (1.0 - beta) * (p.Magnitude / n);

    double x2 = p.X * p.X;
    double y2 = p.Y * p.Y;
    double z2 = p.Z * p.Z;

    double da = 0.0;
    double db = 0.0;
    double dc = 0.0;
```

```
double s = 0.0;
double dSdA = 1.0;

do
{
  alpha -= (s / dSdA);

  da = 1.0 + (alpha * _oneOverRadiiSquared.X);
  db = 1.0 + (alpha * _oneOverRadiiSquared.Y);
  dc = 1.0 + (alpha * _oneOverRadiiSquared.Z);

  double da2 = da * da;
  double db2 = db * db;
  double dc2 = dc * dc;

  double da3 = da * da2;
  double db3 = db * db2;
  double dc3 = dc * dc2;

  s = x2 / (_radiiSquared.X * da2) +
      y2 / (_radiiSquared.Y * db2) +
      z2 / (_radiiSquared.Z * dc2) - 1.0;

  dSdA = -2.0 *
      (x2 / (_radiiToTheFourth.X * da3) +
       y2 / (_radiiToTheFourth.Y * db3) +
       z2 / (_radiiToTheFourth.Z * dc3));
}
while (Math.Abs(s) > 1e-10);

return new Vector3D(
    p.X / da,
    p.Y / db,
    p.Z / dc);
}

// ...
private readonly Vector3D _radiiToTheFourth;
}
```

Listing 2.9. Scaling a point to the surface along the geodetic surface normal.

Patrick Says ○○○○

I was assured this method converges quickly, but, for curiosity's sake, I ran a few tests to see how quickly. I created a 256×128 grid of points, each with a random height up to 10% of the semiminor axis above or below the surface. Then, I tested how many iterations ScaleToGeodeticSurface took to converge on three different ellipsoids, shown in Figure 2.11, tracking the minimum, maximum, and average number of iterations. The results, shown in Table 2.1, are encouraging. For Earth, all points converted in one or two iterations. As the ellipsoid becomes more oblate, more iterations are generally necessary, but never an impractical number.

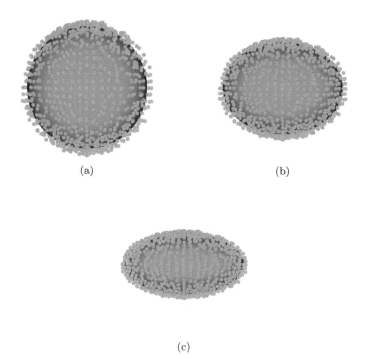

Figure 2.11. Ellipsoids used for `ScaleToGeodeticSurface` testing. (a) An ellipsoid with the oblateness of Earth. (b) Semimajor axis $= 1$ and semiminor axis $= 0.75$. (c) Semimajor axis $= 1$ and semiminor axis $= 0.5$. The images do not show the full resolution 256×128 point grid.

Ellipsoid	Minimum	Maximum	Average
Earth	1	2	1.9903
Semimajor axis: 1 Semiminor axis: 0.75	1	4	3.1143
Semimajor axis: 1 Semiminor axis: 0.5	1	4	3.6156

Table 2.1. Minimum, maximum, and average number of `ScaleToGeodetic Surface` iterations for ellipsoids of different oblateness.

Arbitrary WGS84 points to geographic. Using our function to scale a point to the geodetic surface and the function to convert a surface point from WGS84 to geographic coordinates, it is straightforward to convert an arbitrary point from WGS84 to geographic coordinates.

```
public  Geodetic3D  ToGeodetic3D ( Vector3D  position )
{
    Vector3D  p = ScaleToGeodeticSurface ( position );
    Vector3D  h = position  − p;
    double  height  = Math . Sign ( h . Dot ( position )) ∗ h . Magnitude ;
    return  new  Geodetic3D ( ToGeodetic2D ( p ) ,  height );
}
```

Listing 2.10. Converting a WGS84 point to geographic coordinates.

Given $r = (x, y, z)$, first, we scale it to the geodetic surface, producing r_s. A height vector, $\mathbf{h}$, is then computed:

$$\mathbf{h} = r - r_s.$$

The height of r above or below the ellipsoid is

$$h = \text{sign} \left(\mathbf{h} \cdot \mathbf{r} - \mathbf{0} \right) \| \mathbf{h} \|.$$

Finally, the surface point, r_s, is converted to geographic coordinates as done previously, and the resulting longitude and latitude are paired with h to create the final geographic coordinate, (λ, ϕ, h).

Given our implementations of `ScaleToGeodeticSurface` and `ToGeodetic2D`, this only requires a few lines of code, as shown in Listing 2.10.

2.4 Curves on an Ellipsoid

Many times in virtual globes, we have two endpoints on the surface of an ellipsoid that we wish to connect. Simply connecting the endpoints with a line results in a path that cuts under the ellipsoid, as shown in Figure 2.12(a). Instead, we wish to connect the endpoints with a path that approximately follows the curvature of the ellipsoid by subsampling points, as shown in Figures 2.12(b) and 2.12(c).

There are a variety of path types, each with different properties. A *geodesic curve* is the shortest path connecting two points. *Rhumb lines* are paths of constant bearing; although they are not the shortest path, they are widely used in navigation because of the simplicity of following a constant bearing.

Let's consider a simple way to compute a curve based on intersecting a plane and the ellipsoid surface, as shown in Figure 2.13(a). On a sphere, a *great circle* is the intersection of a plane containing the sphere's center and the sphere's surface. The plane cuts the sphere into two equal halves. When a plane containing the two endpoints and the sphere's center is intersected with the sphere's surface, two *great arcs* are formed: a *minor arc*, which is

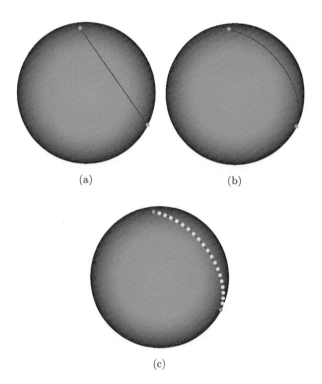

(a) (b)

(c)

Figure 2.12. (a) Connecting two endpoints on the ellipsoid surface with a line results in a path that travels under the ellipsoid. The line is drawn here without depth testing. (b) Subsampling points along the surface approximates a path between the endpoints. (c) Subsampled points shown in yellow.

the shortest path on the sphere between the points, and a *major arc*, the longer path along the great circle.

Using the same plane-intersection technique on an ellipsoid does not result in a geodesic curve, but for near-spherical ellipsoids, such as Earth's oblate spheroid, it produces a reasonable path for rendering purposes, especially when the endpoints are close together.

The algorithm for computing such a curve is straightforward, efficient, and easy to code. Given two endpoints, p and q, and a granularity, γ, we wish to compute a path on an ellipsoid, (a, b, c), centered at the origin, by subsampling points with γ angular separation.

As shown in Figures 2.13(c) to 2.13(e), as γ decreases, the subsampled points become closer together, and thus better approximate the curve. Except in the limit when γ approaches zero, the line segments will always

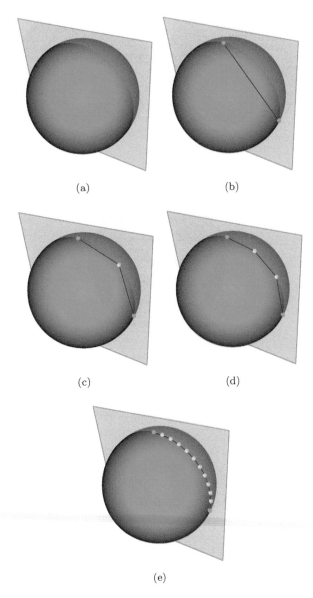

(a)   (b)

(c)   (d)

(e)

Figure 2.13. (a) Our curves are determined by the intersection of a plane and the ellipsoid. (b) The plane contains the ellipsoid's center and the line endpoints. (c)–(e) As the granularity decreases, the subsampled points better approximate a curve on the ellipsoid surface.

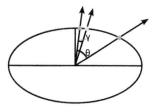

Figure 2.14. θ is the angle between the endpoints. The granularity, γ, is the angular separation between points on the curve.

cut under the ellipsoid; in Section 7.2, we present a rendering strategy that allows the line segments to still win the depth test against the ellipsoid.

To compute the curve between p and q, first create vectors from the ellipsoid center to the endpoints and take the cross product to determine the plane's normal, $\hat{\mathbf{n}}$:

$$\mathbf{p} = p - 0, \mathbf{q} = q - 0, \mathbf{m} = \mathbf{p} \times \mathbf{q}, \hat{\mathbf{n}} = \frac{\mathbf{m}}{\|\mathbf{m}\|}.$$

Next, determine the angle between vectors $\mathbf{p}$ and $\mathbf{q}$, θ, shown in Figure 2.14:

$$\hat{\mathbf{p}} = \frac{\mathbf{p}}{\|\mathbf{p}\|},$$
$$\hat{\mathbf{q}} = \frac{\mathbf{q}}{\|\mathbf{q}\|},$$
$$\theta = \arccos \hat{\mathbf{p}} \cdot \hat{\mathbf{q}}.$$

The number of subsampled points, s, is based on the granularity and the angle between endpoints:

$$n = \lfloor \frac{\theta}{\gamma} \rfloor - 1,$$
$$s = \max(n, 0)$$

Each subsampled point is computed by rotating $\mathbf{p}$ around the plane normal, $\hat{\mathbf{n}}$, and scaling the resulting vector to the geocentric surface as described in Section 2.3.2. Using geocentric scaling, as opposed to geodetic scaling, keeps the point in the plane, and thus on the desired curve.

As shown in the implementation of Ellipsoid .ComputeCurve in Listing 2.11, a for loop is used to compute each subsampled point using an angle, $\phi = i\gamma$, where i is the index of the point.

```
public class Ellipsoid
{
  public IList<Vector3D> ComputeCurve(
      Vector3D p,
      Vector3D q,
      double granularity)
  {
    Vector3D normal = p.Cross(q).Normalize();
    double theta = p.AngleBetween(q);
    int s = Math.Max((int)(theta / granularity) - 1, 0);
    List<Vector3D> positions = new List<Vector3D>(2 + s);

    positions.Add(p);
    for (int i = 1; i <= s; ++i)
    {
      double phi = (i * granularity);
      Vector3D rotated = p.RotateAroundAxis(normal, phi);
      positions.Add(ScaleToGeocentricSurface(rotated));
    }
    positions.Add(q);

    return positions;
  }

  // ...
}
```

Listing 2.11. Computing a curve on the ellipsoid between two surface points.

Try This ○○○○

> Run Chapter02Curves and experiment with the ellipsoid oblateness and curve granularity. How can we implement LOD for a curve?

Try This ○○○○

> Modify **Ellipsoid.ComputeCurve** to create curves at a constant height above or below the ellipsoid.

Question ○○○○

> Instead of using an angular granularity and rotating around the plane's normal, an implementation could simply use linear interpolation to subsample points along the line connecting the endpoints, then call **ScaleToGeocentricSurface** for each point. What are the advantages and disadvantages of this approach?

2.5   Resources

Vallado and McClain cover WGS84 and an incredible number of other topics in depth [171]. Precision concerns in virtual globes, such as those described in Section 2.2.3, with a focus on visualizing space assets, are described by Giesecke [62]. Luebke et al. cover georeferencing in their excellent book on LOD [107].

3

Renderer Design

Some graphics applications start off life with OpenGL or Direct3D calls sprinkled throughout. For small projects, this is manageable, but as projects grow in size, developers start asking, How can we cleanly manage OpenGL state?; How can we make sure everyone is using OpenGL best practices?; or even, How do we go about supporting both OpenGL and Direct3D?

The first step to answering these questions is abstraction—more specifically, abstracting the underlying rendering API such as OpenGL or Direct3D using interfaces that make most of the application's code API-agnostic. We call such interfaces and their implementation a renderer. This chapter describes the design behind the renderer in OpenGlobe. First, we pragmatically consider the motivation for a renderer, then we look at the major components of our renderer: state management, shaders, vertex data, textures, and framebuffers. Finally, we look at a simple example that renders a triangle using our renderer.

If you have experience using a renderer, you may just want to skim this chapter and move on to the meat of virtual globe rendering. Examples in later chapters build on our renderer, so some familiarity with it is required.

This chapter is not a tutorial on OpenGL or Direct3D, so you need some background in one API. Nor is this a description of how to wrap every OpenGL call in an object-oriented wrapper. We are doing much more than wrapping functions; we are raising the level of abstraction.

Our renderer contains quite a bit of code. To keep the discussion focused, we only include the most important and relevant code snippets in these pages. Refer to the code in the OpenGlobe.Renderer project for the full implementation. In this chapter, we are focused on the organization of the public interfaces and the design trade-offs that went into them; we are not concerned with minute implementation details.

Throughout this chapter, when we refer to GL, we mean OpenGL 3.3 core profile specifically. Likewise, when we refer to D3D, we mean Direct3D

11 specifically. Also, we define client code as code that calls the renderer, for example, application code that uses the renderer to issue draw commands.

Finally, software design is often subjective, and there is rarely a single best solution. We want you to view our design as something that has worked well for us in virtual globes, but as only one of a myriad approaches to renderer design.

3.1 The Need for a Renderer

Given that APIs such as OpenGL and Direct3D are already an abstraction, it is natural to ask, why build a renderer layer in our engine at all? Aren't these APIs sufficiently high level enough to use directly throughout our engine?

Many small projects do scatter API calls throughout their code, but as projects get larger, it is important that they properly abstract the underlying API for many reasons:

- *Ease of development.* Using a renderer is almost always easier and more concise than calling the underlying API directly. For example, in GL, the process of compiling and linking a shader and retrieving its uniforms can be simplified to a single constructor call on a shader program abstraction.

  Since most engines are written in object-oriented languages, a renderer allows us to present the procedural GL API using object-oriented constructs. For example, constructors invoke glCreate* or glGen*, and destructors invoke glDelete*, allowing the C# garbage collector to handle resource lifetime management, freeing client code from having to explicitly delete renderer resources.[1]

  Besides conciseness and object orientation, a renderer can simplify development by minimizing or completely eliminating error-prone global states, such as the depth and stencil tests. A renderer can group these states into a coarse-grained render state, which is provided per draw call, eliminating the need for global state.

  When using Direct3D 9, a renderer can also hide the details of handling a "lost device," where the GPU resources are lost due to a user changing the window to or from full screen, a laptop's cover opening/closing, etc. The renderer implementation can shadow a copy of

---

[1]In C++, similar lifetime management can be achieved with smart pointers. Our renderer abstractions implement IDisposable, which gives client code the option to explicitly free an object's resources in a deterministic manner instead of relying on the garbage collector.

GPU resources in system memory, so it can restore them in response
to a lost device, without any interaction from client code.

- *Portability.* A renderer greatly reduces, but does not eliminate, the
  burden of supporting multiple APIs. For example, an engine may
  want to use Direct3D on Windows, OpenGL on Linux, OpenGL ES
  on mobile devices,[2] and LibGCM on PlayStation 3. To support different APIs on different platforms, a different renderer implementation
  can be swapped in, while the majority of the engine code remains
  unchanged. Some renderer implementations, such as a GL renderer
  and GL ES renderer or a GL 3.x renderer and GL 4.x renderer, may
  even share a good bit of code.

  A renderer also makes it easier to migrate to new versions of an
  API or new GL extensions. For example, when all GL calls are isolated, it is generally straightforward to replace global-state selectors
  with direct-state access (EXT_direct_state_access [91]). Likewise, the
  renderer can decide if an extension is available or not and take appropriate action. Supporting some new features or extension requires
  exposing new or different public interfaces; for example, consider how
  one would migrate from GL uniforms to uniform buffers.

- *Flexibility.* A renderer allows a great deal of flexibility since a renderer's implementation can be changed largely independent of client
  code. For example, if it is more efficient to use GL display lists[3] than
  vertex buffer objects (VBOs) on certain hardware, that optimization
  can be made in a single location. Likewise, if a bug is found that was
  caused by a misunderstanding of a GL call or by a driver bug, the fix
  can be made in one location, usually without impacting client code.

  A renderer helps new code plug into an engine. Without a renderer, a
  call to a virtual method may leave GL in an unknown state. The implementor of such a method may not even work for the same company
  that developed the engine and may not be aware of the GL conventions used. This problem can be avoided by passing a renderer to the
  method, which is used for all rendering activities. A renderer enables
  the flexibility of engine "plug-ins" that are as seamless as core engine
  code.

- *Robustness.* A renderer can improve an engine's robustness by providing statistics and debugging aids. In particular, it is easy to count

---

[2]With ARB_ES2_compatibility, OpenGL 3.x is now a superset of OpenGL ES 2.0 [19].
This simplifies porting and sharing code between desktop OpenGL and OpenGL ES.

[3]Display lists were deprecated in OpenGL 3, although they are still available through
the compatibility profile.

the number of draw calls and triangles drawn per frame when GL commands are isolated in a renderer. It can also be worthwhile to have an option to log underlying GL calls for later debugging. Likewise, a renderer can easily save the contents of the framebuffer and textures, or show the GL state at any point in time. When run in debug mode, each GL call in the renderer can be followed by a call to `glGetError` to get immediate feedback on errors.[4]

Many of these debugging aids are also available through third-party tools, such as BuGLe,[5] GLIntercept,[6] and gDEBugger.[7] These tools track GL calls to provide debugging and performance information, similar to what can be done in a renderer.

- *Performance.* At first glance, it may seem that a renderer layer can hurt performance. It does add a lot of virtual methods calls. However, considering the amount of work the driver is likely to do, virtual call overhead is almost never a concern. If it were, virtual calls aren't even required to implement a renderer unless the engine supports changing rendering APIs at runtime, an unlikely requirement. Therefore, a renderer can be implemented with plain or inline methods if desired.

  A renderer can actually help performance by allowing optimizations to be made in a single location. Client code doesn't need to be aware of GL best practices; only the renderer implementation does. The renderer can shadow GL state to eliminate redundant state changes and avoid expensive calls to `glGet*`. Depending on the level of abstraction chosen for the renderer, it can also optimize vertex and index buffers for the GPU's caches and select optimal vertex formats with proper alignment for the target hardware. Renderer abstractions can also make it easier to sort by state, a commonly used optimization.

  For engines written in a managed language like Java or C#, such as OpenGlobe, a renderer can improve performance by minimizing the managed-to-native-code round trip overhead. Instead of calling into native GL code for every fine-grained call, such as changing a uniform or a single state, a single coarse-grained call can pass a large amount of state to a native C++ component that does several GL calls.

- *Additional functionality.* A renderer layer is the ideal place to add functionality that isn't in the underlying API. For example, Sec-

---

[4]If ARB_debug_output is supported, calls to `glGetError` can be replaced with a callback function [93].

[5]http://sourceforge.net/projects/bugle/

[6]http://glintercept.nutty.org/

[7]http://www.gremedy.com/

tion 3.4.1 introduces additional built-in GLSL constants that are not part of the GLSL language, and Section 3.4.5 introduces GLSL uniforms that are not built into GLSL but are still set automatically at draw time by the renderer. A renderer doesn't just wrap the underlying API; it raises the level of abstraction and provides additional functionality.

Even with all the benefits of a renderer, there is an important pitfall to watch for:

- *A false sense of portability.* Although a renderer eases supporting multiple APIs, it does not completely eliminate the pain. David Eberly explains his experience with Wild Magic: "After years of maintaining an abstract rendering API that hides DirectX, OpenGL, and software rendering, the conclusion is that each underlying API suffers to some extent from the abstraction" [45]. No renderer is a "one size fits all" solution. We freely admit that the renderer described in this chapter is biased to OpenGL as we've not implemented it with Direct3D yet.

  One prominent concern is that having both GL and D3D implementations of the renderer requires us to maintain two versions of all shaders: a GLSL version for the GL renderer and an HLSL version for the D3D renderer. Given that shader languages are so similar, it is possible to use a tool to convert between languages, even at runtime. For example, HLSL2GLSL,[8] a tool from AMD, converts D3D9 HLSL shaders to GLSL. A modified version of this tool, HLSL2GLSLFork,[9] maintained by Aras Pranckevičius, is used in Unity 3.0. The Google ANGLE[10] project translates in the opposite direction, from GLSL to D3D9 HLSL.

  To avoid conversions, shaders can be written in NVIDIA's Cg, which supports both GL and D3D. A downside is that the Cg runtime is not available for mobile platforms at the time of this writing.

  Ideally, using a renderer would avoid the need for multiple code paths in client code. Unfortunately, this is not always possible. In particular, if different generations of hardware are supported with different renderers, client code may also need multiple code paths. For example, consider rendering to a cube map. If a renderer is implemented using GL 3, geometry shaders will be available, so the cube map can be rendered in a single pass. If the renderer is implemented with

---

[8] http://sourceforge.net/projects/hlsl2glsl/
[9] http://code.google.com/p/hlsl2glslfork/
[10] http://code.google.com/p/angleproject/

an older GL version, each cube-map face needs to be rendered in a separate pass.

A renderer is such an important piece of an engine that most game engines include a renderer of some sort, as do applications like Google Earth. It is fair to ask, if a renderer is so important, why does everyone roll their own? Why isn't there one renderer that is in widespread use? Because different engines prefer different renderer designs. Some engines want low-level, nearly one-to-one mappings between renderer calls and GL calls, while other engines want very high-level abstractions, such as effects. A renderer's performance and features tend to be tuned for the application it is designed for.

Patrick Says ○○○○

When writing an engine, consider using a renderer from the start. In my experience, taking an existing engine with GL calls scattered throughout and refactoring it to use a renderer is a difficult and error-prone endeavor. When we started on Insight3D, one of my first tasks was to replace many of the GL calls in the existing codebase we were leveraging with calls to a new renderer. Even with all the debug code I included to validate GL state, I injected my fair share of bugs.

Although developing software by starting with solid foundations and building on top is much easier than retrofitting a large codebase later, do not fall into the trap of doing architecture for architecture's sake. A renderer's design should be driven by actual use cases.

3.2   Bird's-Eye View

A renderer is used to create and manipulate GPU resources and issue rendering commands. Figure 3.1 shows our renderer's major components. A small amount of render state configures the fixed-function components of the pipeline for rendering. Given that we are using a fully shader-based design, there isn't much render state, just things like depth and stencil testing. The render state doesn't include legacy fixed-function states that can be implemented in shaders like per-vertex lighting and texture environments.

Shader programs describe the vertex, geometry, and fragment shaders used to execute draw calls. Our renderer also includes types for communicating with shaders using vertex attributes and uniforms.

A vertex array is a lightweight container object that describes vertex attributes used for drawing. It receives data for these attributes through

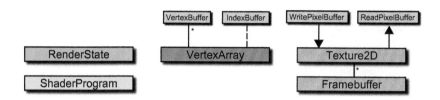

Figure 3.1. Our renderer's major components include render states; shader programs; vertex arrays that reference vertex and index buffers; 2D textures, whose data are manipulated with pixel buffers; and framebuffers, which contain textures.

one or more vertex buffers. An optional index buffer indexes into the vertex buffers to select vertices for drawing.

Two-dimensional textures represent images in driver-controlled memory that, when combined with a sampler that describes filtering and wrapping behavior, are accessible to shaders. Data are transferred to and from textures via write and read pixel buffers, respectively. Pixel buffers are channels for transferring data; ultimately, image data are stored in the texture itself.

Finally, framebuffers are lightweight containers of textures that enable render to texture techniques. A framebuffer can contain multiple textures as color attachments and a texture as a depth or depth/stencil attachment.

A static class named Device is used to create the render objects shown in Figure 3.2. It is helpful to think of Device as a factory for creating objects, but not for issuing rendering commands. A Context is used for issuing rendering commands. This distinction is similar to ID3D11Device and ID3D11DeviceContext in Direct3D.

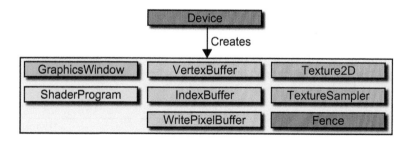

Figure 3.2. Device is a static class that creates renderer objects that are shared between contexts, except for GraphicsWindow, which contains a context.

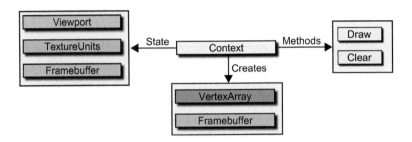

Figure 3.3. Context is used to issue rendering commands, **Draw** and **Clear** and create objects that are not shared between contexts. Rendering methods rely on context-level state, such as the framebuffer, and arguments passed to the methods describing the render state, etc.

A Context is part of a GraphicsWindow, which is created using Device .CreateWindow. A graphics window represents the canvas for drawing and contains the context and events for rendering, window resizing, and input handling. The objects created using Device can be shared among multiple contexts, except, of course, GraphicsWindow. For example, client code can create two different windows that use the same shaders, vertex/index buffers, textures, etc. This chapter describes these types in detail, with the exception of fences, which are described in Section 10.4 along with OpenGL multithreading.

As shown in Figure 3.3, in addition to issuing render commands, a context is used to create certain renderer objects and contains some state. Objects created by a Context can only be used in that context, unlike objects created by the Device, which can be used in any context. Only two objects fall into this nonshareable category: vertex arrays and framebuffers, both lightweight containers.

Patrick Says ○○○○

As much as an interface designer tries not to let implementation details (e.g., the underlying API) influence public interfaces, here is a case where it does. Vertex arrays and framebuffers cannot be shared among contexts in our renderer because GL does not allow them to be shared. Since these are lightweight containers, the GL overhead for making them shareable might be significant relative to the object itself.

A context contains some state: the viewport transformation, the framebuffer to render to, and the textures and samplers to render with. All the other information needed for rendering, such as a shader program and vertex array, are passed to Context.**Draw**. Our design uses only context-level state where appropriate to simplify state management for client code. This organization is much different from the state machine used by OpenGL.

The relevant subsections of the Device and Context interfaces are shown in Listings 3.1 and 3.2, respectively.[11]

```
public static class Device
{
  public static GraphicsWindow CreateWindow(int width,
                                            int height);
  public static ShaderProgram CreateShaderProgram(
      string vertexShaderSource,
      string geometryShaderSource,
      string fragmentShaderSource);
  public static VertexBuffer CreateVertexBuffer(
      BufferHint usageHint, int sizeInBytes);
  public static IndexBuffer CreateIndexBuffer(
      BufferHint usageHint, int sizeInBytes);
  public static WritePixelBuffer CreateWritePixelBuffer(
      PixelBufferHint usageHint, int sizeInBytes);
  public static Texture2D CreateTexture2D(
      Texture2DDescription description);
  public static TextureSampler CreateTexture2DSampler(/* ... */);
  public static Fence CreateFence();
  // ...
}
```

Listing 3.1. Device interface.

```
public abstract class Context
{
  public abstract VertexArray CreateVertexArray();
  public abstract Framebuffer CreateFramebuffer();

  public abstract TextureUnits TextureUnits { get; }
  public abstract Rectangle Viewport { get; set; }
  public abstract Framebuffer Framebuffer { get; set; }

  public abstract void Clear(ClearState clearState);
  public abstract void Draw(PrimitiveType primitiveType,
      int offset, int count, DrawState drawState,
      SceneState sceneState);
  // ...
}
```

Listing 3.2. Context interface.

---

[11]Some members of Context use C# properties. These are signified with { **get** ; } after the member's name for read-only properties or { **get; set;** } for read/write properties. In C++ and Java, properties would be implemented with explicit get and/or set methods.

3.2.1 Code Organization

Our renderer is implemented in the OpenGlobe.Renderer.dll assembly. All publicly exposed types are in the `OpenGlobe.Renderer` namespace. All types for the OpenGL 3.3 implementation are in the `OpenGlobe.Renderer.GL3x` namespace and also have a `GL3x` appended to their name. Since these types are an implementation detail, they are defined with internal scope so they are not accessible outside of the renderer assembly.[12] For example, Context is a public type in the `OpenGlobe.Renderer` namespace, and ContextGL3x is an internal type in the `OpenGlobe.Renderer.GL3x` namespace.

Some types in `OpenGlobe.Renderer` do not rely on the underlying API, so they do not have a GL3x counterpart. For example, RenderState is just a container of render-state parameters and does not require any GL calls. Types that do rely on the underlying API are defined as abstract classes with one or more virtual or abstract members.[13] The corresponding GL3x type inherits from this class and overrides the abstract and possibly virtual methods.

3.3 State Management

A major design decision for a renderer is how render state is managed. Render state configures the fixed-function components of the pipeline for rendering and affects draw calls. It includes things like the scissor test, depth test, stencil test, and blending.

3.3.1 Global State

In raw GL, render states are global states that can be set any time, as long as a valid context is current in the calling thread. For example, the depth test can be enabled with a call like:

```
GL.Enable(EnableCap.DepthTest);
```

The simplest way to expose this through a renderer would be to mirror the GL design: provide `Enable` and `Disable` methods and an enumeration defining states that can be enabled and disabled. This design still has the fundamental problem of global state. At any given point in time, is the depth test enabled or disabled? What if a virtual method is called? How did it leave the depth test state? I don't know. Do you?

---

[12]In C++, this is similar to not exporting the class from the dll.
[13]An abstract method in C# is equivalent to a pure virtual function in C++.

One approach to managing global state is to guarantee a set of states before a method is called, and require the called method to restore any states it changed. For example, our convention could be that the depth test is always enabled, so if a method wants the depth test disabled, it must do something like the following:

```
public virtual void Render()
{
  GL.Disable(EnableCap.DepthTest);
  // ... Draw call
  GL.Enable(EnableCap.DepthTest);
}
```

The obvious problem here is that this can lead to state thrashing, where every method sets and restores the same state. For example, if the above method was called 10 times, it would disable and enable the depth test 10 times, when it only needed to do it once. The driver may optimize away state changes if the state doesn't actually change from draw call to draw call, but there is still some driver overhead associated with calling into GL. This is peanuts compared to the real problem: this design requires the person implementing the method to know what the incoming state is and remember to restore it.

Let's pretend that the developer is able to remember what the incoming state is. We can surround calls to their method with push and pop attribute GL calls, like the following:

```
// ... Set initial states
GL.PushAttrib(AttribMask.AllAttribBits);
GL.PushClientAttrib(ClientAttribMask.ClientAllAttribBits);
Render();
GL.PopClientAttrib();
GL.PopAttrib();

// ...

public virtual void Render()
{
  GL.Disable(EnableCap.DepthTest);
  // ... Draw call
  // No need to restore
}
```

With this design, Render is not required to restore any states because the push and pop attribute saved and restored the states. Since it is not known what states Render will change, all attributes are pushed and popped, which is likely to be overkill. To add to the fun, these methods were deprecated in OpenGL 3. Instead of pushing and popping state, we can "restore" the state by explicitly setting the entire state before each call to `Render`, for example:

```
GL.Enable(EnableCap.DepthTest);
// ... Set other states
Render();
```

As before, this still leads to redundant state changes and requires the person implementing `Render` to know what the incoming state is.

Patrick Says ○○○○

> I've used this approach where the incoming state is well defined and must be preserved. It was sometimes painful because I always had to double check what the incoming state was before writing new code, for example, "This object is translucent so depth writes are disabled, or are they?"

The final evil that comes from global render state is developers can write seemingly harmless code like this:

```
public virtual void Render()
{
  GL.ColorMask(false, false, false, false);
  GL.DepthMask(false);
  // ... Draw call

  GL.ColorMask(true, true, true, true);
  // ... Another draw call
}
```

At first glance, it appears obvious that depth writes are disabled for the second draw call. How obvious would it be if several other state changes were added to this function? What if a developer comments out the call to `GL.DepthMask`? Did they intend that for just the first draw call or every draw call? Whoops. How much easier would it be if, instead, the render state were completely defined for each draw call?

3.3.2   Defining Render State

Just because the GL API uses global state doesn't mean that a renderer has to expose global state. Global state is an implementation detail. A renderer can present render state using any abstraction it pleases. One approach that eliminates global state and enables sorting by state is to group all render states into a single object that is passed to a draw call. The draw call then does the actual GL calls to set the states before issuing the GL draw call. In this design, there is never any question about what the current state is; there is no current state. Each draw call has its own render state, which may or may not change from call to call.

We define a RenderState class with the following properties:

```csharp
public class RenderState
{
  public PrimitiveRestart PrimitiveRestart { get; set; }
  public FacetCulling FacetCulling { get; set; }
  public RasterizationMode RasterizationMode { get; set; }
  public ScissorTest ScissorTest { get; set; }
  public StencilTest StencilTest { get; set; }
  public DepthTest DepthTest { get; set; }
  public DepthRange DepthRange { get; set; }
  public Blending Blending { get; set; }
  public ColorMask ColorMask { get; set; }
  public bool DepthMask { get; set; }
}
```

Listing 3.3. RenderState Properties.

The types for these properties are all API-agnostic and are defined in `OpenGlobe.Renderer`. They are exactly what you would expect them to be

```csharp
public enum DepthTestFunction
{
  Never ,
  Less ,
  Equal ,
  LessThanOrEqual ,
  Greater ,
  NotEqual ,
  GreaterThanOrEqual ,
  Always
}

public class DepthTest
{
  public DepthTest ()
  {
    Enabled = true;
    Function = DepthTestFunction.Less;
  }

  public bool Enabled { get; set; }
  public DepthTestFunction Function { get; set; }
}
```

Listing 3.4. DepthTest state.

given their names. For example, the DepthTest consists of an **Enabled** boolean property and a **DepthTestFunction** enumeration that defines the comparison function used when the depth test is enabled (see Listing 3.4).

When default constructed, RenderState properties match the default GL states, with two exceptions. The depth test, DepthTest, is enabled since that is the common case in our engine, and facet culling, Render State.**FacetCulling.Enabled**, is enabled instead of disabled.

All other RenderState properties are similar to DepthTest. It is worth mentioning that the stencil test, RenderState.**StencilTest**, has separate front and back facing state, and blending, RenderState.**Blending**, has separate RGB and alpha blend factors. These objects are just containers; they store state but do not actually set the GL global state.

Client code simply allocates a RenderState and then sets any properties that have inappropriate defaults. For example, the following code defines the render state for billboards (see Chapter 9):

```
RenderState renderState = new RenderState();
renderState.FacetCulling.Enabled = false;
renderState.Blending.Enabled = true;
renderState.Blending.SourceRGBFactor =
    SourceBlendingFactor.SourceAlpha;
renderState.Blending.SourceAlphaFactor =
    SourceBlendingFactor.SourceAlpha;
renderState.Blending.DestinationRGBFactor =
    DestinationBlendingFactor.OneMinusSourceAlpha;
renderState.Blending.DestinationAlphaFactor =
    DestinationBlendingFactor.OneMinusSourceAlpha;
```

Our RenderState and related types are similar to D3D state objects, which group state into coarse-grained objects: ID3D11BlendState, ID3D11 DepthStencilState, and ID3D11RasterizerState. The main difference is that the D3D types are immutable and, therefore, cannot be changed once they are created. Immutable types allow some optimizations, but they also reduce the flexibility to client code. For example, with a mutable render state, an object can determine its depth write property right before rendering based on its alpha value. With an immutable render state, the object either needs to create a new render state if its depth write property changes or keep two render states that are selected based on its alpha.

The other difference between D3D state objects and our RenderState is that D3D state objects are still assigned to global state using methods such as ID3D11DeviceContext::**OMSetBlendState**, whereas we completely eliminate global render state by passing objects directly to draw calls.

When I first designed render states for Insight3D, I used an immutable type that was created based on a "template," which defined the actual states. The template was passed to a global factory, which either created a new render state or returned one from its cache. The benefit was that since all render states were known, it was possible to use bucket sort to sort by state (see Section 3.3.6). The problem was, client code was always releasing render states and asking for new ones. I ultimately decided the flexibility of a mutable render state outweighed the performance gains of an immutable one and switched the implementation to use a mutable render state that is sorted at draw time with a comparison sort (e.g., `std::sort`). Ericson describes a similar flexibility versus performance trade-off for state sorting in God of War III [47].

○○○○ Patrick Says

3.3.3   Syncing GL State with Render State

So far, we've defined RenderState as a container object for render states that affect the fixed-function configuration of the pipeline during draw calls. In order to apply a RenderState, it is passed to a draw call, like the following:

```
RenderState renderState = new RenderState();
// ... Set states
context.Draw(PrimitiveType.Triangles, renderState, /* ... */);
```

Of course, a RenderState does not need to be allocated before every draw call. An object can allocate one or more RenderStates at construction time and set their properties when needed. The same RenderState can also be used with different contexts.

The implementation of ContextGL3x.**Draw** is exactly what you would expect. It uses fine-grained GL calls to sync the GL state with the state passed in. The context keeps a "shadowed" copy of the current state, **_renderState**, so it can avoid GL calls for state that does not need to change. For example, to set the depth test state, Context.**Draw** calls **ApplyDepthTest**:

```
private void ApplyDepthTest(DepthTest depthTest)
{
  if (_renderState.DepthTest.Enabled != depthTest.Enabled)
  {
    Enable(EnableCap.DepthTest, depthTest.Enabled);
    _renderState.DepthTest.Enabled = depthTest.Enabled;
  }
```

```
  if (depthTest.Enabled)
  {
    if (_renderState.DepthTest.Function != depthTest.Function)
    {
      GL.DepthFunc(TypeConverterGL3x.To(depthTest.Function));
      _renderState.DepthTest.Function = depthTest.Function;
    }
  }
}

protected static void Enable(EnableCap enableCap, bool enable)
{
  if (enable)
  {
    GL.Enable(enableCap);
  }
  else
  {
    GL.Disable(enableCap);
  }
}
```

To set the depth function, our renderer's enumeration is converted to the GL value using TypeConverterGL3x.**To**. This can be implemented with a series of if ... else statements, a switch statement, or a table lookup. A robust implementation should verify the enumeration and assert or throw an exception if appropriate. Unless a renderer's enumeration values match the GL values, an enumeration cannot be cast directly to the GL type.

The majority of the code in **OpenGlobe.Renderer.GL3x**.ContextGL3x sets GL states just like the above snippet. When shadowing state, it is important that the shadowed copy never becomes out of sync with the GL state. When the context is constructed, the GL state should be synced with the shadowed state (see ContextGL3x.**ForceApplyRenderState**), and the shadowed state should always be updated when a GL state change is made.

Try This ○○○○

If several draw calls are made with the same render state, as done when rendering terrain with geometry clipmapping in Chapter 13, it may be worth avoiding all the fine-grained if statements that compare individual properties of the shadowed state with the render state passed in. Implement a quick accept, coarse-grained check by remembering the last render state instance used and its "version," which is just an integer that is incremented every time a property of the render state changes. ContextGL3x. **Draw** can skip all fine-grained checks if the passed-in render state reference equals the shadowed render state reference and their versions match, indicating that the same render state was used in the last draw call.

3.3.4 Draw State

Although we've described Context.**Draw** as taking a RenderState, it actually takes a higher-level container object called DrawState. In order to issue a draw call, the render state is needed to configure the fixed-function pipeline but another state is required to configure other parts of the pipeline. Notably, a shader program that is executed in response to the draw called is required (see Section 3.4), as is a vertex array referencing vertex buffers and an optional index buffer (see Section 3.5).

In GL, these are global states; the shader program is specified using glUseProgram,[14] and the vertex array is specified using glBind VertexArray, both before issuing a draw call. In D3D, these are also global state, bound using ID3D11DeviceContext methods; shader stages are bound using *SetShader methods such as VSSetShader and PSSetShader, and vertex-array state is bound using IA* methods such as IASet VertexBuffers and IASetIndexBuffer.

These global states lead to the same problems as global render states. The solution is the same: group them into a container that is passed to a draw call. DrawState is such a container; it includes the render state, shader program, and vertex array used for drawing, as shown in Listing 3.5.

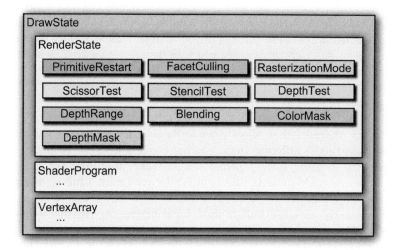

Figure 3.4. DrawState and RenderState properties. A DrawState object is passed to a draw call, which eliminates the need for global render, shader, and vertex-array states.

---

[14]When ARB_separate_shader_objects is supported, **glBindProgramPipeline** can be used to bind a pipeline program instead [87].

```
public class DrawState
{
    // ... Constructors
    public RenderState RenderState { get; set; }
    public ShaderProgram ShaderProgram { get; set; }
    public VertexArray VertexArray { get; set; }
}
```

Listing 3.5. DrawState properties.

This also enables sorting by shader, as explained in Section 3.3.6. A diagram showing DrawState properties is shown in Figure 3.4. A draw call does more than just call `glUseProgram` for the shader program and `glBindVertexArray` for the vertex array, but let's save that discussion until Sections 3.4 and 3.5, respectively.

Patrick Says ○○○○

> When I first designed Context.**Draw**, it did take just a Render State object. Client code was responsible for "binding" a shader program and vertex array to the context before calling **Draw**. Then I realized these were unnecessary global states, with the same problems associated with global render states, so I combined them with RenderState into a higher-level container named DrawState. When abstracting any API, it is important to remember that you don't need to let that API's implementation details show through in your design. This is a great ideal to strive for, but sometimes easier said than done.

3.3.5 Clear State

A number of states that affect draw calls also affect clearing the framebuffer. Clearing is different than drawing though, because geometry does not need to go through the pipeline and a shader program is not executed. Although one could clear the framebuffer by rendering a full screen quad, this is not a good practice because it doesn't take advantage of fast clears, which properly initialize buffers used for compression and hierarchical z-culling [136].

In GL, clears are affected by a number of states, including the scissor test and color, depth, and stencil masks. Clears also rely on color, depth, and stencil clear-value states set with `glClearColor`, `glClearDepth`, and `glClearStencil`, respectively. Once the states are configured, one or more buffers are cleared by calling `glClear`.

D3D has a different design that does not rely on global state; ID3D11 DeviceContext::**ClearRenderTargetView** clears a render target (e.g., a color

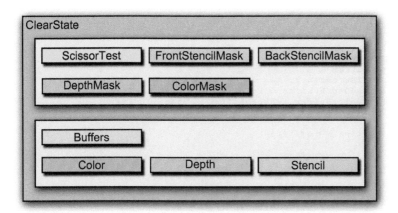

Figure 3.5. ClearState properties. Similar to how DrawState is used for draw calls, a ClearState object is passed to clear calls, which eliminates the need for related global states.

buffer) and ID3D11DeviceContext::**ClearDepthStencilView** clears a depth and/or stencil buffer. The clear values are passed as arguments instead of global state. Separate clear methods for the depth and stencil buffers are not provided; clearing them together is more efficient because they are usually stored in the same buffer [136].

Our renderer design for clears is similar to D3D, even though it is implemented using GL. It uses a container object, ClearState, shown in Listing 3.6 and Figure 3.5, to encapsulate the state needed for clearing. One of its members, **ClearBuffers**, is a bitmask that defines which buffers are cleared. To clear one or more buffers in the framebuffer, a ClearState object is created and passed to Context.**Clear**. For example, the following client code clears only the depth and stencil buffers:

```
ClearState clearState = new ClearState();
clearState.Buffers = ClearBuffers.DepthBuffer |
                     ClearBuffers.StencilBuffer;
context.Clear(clearState);
```

The ContextGL3x.**Clear** implementation is a series of straightforward GL calls based on the ClearState passed in. Similar to shadowing render state, the color, depth, and stencil clear values are shadowed to avoid unnecessary GL calls.

```
[Flags]
public enum ClearBuffers
{
   ColorBuffer = 1,
   DepthBuffer = 2,
   StencilBuffer = 4,
   ColorAndDepthBuffer = ColorBuffer | DepthBuffer,
   All = ColorBuffer | DepthBuffer | StencilBuffer
}

public class ClearState
{
   // ... Constructor

   public ScissorTest ScissorTest { get; set; }
   public ColorMask ColorMask { get; set; }
   public bool DepthMask { get; set; }
   public int FrontStencilMask { get; set; }
   public int BackStencilMask { get; set; }

   public ClearBuffers Buffers { get; set; }
   public Color Color { get; set; }
   public float Depth { get; set; }
   public int Stencil { get; set; }
}
```

Listing 3.6. ClearState properties.

It is important to note that the scissor test does not affect clears in D3D, but it does affect clears in GL. As mentioned in Section 3.1, designing a renderer for portability has some challenges. In this case, if both GL and D3D are supported, the scissor test may be removed from ClearState and always disabled for clears. Alternatively, the D3D implementation could assert or throw an exception if the scissor test is enabled, although this forces client code to know which renderer it is using.

3.3.6   Sorting by State

A common optimization is to sort by state. Doing so avoids wasted CPU driver overhead and helps exploit the parallelism of the GPU (see Section 10.1.2) by minimizing pipeline stalls [52]. The most expensive state changes are those that require large-scale reconfiguration of the pipeline, such as changing shaders, depth test state, or blending state.

State sorting is not done by our renderer itself but, instead, can be implemented on top of it by sorting by DrawState and issuing Context .Draw calls in an order that minimizes changing expensive states. It is also common to sort by texture or combine multiple textures in a single texture, as discussed in Section 9.2.

One approach to rendering a scene with state sorting is to use three passes:

1. First, hierarchical culling determines a list of visible objects.

2. Next, the visible objects are sorted by state.

3. Finally, the visible objects are drawn in sorted order.

Of course, there are many variations. If all the states are known before traversing the scene, the first two passes can be combined, and state sorting basically comes for free. During initialization, a list of sorted buckets is allocated with one bucket for each possible state. Each frame is rendered in two passes:

1. Hierarchical culling determines the visible objects and drops each into a bucket based on its state. Mapping an object to a bucket based on its state can be done in $\mathcal{O}(1)$ time if each state is given a unique index.

2. Next, the already sorted buckets are traversed and objects in non-empty buckets are drawn. After a bucket is rendered, it is cleared in preparation for the next frame.

This design assumes all possible states are known at initialization time and that the number of possible states is on the order of the number of objects. That is, this design is not ideal if there are 100,000 possible states and only ten objects.

For some scenes, hierarchical culling is not required. If the set of states is known in advance and objects do not change state, the scene can be rendered in a single pass that simply draws the objects in sorted order, culling or not culling individually along the way.

Regardless of when state sorting occurs, a sort order needs to be defined. This can easily be done with our DrawState object using a comparison method such as CompareDrawStates shown in Listing 3.7. This methods returns -1 when left $<$ right, 1 when left $>$ right, and 0 when left $=$ right. To sort by state, CompareDrawStates can be provided to List <T>.Sort, similar to how a function object is passed to std::sort in C++.[15]

A comparison method should compare the most expensive state first, then compare less expensive states until an ordering can be determined. The method CompareDrawStates sorts first by shader, then by depth test enabled. It should then move on, comparing other render states such as the depth compare function and blending states. The problem with CompareDraw States's implementation is it gets long as more states are compared, and the large number of branches may hurt performance when used for sorting every frame.

---

[15]To be exact, the function object used by std::sort requires only a strict weak ordering, whereas the comparison method shown here is a full ordering.

```
private static int CompareDrawStates(DrawState left,
                                     DrawState right)
{
  // Sort by shader first
  int leftShader = left.ShaderProgram.GetHashCode();
  int rightShader = right.ShaderProgram.GetHashCode();

  if (leftShader < rightShader)
  {
    return -1;
  }
  else if (leftShader > rightShader)
  {
    return 1;
  }

  // Shaders are equal, compare depth test enabled
  int leftEnabled =
      Convert.ToInt32(left.RenderState.DepthTest.Enabled);
  int rightEnabled =
      Convert.ToInt32(right.RenderState.DepthTest.Enabled);

  if (leftEnabled < rightEnabled)
  {
    return -1;
  }
  else if (rightEnabled > leftEnabled)
  {
    return 1;
  }

  // ... Continue comparing other states in order of most to ←
      least expensive...
  return 0;
}
```

Listing 3.7. A comparison method for sorting by DrawState. First it sorts by shader, then by depth test enabled. A complete implementation would sort by additional states.

In order to write a more concise and efficient comparison method, RenderState can be changed to store all the sortable states in one or more bitmasks. Each bitmask stores the most expensive states in the most significant bits and the less expensive states in the least significant bits. The large number of individual render-state comparisons in the comparison method is then replaced with a few comparisons using the bitmasks. This can also significantly reduce the amount of memory used by RenderState, which helps cache performance.

Try This ○○○○

> Modify RenderState so a bitmask can be used for state sorting.

The render state used in Insight3D uses a bitmask to store state with a
bit ordering such that states are sorted according to Forsyth's advice [52].
This has the advantage of using very little memory and enabling a simple and efficient functor for sorting. Debugging can be tricky, though,
because it is not obvious what the current render state is given that individual states are jammed into a bitmask. When using a design like
this, it is worth writing debug code to easily visualize the states in the
bitmask.

○○○○ Patrick Says

It is also common to sort by things other than state. For example, all
opaque objects are usually rendered before all translucent objects. Opaque
objects may be sorted by depth near to far to take advantage of z-buffer
optimizations, as discussed in Section 12.4.5, and translucent objects may
be sorted far to near for proper blending. State sorting is also not fully
compatible with multipass rendering because each pass depends on the
results of the previous pass, making it hard to reorder draw calls.

3.4   Shaders

In our renderer, shader support involves more interfaces and code than
any of the other major renderer components, and rightfully so: shaders
are at the center of modern 3D engine design. Even mobile devices supporting OpenGL ES 2.0 have very capable vertex and fragment shaders.
Today's desktops have a highly programmable pipeline, including programmable vertex, tessellation control, tessellation evaluation, geometry,
and fragment-shading stages. In D3D, the tessellation stages are called
hull and domain, respectively, and a fragment shader is called a pixel
shader. We use the GL terminology. This section focuses on abstracting the programmable stages supported by OpenGL 3.x and Direct3D 10
class hardware: vertex, geometry, and fragment stages.

Add support for tessellation control and evaluation shaders to the renderer. How should client code react if tessellation is not supported by
the hardware?

○○○○ Try This

3.4.1 Compiling and Linking Shaders

To begin, let's consider compiling and linking shaders with a simple example:

```
string vs = // ...
string fs = // ...
ShaderProgram sp = Device.CreateShaderProgram(vs, fs);
```

Two strings, vs and fs, contain the vertex and fragment-shader source code. A geometry shader is optional and not shown here. The source could be a hard-coded string, a procedurally generated string from different code snippets, or a string read from disk. Most of our examples store shader source in .glsl files, separate from C# source files. These files are included in the C# project and marked as an embedded resource so they are compiled into the assembly.[16] At design time, shaders are in individual files, but at runtime, they are embedded inside the .dll or .exe file. A shader's source is retrieved at runtime with a helper function that retrieves a string given a resource's name:

```
string vs = EmbeddedResources.GetText(
    "OpenGlobe.Scene.Globes.RayCasted.Shaders.GlobeVS.glsl");
```

Patrick Says ○○○○

I've organized shaders many ways over the years, and I am convinced that embedded resources is the most convenient. Since shaders are part of the assembly, this approach does not require additional files at runtime but still provides the convenience of individual source files at design time. Unlike hard-coded strings and shaders generated procedurally at runtime, shaders embedded as resources can easily be authored in third party tools. There is still something to be said for the potential performance advantages of procedurally generated shaders though, which I've used in a few places in Insight3D. For hardware that supports ARB_shader_subroutine, there should be less of a need to procedurally generate shaders since different shader subroutines can be selected at runtime, similar to virtual calls [20].

---

[16]In C++ on Windows, shader source files can be embedded as user-defined resources and accessed with FindResource and related functions.

Strings containing shader source are passed to Device.Create
ShaderProgram to create a shader program, fully constructed and ready to be
used for rendering (of course, the user may want to set uniforms first). As
mentioned in Section 3.3.4, a shader program is part of DrawState. It does
not need to be bound to the context like in vanilla GL or D3D;[17] instead,
a shader program is specified to each draw call, as part of DrawState, to
avoid global state. Note that clearing the framebuffer does not invoke a
shader, so a shader program is not part of ClearState.

Our GL implementation of a ShaderProgram is in ShaderProgram
GL3x, which uses a helper class for shader objects, ShaderObject
GL3x. A shader object represents one programmable stage of the pipeline
and is a GL renderer implementation detail. It is not part of the renderer
abstraction, which represents all stages together in one shader program. It
can be useful to expose interfaces for individual pipeline stages, especially
when using ARB_separate_shader_objects, but we did not have the need.

The ShaderObjectGL3x constructor calls glCreateShader, glShader
Source, glCompileShader, and glGetShader to create and compile a shader
object. If compilation fails, the renderer throws a custom CouldNotCreate
VideoCardResourceException exception. This is one of the benefits of a
renderer abstraction layer: it can convert error return codes into object-
oriented exceptions.

The ShaderProgramGL3x constructor creates shader objects, which are
then linked into a shader program using glCreateProgram, glAttachShader,
glLinkProgram, and glGetProgram. Similar to a compiler error, a link error
is converted into an exception.

All of these GL calls show another benefit of a renderer: conciseness.
A single call to Device.CreateShaderProgram in client code constructs a
ShaderProgramGL3x, which makes many GL calls, including proper error
handling, on the client's behalf.

Besides adding object-oriented constructs and making client code more
concise, a renderer allows adding useful building blocks for shader authors.
Our renderer has built-in constants, which are constants available to all
shaders but not defined in GLSL by default. These constants are always
prefixed with og_. A few built-in constants are shown in Listing 3.8. The
values for these GLSL constants are computed using C#'s Math constants
and constants in OpenGlobe's static Trig class.

Shader authors can simply use these constants, without explicitly declaring them. This is implemented in ShaderObjectGL3x by providing an array
of two strings to glShaderSource: one with the built-in constants and another with the actual shader source. In GLSL, a #version directive must

---

[17]In D3D, "shader programs" do not exist; instead, individual shader stages are bound
to the context. This is also possible in GL, with ARB_separate_shader_objects.

```
float  og_pi = 3.14159265358979;
float  og_oneOverPi = 0.318309886183791;
float  og_twoPi = 6.28318530717959;
float  og_halfPi = 1.5707963267949;
```

Listing 3.8. Selected GLSL constants built into our renderer.

be the first source line and cannot be repeated, so we include #version 330 as the first line of the built-in constants and comment out the #version line if it was provided in the actual source.

Try This ○○○○

In D3D, shader linking is not required. Instead, shaders are compiled (potentially offline) with **D3D10CompileShader**, and objects for individual stages are created with ID3D11DeviceContext calls like **CreateVertexShader** and **CreatePixelShader**, then bound to individual stages for rendering with calls such as **VSSetShader** and **PSSetShader**. This is also possible in GL with ARB_get_program_binary [103] and ARB_separate_shader_objects.

Use these extensions to modify the renderer to support offline compilation and to treat shader stages as individual objects and properties of DrawState. In OpenGL, the binary shader may be rejected if it was compiled for different hardware or even compiled using a different driver version, so a fallback method must be provided.

Try This ○○○○

Since 3D engines are designed for use by multiple applications, it would be useful for applications to define their own built-in constants that are available for all shaders. Modify the renderer to support this.

Try This ○○○○

Built-in constants are useful, but built-in reusable functions are even more useful. How would you add support for this to the renderer? Would it work the same way as built-in constants? Or is a #include mechanism preferred? What are the trade-offs?

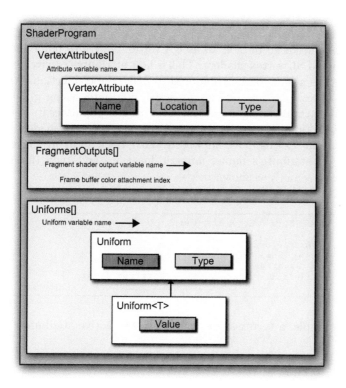

Figure 3.6. ShaderProgram properties. A shader program contains a collection of attribute variables, a collection of fragment-shader output variables, and a collection of uniforms, all indexable by name.

After compiling and linking, a few collections, shown in Listing 3.9 and Figure 3.6, are populated based on the shader program's inputs and outputs. We consider these collections separately in the following sections.

```
public abstract class ShaderProgram : Disposable
{
  public abstract ShaderVertexAttributeCollection
    VertexAttributes { get; }
  public abstract FragmentOutputs FragmentOutputs { get; }
  public abstract UniformCollection Uniforms { get; }

  // ...
}
```

Listing 3.9. Common ShaderProgram properties.

3.4.2 Vertex Attributes

Client code needs to be able to assign vertex buffers to named vertex attributes in GLSL vertex shaders. This is how attributes receive data. Each attribute has a numeric location used to make the connection. After linking, ShaderProgramGL3x queries its active attributes using `glGetProgram`, `glGetActiveAttrib`, and `glGetAttribLocation` and populates its publicly exposed collection of attributes used in the vertex shader: `Vertex Attributes` in Listing 3.9 and Figure 3.6. Each item in the collection includes an attribute's name, numeric location, and type (for example, Float, FloatVector4), as shown below:

```
public class ShaderVertexAttribute
{
  // ... Constructor
  public string Name { get; }
  public int Location { get; }
  public ShaderVertexAttributeType DataType { get; }
}
```

For example, a vertex shader may have two active attributes:

```
in vec4 position;
in vec3 normal;
```

The associated shader program would find and expose two attributes:

	Name	Location	Type
sp.VertexAttributes["position"]	"position"	0	FloatVector4
sp.VertexAttributes["normal"]	"normal"	1	FloatVector3

This collection of attributes is passed to Context.`CreateVertex Array` (see Section 3.5.3) to connect vertex buffers to attributes. Client code can rely on the shader program to create this collection, but sometimes, it is convenient to not have to create a shader program before creating a vertex array.

To achieve this, the attributes, including their location, need to be known in advance. Client code can use this information to explicitly create a ShaderVertexAttributeCollection instead of relying on the one provided by a shader program.

Our renderer includes built-in constants, shown in Listing 3.10, that can be used to explicitly specify a vertex attribute's location directly in the

```
#define  og_positionVertexLocation           0
#define  og_normalVertexLocation             1
#define  og_textureCoordinateVertexLocation  2
#define  og_colorVertexLocation              3
```

Listing 3.10. Built-in constants for specifying vertex attribute locations.

vertex-shader source. The attribute declarations in the vertex shader would look like the following:

```
layout(location = og_positionVertexLocation) in vec4 position;
layout(location = og_normalVertexLocation) in vec3 normal;
```

Instead of relying on a shader program to construct a collection of vertex attributes, client code can do it with code like the following:

```
ShaderVertexAttributeCollection vertexAttributes =
    new ShaderVertexAttributeCollection();
vertexAttributes.Add(new ShaderVertexAttribute(
    "position", VertexLocations.Position,
    ShaderVertexAttributeType.FloatVector4, 1));
vertexAttributes.Add(new ShaderVertexAttribute(
    "normal", VertexLocations.Normal,
    ShaderVertexAttributeType.FloatVector3, 1));
```

Here, a static class VertexLocations contains the constants for the vertex location for position and normal. The constants in VertexLocations are the C# equivalent of the GLSL constants; in fact, VertexLocations is used to procedurally create the GLSL constants.

The client code still needs to know about the attributes used in a shader program, but it doesn't have to create the shader program first. By specifying locations as built-in constants, locations for certain vertex attributes, such as position or normals, can be consistent across an application.

In Section 3.5.3, we will see how exactly the attribute collection is used to connect vertex buffers to attributes.

> Our renderer does not support array vertex attributes. Add support for them.

○○○○ Try This

3.4.3   Fragment Outputs

Similar to how vertex attributes defined in a vertex shader need to be connected to vertex buffers in a vertex array, fragment-shader output variables
need to be attached to framebuffer color attachments when multiple color
attachments are used. Thankfully, the process is even simpler than that
for vertex attributes.

Most of our fragment shaders just have a single output variable, usually
declared as out vec3 **fragmentColor**; or out vec4 **fragmentColor**;, depending
on if an alpha value is written. Usually, we rely on the fixed function to
write depth, but some fragment shaders write it explicitly by assigning to
gl_FragDepth, such as the shader for GPU ray casting a globe in Section 4.3.

When a fragment shader has multiple output variables, they need to
be connected to the appropriate framebuffer color attachments. Similar to
vertex attribute locations, client code can do this by asking a shader program for its output variable locations or by explicitly assigning locations
in the fragment-shader source. For the former, ShaderProgram exposes a
FragmentOutputs collection, shown earlier in Listing 3.9 and Figure 3.6.
This maps an output variable's name to its numeric location. It is implemented in FragmentOutputsGL3x using **glGetFragDataLocation**.

For example, consider a fragment shader with two output variables:

```
out vec4 dayColor;
out vec4 nightColor;
```

Client code can simply ask the shader program for the location of an
output variable given its name and assign a texture to that color attachment
location:

```
framebuffer.ColorAttachments[sp.FragmentOutputs("dayColor")] = ↵
    dayTexture;
```

Section 3.7 goes into more detail on framebuffers and their color attachments. Alternatively, a fragment shader can explicitly declare its output
variable locations with the same syntax used to declare vertex attribute
locations:

```
layout(location = 0) out vec4 dayColor;
layout(location = 1) out vec4 nightColor;
```

Now client code can know an output variable's location before creating a shader program.

3.4.4 Uniforms

Vertex buffers provide vertex shaders with data that can vary every vertex. Typical vertex components include positions, normals, texture coordinates, and colors. These types of data usually vary per vertex, but other types of data vary less frequently and would be wasteful to store per vertex. Uniform variables are a way to provide data to any shader stage that vary only up to every primitive. Since uniforms are set before a draw call, completely separate from vertex data, they are typically the same for many primitives. Common uses for uniforms include the model-view-projection transformation; light and material properties, such as diffuse and specular components; and application-specific values, such as the position of the sun or the viewer's altitude.

A shader program presents a collection of active uniforms called Uniforms, shown in Listing 3.9 and Figure 3.6, so client code can inspect uniforms and change their values. A shader-authoring tool may want to know what uniforms a shader program uses, so it can present text boxes, color selectors, etc., to modify their values. Most of our example code already knows what uniforms are used and simply needs to set their values, either once after creating a ShaderProgram or before every draw call. This section describes how a shader program in our renderer presents uniforms and makes GL calls to modify them. The following section discusses uniforms that are set automatically by the renderer.

Our renderer supports uniform data types defined by Uniform Type, which include the following:

- float, int, and bool scalars.

- float, int, and bool 2D, 3D, and 4D vectors. In GLSL, these are types like vec4, ivec4, and bvec4, which correspond to Vector4F, Vector4I, and Vector4B, respectively, in OpenGlobe.

- $2 \times 2, 3 \times 3, 4 \times 4, 2 \times 3, 2 \times 4, 3 \times 2, 3 \times 4, 4 \times 2$, and 4×3 float matrices, where an n x m matrix has n columns and m rows. In GLSL, these are types like mat4 and mat2x3, which correspond to Matrix4F and Matrix23, respectively.

- Sampler uniforms used to refer to textures (see Section 3.6), each of which is really just a scalar int.

Our most commonly used uniforms are 4×4 matrices, floating-point vectors, and samplers.

```
public class Uniform
{
    // ... Protected constructor

    public string Name { get; }
    public UniformType DataType { get; }
}

public abstract class Uniform<T> : Uniform
{
    protected Uniform(string name, UniformType type)
      : base(name, type)
    {
    }

    public abstract T Value { set; get; }
}
```

Listing 3.11. Each uniform is accessed by client code using an instance of Uniform<T>.

Try This ○○○○

OpenGL also includes unsigned ints and array uniforms. Add support to the renderer for them.

Try This ○○○○

OpenGL 4.1 includes support for uniform types with double precision. Double precision is a "killer feature" for virtual globes, as described in Chapter 5. Add support to the renderer for double-precision uniforms. Since OpenGL 4.x requires different hardware than 3.x, how should this be designed?

After a shader program is linked, ShaderProgramGL3x uses gl GetProgram, glGetActiveUniform, glGetActiveUniforms, and glGetUniform Location to populate its collection of uniforms. Each uniform in the collection is an instance of Uniform<T> shown in Listing 3.11 and Figure 3.7.

Let's consider a shader with three uniforms:

```
uniform mat4 modelViewPerspectiveMatrix;
uniform vec3 sunPosition;
uniform sampler2D diffuseTexture;
```

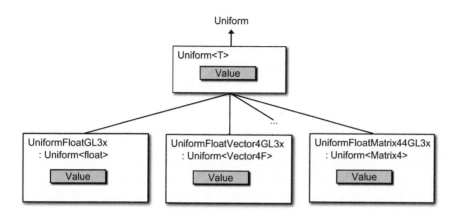

Figure 3.7. Client code sets uniform values using Uniform<T> since different types of uniforms need different data types for their **Value** properties. For each possible **T**, there is a GL implementation that calls the right GL function (e.g., `glUniform1f`, `glUniform4f`).

Once client code creates a ShaderProgram for a shader with these uniforms, the **Uniforms** collection will contain three Uniform objects:

	Name	**Type (UniformType)**
Uniforms["modelView PerspectiveMatrix"]	"modelView PerspectiveMatrix"	FloatVector4
Uniforms["sunPosition"]	"sunPosition"	FloatVector3
Uniforms["diffuseTexture"]	"diffuseTexture"	Sampler2D

The **Uniforms** collection presents uniforms through their base class, Uniform, so client code needs to cast a uniform to the appropriate Uniform <T> type to access its **Value** property. For example, the three uniforms in the above example would be assigned values with code like the following:

```
((Uniform<Matrix4F >)sp.Uniforms["modelViewPerspectiveMatrix"]).
    Value = new Matrix4F(/* ... */);
((Uniform<Vector4F >)sp.Uniforms["sunPosition"]).Value =
    new Vector4F(/* ... */);
// Texture unit zero
((Uniform<int >)sp.Uniforms["diffuseTexture"]).Value = 0;
```

If the uniform is set before every draw call, the string lookup and cast should be done just once and the Uniform<T> cached in client code for use before every draw call:

```
Uniform<Matrix4F> u =
    (Uniform<Matrix4F>)sp.Uniforms["modelViewPerspectiveMatrix"];

// ...

while (/* ... */)
{
  u.Value = // ...
  context.Draw(/* ... */);
}
```

If the client code does not know what Uniform<T> to cast to, it can check the `Type` property and then make the appropriate cast.

Our GL implementation implements each Uniform<T> in a different class. The inheritance relationship between Uniform<T> and a few GL classes is shown in Figure 3.7. The naming convention is straightforward: Uniform<float> is implemented in UniformFloatGL3x, and so on. The code ShaderProgram GL3x.`CreateUniform` uses the uniform type returned by `glGetActiveUniform` to create the appropriate Uniform<T>-derived type.

Each GL class stores a copy of the uniform's value and makes the appropriate `glUniform*` call to sync the value with GL. When a user assigns a value to the `Value` property, the GL call is not made immediately. Instead, a delayed technique is used [32]. The code ShaderProgramGL3x keeps a collection of all of its uniforms and a separate list of "dirty" uniforms. When client code sets a uniform's `Value` property, the uniform saves the value and adds itself to the shader program's "dirty list" if it wasn't already added.

When a draw call is made, the shader program that is part of the DrawState is bound to the context with `glUseProgram`, then the program is "cleaned." That is, a `glUniform*` call is made for each uniform in the dirty list, and then the dirty list is cleared. If most of a program's uniforms change from draw call to draw call, this implementation could be changed to not use a dirty list and, instead, just iterate through the collection of all uniforms, making a call to `glUniform*` for each. A highly tuned engine could do timings at runtime and use the more efficient approach.

The benefit of this delayed technique is that no redundant calls to `glUniform*` are made if client code redundantly sets a uniform's `Value` property. A context doesn't even need to be current when setting the property. It also simplifies GL state management since a call to `glUniform*` requires the corresponding shader program be bound.[18]

Direct3D 9 has constants that are similar to GL uniforms. A shader's constants can be queried and set using ID3DXConstantTable. D3D 10 replaced constants with constant buffers, which are buffers containing con-

---

[18] Using ARB_separate_shader_objects also simplifies state management since the program is passed directly to `glProgramUniform*` instead of needing to be bound.

stants that should be grouped together based on their expected update frequency. Buffers can improve performance by reducing per-constant overhead. OpenGL introduced a similar feature called uniform buffers.

> Our renderer includes support for uniform buffers. See Uniform Block∗ and their implementations in UniformBlock∗GL3x, as well as ShaderProgramGL3x.**FindUniformBlocks**. When these were originally coded, drivers were not stable enough to allow their use for examples throughout the book. Now that stable drivers are available, how would you organize uniforms into uniform buffers? Given that D3D 10 dropped support for constants, should our renderer drop support for uniforms and only expose uniform buffers?

◯◯◦◦ **Try This**

3.4.5 Automatic Uniforms

Uniforms are a convenient way to provide shaders with variables that are constant across one or more draw calls. Thus far, our renderer requires client code to explicitly assign every active uniform of every shader using Uniform<T>. Some uniforms, like the model-view matrix and the camera position, will be the same across many draw calls or even the entire frame. It is inconvenient for client code to have to explicitly set these uniforms for every shader, or even at all. Wouldn't it be great if the renderer defined a collection of uniforms that are automatically set before a draw call? Wouldn't it be even better if client code could add new uniforms to this collection? After all, one of the reasons to use a renderer in the first place is to add functionality above the underlying API (see Section 3.1).

Our renderer defines a number of these uniforms and a framework for adding new ones. We call these automatic uniforms; they go by different names in various engines: preset uniforms, engine uniforms, engine variables, engine parameters, etc. The idea is always the same: a shader author simply declares a uniform (e.g., mat4 **og_modelViewMatrix**) and it is automatically set, without the need to write any client code.

Similar to the built-in GLSL constants of Section 3.4.5, our automatic uniform names are prefixed with **og_**. Our implementation is similar to Cozzi's [31], although we divide automatic uniforms into two types:

- *Link automatic.* Uniforms that are set just once after a shader is compiled and linked.

- *Draw automatic.* Uniforms that need to be set as often as every draw call.

The vast majority of our automatic uniforms are draw automatic. Link automatics are obviously less useful but have less overhead since they do not need to be set per draw call. An automatic uniform implements either LinkAutomaticUniform or DrawAutomaticUniform and DrawAutomatic UniformFactory. These abstract classes are shown in Listing 3.12.

A link automatic simply needs to provide its name and know how to assign its value to a Uniform. The Device stores a collection of link automatics. After a shader is compiled and linked, ShaderProgram.Initialize AutomaticUniforms iterates through the shader's uniforms, and if any uniforms match the name of a link automatic, the abstract Set method is called to assign a value to the uniform.

The only link automatic in our renderer is og_textureN, which assigns a sampler uniform to texture unit N. For example, sampler2D og_texture1 defines a 2D sampler that is automatically set to texture unit 1. Listing 3.13 shows how this would be implemented.

Draw automatics are slightly more involved than link automatics. Similar to link automatics, the Device stores a collection of all draw automatic factories. Unlike with link automatics, when ShaderProgram.Initialize Automatic Uniforms comes across a draw automatic, it cannot simply assign a value because the value can vary from draw call to draw call. Instead, the uniform factory creates the actual draw automatic uniform, which is stored in a collection of draw automatics for the particular shader. Before each draw call, the shader is "cleaned"; it iterates through its draw automatic uniforms and calls the abstract Set method to assign their values.

Automatic uniforms are implemented independently of the GL renderer, except that ShaderProgramGL3x needs to call the protected Initial

```
public abstract class LinkAutomaticUniform
{
  public abstract string Name { get; }
  public abstract void Set(Uniform uniform);
}

public abstract class DrawAutomaticUniformFactory
{
  public abstract string Name { get; }
  public abstract DrawAutomaticUniform Create(Uniform uniform);
}

public abstract class DrawAutomaticUniform
{
  public abstract void Set(Context context, DrawState drawState,
      SceneState sceneState);
}
```

Listing 3.12. Abstract classes for automatic uniforms.

```
internal  class  TextureUniform1  :  LinkAutomaticUniform
{
  public  override  string  Name
  {
    get { return "og_texture1"; }
  }

  public  override  void  Set(Uniform  uniform)
  {
    ((Uniform<int>)uniform).Value = 1;
  }
}
```

Listing 3.13. An example link automatic uniform.

izeAutomaticUniforms after compiling and linking and SetDrawAuto
maticUniforms when it is cleaned. Individual automatic uniforms, such
as those in Listings 3.13 and 3.15, are implemented without any knowledge
of the underlying API.

Given the signature of DrawAutomaticUniform.Set in Listing 3.12, a draw
automatic can use the Context and the DrawState to determine the value to
assign to its uniform. But how does this help with uniforms like the model-
view matrix and the camera's position? It doesn't, of course. The third
argument to Set is a new type called SceneState that encapsulates scene-
level state like transformations and the camera, as shown in Listing 3.14.
Similar to DrawState, a SceneState is passed to a draw call, which ultimately
makes its way to an automatic uniform's Set method. An application
can use one SceneState for all draw calls or use different SceneStates as
appropriate.

```
public class Camera
{
  // ...

  public Vector3D Eye { get; set; }
  public Vector3D Target { get; set; }
  public Vector3D Up { get; set; }
  public Vector3D Forward { get; }
  public Vector3D Right { get; }

  public double FieldOfViewX { get; }
  public double FieldOfViewY { get; set; }
  public double AspectRatio { get; set; }

  public double PerspectiveNearPlaneDistance { get; set; }
  public double PerspectiveFarPlaneDistance { get; set; }

  public double Altitude(Ellipsoid shape);
}

public class SceneState
{
```

```
// ...
    public float DiffuseIntensity { get; set; }
    public float SpecularIntensity { get; set; }
    public float AmbientIntensity { get; set; }
    public float Shininess { get; set; }

    public Camera Camera { get; set; }
    public Vector3D CameraLightPosition { get; }

    public Matrix4D ComputeViewportTransformationMatrix(Rectangle
        viewport, double nearDepthRange, double farDepthRange);
    public static Matrix4D ComputeViewportOrthographicMatrix(
        Rectangle viewport);
    public Matrix4D OrthographicMatrix { get; }
    public Matrix4D PerspectiveMatrix { get; }
    public Matrix4D ViewMatrix { get; }
    public Matrix4D ModelMatrix { get; set; }
    public Matrix4D ModelViewPerspectiveMatrix { get; }
    public Matrix4D ModelViewOrthographicMatrix { get; }
    public Matrix42 ModelZToClipCoordinates { get; }
}
```

Listing 3.14. Selected Camera and SceneState members used to implement draw automatic uniforms.

Let's firm up our understanding of draw automatics by looking at the implementation for og_wgs84Height in Listing 3.15. This floating-point uniform is the height of the camera above the WGS84 ellipsoid (see Section 2.2.1), basically the viewer's altitude, neglecting terrain. One use for this is to fade in/out layers of vector data.

The factory implementation in Listing 3.15 is straightforward, as is the implementation for all draw automatic factories; one property simply returns the uniform's name, and the other method actually creates the draw automatic uniform. The automatic uniform itself, Wgs84Height Uniform, asks the scene state's camera for the height above the WGS84 ellipsoid and assigns it to the uniform.

```
internal class Wgs84HeightUniformFactory :
        DrawAutomaticUniformFactory
{
    public override string Name
    {
        get { return "og_wgs84Height"; }
    }

    public override DrawAutomaticUniform Create(Uniform uniform)
    {
        return new Wgs84HeightUniform(uniform);
    }
}

internal class Wgs84HeightUniform : DrawAutomaticUniform
{
```

```
public Wgs84HeightUniform(Uniform uniform)
{
    _uniform = (Uniform<float >)uniform;
}

public override void Set(Context context , DrawState drawState ,
                         SceneState sceneState)
{
    _uniform.Value =
        (float)sceneState.Camera.Height(Ellipsoid.Wgs84);
}

private Uniform<float> _uniform;
}
```

Listing 3.15. Implementation for `og_wgs84Height` draw automatic uniform.

Since automatic uniforms are found based on name only, it is possible that a shader author may declare an automatic uniform with the wrong data type (e.g., int `og_wgs84Height` instead of float `og_wgs84Height`), resulting in an exception at runtime when a cast to Uniform<T> fails. Add more robust error handling that reports the data type mismatch or only detects an automatic uniform if its name and data type match.

Try This ○○○○

Without caching camera and scene-state values, automatic uniforms can introduce CPU overhead by asking the camera or scene state to redundantly compute things. For example, in Wgs84HeightUniform, the call to Camera.**Height** is likely to be redundant for each shader using the `og_wgs84Height` automatic uniform. Add caching to SceneState and Camera to avoid redundant computations.

Try This ○○○○

How can uniform buffers, UniformBlock*, simplify automatic uniforms?

Try This ○○○○

Table 3.1 lists the most common automatic uniforms used in examples throughout this book. We've only included uniforms that we actually use in our examples, but many others are useful in the general sense. Draw automatic uniforms don't even have to set a uniform's value before every

GLSL Uniform Name	Source Renderer Property/Method	Description
vec3 og_cameraEye	Camera.Eye	The camera's position in world coordinates.
vec3 og_cameraLightPosition	SceneState. CameraLightPosition	The position of a light attached to the camera in world coordinates.
vec4 og_diffuseSpecular AmbientShininess	DiffuseIntensity, SpecularIntensity, AmbientIntensity, and Shininess SceneState properties	Lighting equation coefficients.
mat4 og_modelViewMatrix	SceneState.ModelViewMatrix	The model-view matrix, which transforms model coordinates to eye coordinates.
mat4 og_modelViewOrtho graphicMatrix	SceneState.Model ViewOrthographicMatrix	The model-view-orthographic projection matrix, which transforms model coordinates to clip coordinates using an orthographic projection.
mat4 og_modelViewPerspect iveMatrix	SceneState. ModelViewPerspectiveMatrix	The model-view-perspective projection matrix, which transforms model coordinates to clip coordinates using a perspective projection.
mat4x2 og_modelZ ToClipCoordinates	SceneState.ModelZ ToClipCoordinates	A 4×2 matrix that transforms a Z component in model coordinates to clip coordinates, which is useful in GPU ray casting (see Section 4.3).
float og_perspective FarPlaneDistance	Camera.Perspective FarPlaneDistance	The distance from the camera to the far plane. Used for a logarithmic depth buffer (see Section 6.4).
float og_perspective NearPlaneDistance	Camera.Perspective NearPlaneDistance	The distance from the camera to the far plane, which is useful for clipping against the near plane when rendering wide lines (see Section 7.3.3).
mat4 og_perspectiveMatrix	SceneState.PerspectiveMatrix	The perspective projection matrix, which transforms eye coordinates to clip coordinates.
mat4 og_viewportOrtho graphicMatrix	SceneState.ComputeViewport OrthographicMatrix()	The orthographic projection matrix for the entire viewport.
mat4 og_viewportTransform ationMatrix	SceneState.ComputeViewport TransformationMatrix()	The viewport transformation matrix, which transforms normalized device coordinates to window coordinates, which is used for rendering wide lines and billboards.
mat4 og_viewport	Context.Viewport	The viewport's left, bottom, width, and height values.
float og_wgs84Height	Camera.Altitude()	The camera's altitude above the WGS84 ellipsoid.
sampler* og_textureN	n/a	A texture bound to texture unit N, $0 \leq N <$ Context.Texture Units.Count.

Table 3.1. Selected renderer automatic uniforms. Many of these replace the deprecated OpenGL built-in uniforms.

draw call. In Insight3D, some draw automatic uniforms set a sampler uniform once and actually bind a texture (see Section 3.6) (e.g., the terrain's depth or silhouette texture) before each draw call. A similar technique can be done with noise textures.

3.4.6   Caching Shaders

Section 3.3.6 discussed the performance benefits of sorting `context.Draw` calls by DrawState. In Listing 3.7, the following code was used to sort by shader:

```
private static int CompareDrawStates(DrawState left,
                                     DrawState right)
{
  int leftShader = left.ShaderProgram.GetHashCode();
  int rightShader = right.ShaderProgram.GetHashCode();

  if (leftShader < rightShader)
  {
    return -1;
  }
  else if (leftShader > rightShader)
  {
    return 1;
  }
  // ... If shaders are the same, sort by other state.
}
```

In order to sort by shader, we need to be able to compare shaders. We are not concerned with the final sorted order, just that shaders that are the same wind up next to each other, in order to minimize the number of actual shader changes required (i.e., calls to `glUseProgram` in the GL implementation). Above, the C# method `GetHashCode` determines the ordering, similar to how a C++ implementation might use the shader's memory address for sorting. The key to sorting by shader is that shaders that are the same need to be the same ShaderProgram instance.

How many unique instances of ShaderPrograms does the following create?

```
string vs = // ...
string fs = // ...
ShaderProgram sp = Device.CreateShaderProgram(vs, fs);
ShaderProgram sp2 = Device.CreateShaderProgram(vs, fs);
```

Even though both shaders are created with the same source, this creates two distinct ShaderProgram instances, just like making similar GL calls would.

```
public class ShaderCache
{
  public ShaderProgram FindOrAdd(
      string key,
      string vertexShaderSource,
      string fragmentShaderSource);
  public ShaderProgram FindOrAdd(
      string key,
      string vertexShaderSource,
      string geometryShaderSource,
      string fragmentShaderSource);
  public ShaderProgram Find(string key);
  public void Release(string key);
}
```

Listing 3.16. ShaderCache interface.

For sorting by shader, we want both shaders to be the same instance so the compare method can group them next to each other.

This calls for a shader cache, which is implemented in ShaderCache. A shader cache simply maps a unique key to a ShaderProgram instance. We use user specified strings as keys, but one could also use integers or even the shader source itself.

The shader cache is part of the renderer, but it is not dependent on a particular API; it only deals with ShaderPrograms. As shown in Listing 3.16, the shader cache supports adding, finding, and releasing shaders.

Reference counting is used to track the number of clients with references to the same shader. When a shader is added to the cache, its reference count is one. Each time it is retrieved from the cache, either via FindOrAdd or Find, its count is increased. When a client is finished with a cached shader, it calls Release, which decreases the count. When the count reaches zero, the shader is removed from the cache.

Our example that created two distinct ShaderPrograms with the same shader source can be rewritten to use a shader cache, so only one Shader Program instance is created, making state sorting work as expected:

```
string vs = // ...
string fs = // ...
ShaderCache cache = new ShaderCache();
ShaderProgram sp = cache.FindOrAdd("example key", vs, fs);
ShaderProgram sp2 = cache.Find("example key");
// sp and sp2 are the same instance
cache.Release("example key");
cache.Release("example key");
```

The call to Find could be replaced with FindOrAdd. The difference is that Find does not require having the shader's source code and will return null if the key is not found, whereas FindOrAdd will create the shader if the key is not found. When procedurally generating shaders, `Find` is useful because the cache can be checked without procedurally generating the entire source.

Client code will usually only want one shader cache, but nothing is stopping clients from creating several. The implementation of ShaderCache is a straightforward use of a Dictionary[19] that maps a string to a ShaderProgram and its reference count. Since multiple threads may want to access a shader cache, each method is protected with a coarse-grained lock to serialize access to the shared cache. See Chapter 10 for more information on parallelism and threads.

One could argue that a coarse-grained lock hurts parallelism here. In particular, since the lock is held while a shader is compiled and linked (e.g., it calls Device.`CreateShaderProgram`), only one shader can be compiled/linked at a time, regardless of how many threads are used. In practice, is this really a problem? How would you change the locking strategy to allow multiple threads to compile/link in parallel, assuming the driver itself is not holding a coarse-grained lock preventing this?

○○○○ Try This

Why not build caching directly into Device.`CreateShaderProgram`? What are the trade-offs between doing so and our design?

○○○○ Question

When I first implemented a shader cache in Insight3D, I created a cache for both shader programs and shader objects. I found the cache for shader programs useful to enable sorting by shader, but the cache for shader objects wasn't all that useful, except for minimizing the number of GL shader objects created. In our renderer, this is not a concern because the concept of shader object doesn't exist; only entire shader programs are exposed.

○○○○ Patrick Says

---

[19]In C++, a `std::map` or `std::hash_map` could be used.

3.5   Vertex Data

A primary source of vertex-shader input is attributes that vary from vertex to vertex. Attributes are generally used for values such as positions, normals, and texture coordinates. For example, a vertex shader may declare the following attributes:

```
in  vec3  position;
in  vec3  normal;
in  vec2  textureCoordinate;
```

To clear up some terminology, a vertex is composed of attributes. Above, a vertex is composed of a position, normal, and texture-coordinate attribute. An attribute is composed of components. Above, the position attribute is composed of three floating-point components, and the texture-coordinate attribute is composed of two (see Figure 3.8).

In our renderer, vertex buffers store data used for attributes, and index buffers store indices used to select vertices for rendering. Both types of buffers are shown in Figure 3.9.

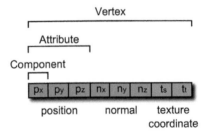

Figure 3.8. A vertex is composed of one or more attributes, which are composed of one or more components.

3.5.1   Vertex Buffers

Vertex buffers are raw, untyped buffers that store attributes in driver-controlled memory, which could be GPU memory, system memory, or both. Clients can copy data from arrays in system memory to a vertex buffer's driver-controlled memory or vice versa. The most common thing our examples do is compute positions on the CPU, store them in an array, and then copy them to a vertex buffer.

Vertex buffers are represented using the abstract class Vertex Buffer, whose GL implementation is VertexBufferGL3x. A vertex buffer is

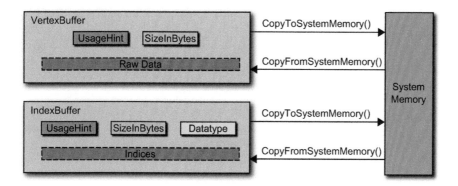

Figure 3.9. Vertex and index buffers. A vertex buffer is an untyped buffer, containing one or more attributes, that is interpreted once assigned to a vertex array. An index buffer is fully typed, having an explicit **Datatype** property. Both buffers have methods to copy their contents to and from system memory.

created using Device.**CreateVertexBuffer**, whose implementation is shown in Listing 3.17. The client only needs to provide two arguments to create a vertex buffer: a usage hint indicating how they intend to copy data to the buffer and the size of the buffer in bytes.

The usage hint is defined by the BufferHint enumeration, which can have three values:

- **StaticDraw.** The client will copy data to the buffer once and draw with it many times. Many vertex buffers in virtual globes should use this hint, which is likely to make the driver store the buffer in GPU memory.[20] Vertex buffers for things like terrain and stationary vector data are usually used for many frames and will benefit significantly from not being sent over the system bus every frame.

- **StreamDraw.** The client will copy data to the buffer once and draw with it at most a few times (e.g., think of a streaming video, where each video frame is used just once). In virtual globes, this hint is useful for real-time data. For example, billboards may be rendered to display the position of every commercial flight in the United States. If position updates are received frequently enough, the vertex buffer storing positions should use this hint.

- **DynamicDraw.** The client will repeatedly copy data to the buffer and draw with it. This does not imply the entire vertex buffer will change

---

[20] An OpenGL driver is also likely to shadow a copy in system memory to handle what D3D calls a lost device where the GPU resources need to be restored.

```
public static class Device
{
  // ...

  public static VertexBuffer CreateVertexBuffer(
      BufferHint usageHint, int sizeInBytes)
  {
    return new VertexBufferGL3x(usageHint, sizeInBytes);
  }
}
```

Listing 3.17. `CreateVertexBuffer` implementation.

```
public abstract class VertexBuffer : Disposable
{
  public virtual void CopyFromSystemMemory<T>(
      T[] bufferInSystemMemory) where T : struct;
  public virtual void CopyFromSystemMemory<T>(
      T[] bufferInSystemMemory,
      int destinationOffsetInBytes) where T : struct;
  public abstract void CopyFromSystemMemory<T>(
      T[] bufferInSystemMemory,
      int destinationOffsetInBytes,
      int lengthInBytes) where T : struct;

  public virtual T[] CopyToSystemMemory<T>() where T : ↵
      struct;
  public abstract T[] CopyToSystemMemory<T>(
      int offsetInBytes, int sizeInBytes) where T : struct;

  public abstract int SizeInBytes { get; }
  public abstract BufferHint UsageHint { get; }
}
```

Listing 3.18. VertexBuffer interface.

repeatedly; the client may only update a subset each frame. In the previous example, this hint is useful if only a subset of aircraft positions change each update.

It is worth experimenting with BufferHint to see which yields the best performance for your scenario. Keep in mind that these are just hints passed to the driver, although many drivers take these hints seriously, and thus, they affect performance. In our GL renderer implementation, BufferHint corresponds to the `usage` argument passed to `glBuffer Data`. See NVIDIA's white paper for more information on setting it appropriately [120].

The VertexBuffer interface is shown in Listing 3.18. Client code copies data from an array or subset of an array to the vertex buffer with a `CopyFromSystemMemory` overload and copies to an array using a `Copy`

ToSystemMemory overload. The implementor (e.g., VertexBufferGL3x) only
has to implement one version of each method, as the virtual overloads
delegate to the abstract overload.

> Add a CopyFromSystemMemory overload that includes a sourceOffset
> InBytes argument, so copying doesn't have to begin at the start of the
> array. When is this overload useful?

○○○○ Try This

Even though a vertex buffer is typeless, that is, it contains raw data
and does not interpret them as any particular data type, the copy methods
use C# generics[21] to allow client code to copy to and from arrays without
casting. This is just syntactic sugar; the buffer itself should be considered to
be raw bytes until it is interpreted using a vertex array (see Section 3.5.3).

A vertex buffer may store just a single attribute type, such as positions,
or it may store multiple attributes, such as positions and normals, either
interleaving each position and normal or storing all positions followed by
all normals. These approaches are shown in Figure 3.10 and are described
in the following three sections.

Separate buffers. One approach is to use a separate vertex buffer for each
attribute type, as shown in Figure 3.10(a). For example, if a vertex shader
requires positions, normals, and texture coordinates, three different vertex buffers are created. The position vertex buffer would be created and
populated with code like the following:

```
Vector3F [] positions = new Vector3F []
{
  new Vector3F (1, 0, 0),
  new Vector3F (0, 1, 0)
};

int sizeInBytes = ArraySizeInBytes.Size (positions);
VertexBuffer positionBuffer = Device.CreateVertexBuffer (
    BufferHint.StaticDraw, sizeInBytes);
positionBuffer.CopyFromSystemMemory (positions);
```

An array of 3D floating point-vectors is copied to a vertex buffer storing positions. Similar code would be used to copy normal and texture-
coordinate arrays into their vertex buffers.

---

[21]C# generics are similar to C++ templates.

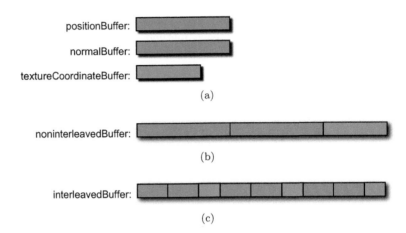

positionBuffer:

normalBuffer:

textureCoordinateBuffer:

(a)

noninterleavedBuffer:

(b)

interleavedBuffer:

(c)

Figure 3.10. (a) Separate buffers for positions, normals, and texture coordinates. Storing each attribute in a separate buffer is the most flexible approach. (b) A single noninterleaved buffer stores positions, followed by normals, and then texture coordinates, requiring only one buffer for multiple attributes. (c) A single interleaved buffer stores all three components, generally resulting in the best performance for static data [14].

The strength of this approach is flexibility. Each buffer can be created with a different BufferHint. For example, if we are rendering billboards whose positions change frequently, we may create a vertex buffer for positions using `StreamDraw` and a vertex buffer for texture coordinates, which do not change, using `StaticDraw`. The flexibility of separate buffers also allows us to reuse a vertex buffer across multiple batches (e.g., multiple batches of billboards may use different vertex buffers for positions but the same vertex buffer for texture coordinates).

Noninterleaved buffers. Instead of storing one attribute per vertex buffer, all attributes can be stored in a single vertex buffer, as shown in Figure 3.10(b). Creating fewer vertex buffers helps reduce per-buffer overhead.

Visualize the vertex buffer as an array of raw bytes that we can copy any data into, since, for example, the data type for positions and textures coordinates may be different. The `CopyFromSystemMemory` overload taking an offset in bytes that describes where to start copying into the vertex buffer should be used to concatenate arrays of attributes in the vertex buffer:

```
Vector3F [] positions = // ...
Vector3F [] normals = // ...
Vector2H [] textureCoordinates = // ...
```

```
VertexBuffer vertexBuffer = Device.CreateVertexBuffer(
    BufferHint.StaticDraw,
    ArraySizeInBytes.Size(positions) +
    ArraySizeInBytes.Size(normals) +
    ArraySizeInBytes.Size(textureCoordinates));

int normalsOffset = ArraySizeInBytes.Size(positions);
int textureCoordinatesOffset =
    normalsOffset + ArraySizeInBytes.Size(normals);

vertexBuffer.CopyFromSystemMemory(positions);
vertexBuffer.CopyFromSystemMemory(normals, normalsOffset);
vertexBuffer.CopyFromSystemMemory(textureCoordinates,
                                  textureCoordinatesOffset);
```

Although noninterleaved attributes use only a single buffer, memory coherence may suffer because a single vertex composed of multiple attributes
will need to be fetched from different parts of memory. We have not used
this approach in practice, and instead, interleave attributes in a single
buffer.

Interleaved buffers. As shown in Figure 3.10(c), a single vertex buffer can
store multiple attributes by interleaving each attribute. That is, if the
buffer contains positions, normals, and texture coordinates, a single position is stored, followed by a normal and then a texture coordinate. The
interleaved pattern continues for each vertex.

A common way to achieve this is to store each component for a vertex
in a struct laid out sequentially in memory:

```
[StructLayout(LayoutKind.Sequential)]
public struct InterleavedVertex
{
    public Vector3F Position { get; set; }
    public Vector3F Normal { get; set; }
    public Vector2H TextureCoordinate { get; set; }
}
```

Then, create an array of these structs:

```
InterleavedVertex[] vertices = new InterleavedVertex[]
{
    new InterleavedVertex()
    {
        Position = new Vector3F(1, 0, 0),
        Normal = new Vector3F(1, 0, 0),
        TextureCoordinate = new Vector2H(0, 0),
    },
    // ...
};
```

Finally, create a vertex buffer that is the same size as the array, in bytes, and copy the entire array to the vertex buffer:

```
VertexBuffer vertexBuffer = Device.CreateVertexBuffer(
    BufferHint.StaticDraw, ArraySizeInBytes.Size(vertices));
vertexBuffer.CopyFromSystemMemory(vertices);
```

When rendering large, static meshes, interleaved buffers outperform their counterparts [14]. A hybrid technique can be used to get the performance advantages of interleaved buffers and the flexibility of separate buffers. Since multiple vertex buffers can be pulled from in the same draw call, a static interleaved buffer can be used for attributes that do not change, and a dynamic or stream vertex buffer can be used for attributes that change often.

GL renderer implementation. Our GL vertex buffer implementation in VertexBufferGL3x makes straightforward use of GL's buffer functions with a target of `GL_ARRAY_BUFFER`. The constructor creates the name for the buffer object with `glGenBuffers`. When the object is disposed, `glDeleteBuffers` deletes the buffer object's name.

The constructor also allocates memory for the buffer with `glBuffer Data` by passing it a null pointer. The object `CopyFromSystem Memory` calls `glBufferSubData` to copy data from the input array to the buffer object. It is likely that the driver is implemented such that the initial call to `glBufferData` does not have much overhead, and instead, the allocation occurs at the first call to `glBufferSubData`. Alternatively, our implementation could delay calling `glBufferData` until the first call to `CopyFrom SystemMemory`, at the cost of some additional bookkeeping.

A call to `glBufferData` and `glBufferSubData` is always preceded with calls to `glBindBuffer`, to ensure that the correct GL buffer object is modified, and `glBindVertexArray`(0), to ensure a GL vertex array is not accidentally modified.

Try This ○○○○

Instead of updating a GL buffer object with `glBufferData` and `gl BufferSubData`, the buffer or a subset of it can be mapped into the application's address space using `glMapBuffer` or `glMapBuffer Range`. It can then be modified using a pointer like any other memory. Add this feature to the renderer. What are the advantages of this approach? What challenges does using a managed language like C# impose?

3.5.2   Index Buffers

Index buffers are represented using the abstract class IndexBuffer, whose
GL implementation is IndexBufferGL3x. The interface and implementation
are nearly identical to vertex buffers. Client code usually looks similar,
like the following example, which copies indices for a single triangle into a
newly created index buffer:

```
ushort[] indices = new ushort[] { 0, 1, 2 };

IndexBuffer indexBuffer = Device.CreateIndexBuffer(
    BufferHint.StaticDraw, indices.Length * sizeof(ushort));
indexBuffer.CopyFromSystemMemory(indices);
```

Unlike vertex buffers, index buffers are fully typed. Indices can be
either unsigned shorts or unsigned ints, as defined by the IndexBufferData
type enumeration. Client code does not need to explicitly declare the data
type for indices in an index buffer. Instead, the generic parameter T of
CopyFromSystemMemory is used to determine the data type.

> What are the design trade-offs between having client code explicitly de-
> clare a data type for the index buffer compared to implicitly determining
> the data type when CopyFromSystemMemory is called?

○○○○ Question

Clients should strive to use the data type that uses the least amount
of memory but still allows indexing fully into a vertex buffer. If 64 K or
fewer vertices are used, unsigned short indices suffice, otherwise unsigned
int indices are called for. In practice, we have not noticed performance
differences between unsigned shorts and unsigned ints, although "size is
speed" in many cases, and using less memory, especially GPU memory, is
always a good practice.

> A useful renderer feature is to trim indices down to the smallest sufficient
> data type. This allows client code to always work with unsigned ints,
> while the renderer only allocates unsigned shorts if possible. Design and
> implement this feature. Under what circumstances is this not desirable?

○○○○ Try This

The only noteworthy GL implementation difference between our index and vertex buffers is that the index buffer uses the `GL_ELEMENT_ARRAY_BUFFER` target instead of `GL_ARRAY_BUFFER`.

3.5.3   Vertex Arrays

Vertex and index buffers simply store data in driver-controlled memory; they are just buffers. A vertex array defines the actual components making up a vertex, which are pulled from one or more vertex buffers. A vertex array also references an optional index buffer that indexes into these vertex buffers. Vertex arrays are represented using the abstract class VertexArray, shown in Listing 3.19 and Figure 3.11.

Each attribute in a vertex array is defined by a VertexBuffer Attribute, shown in Listing 3.20. Attributes are accessed using a zero-

```
public abstract class VertexArray : Disposable
{
  public virtual VertexBufferAttributes Attributes { get; }
  public virtual IndexBuffer IndexBuffer { get; set; }
}
```

Listing 3.19. VertexArray interface.

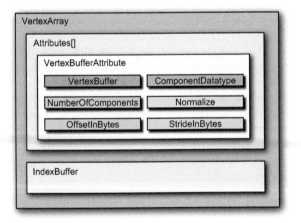

Figure 3.11. A vertex array contains one or more vertex attributes (e.g., position, normal) and an optional index buffer. The **VertexArray** member of DrawState and the arguments passed to Context.**Draw** determine the vertices used for rendering.

```
public enum ComponentDatatype
{
  Byte ,
  UnsignedByte ,
  Short ,
  UnsignedShort ,
  Int ,
  UnsignedInt ,
  Float ,
  HalfFloat ,
}

public class VertexBufferAttribute
{
  // ... Constructors
  public VertexBuffer VertexBuffer { get; }
  public ComponentDatatype ComponentDatatype { get; }
  public int NumberOfComponents { get; }
  public bool Normalize { get; }
  public int OffsetInBytes { get; }
  public int StrideInBytes { get; }
}
```

Listing 3.20. VertexBufferAttribute interface.

based index that corresponds to the shader's attribute location (see Section 3.4.2). Each vertex array supports up to Device.**MaximumNumberOfVertexAttributes** attributes.

An attribute is defined by the vertex buffer containing the data, its component's data type (ComponentDatatype), and its number of components. Unlike indices, which only support unsigned shorts and unsigned ints, a vertex component supports these signed and unsigned integral types, as well as bytes, floats, and half-floats. The latter are useful for saving memory for texture coordinates in the range $[0, 1]$ compared to floats.

If each attribute type is in a separate vertex buffer, it is straightforward to create a vertex array referencing each component. A VertexBuffer Attribute referencing the appropriate vertex buffer is created for each component as shown below:

```
VertexArray va = window.Context.CreateVertexArray();
va.Attributes[0] = new VertexBufferAttribute(
    positionBuffer , ComponentDatatype.Float , 3);
va.Attributes[1] = new VertexBufferAttribute(
    normalBuffer , ComponentDatatype.Float , 3);
va.Attributes[2] = new VertexBufferAttribute(
    textureCoordinatesBuffer , ComponentDatatype.HalfFloat , 2);
```

Positions and normals are each composed of three floating-point components (e.g., x, y, and z), and texture coordinates are composed of two floating-point components (e.g., s and t). If a single vertex buffer contains multiple attribute types, the `OffsetInBytes` and `StrideInBytes` VertexBufferAttribute properties can be used to select attributes from the appropriate parts of the vertex buffer. For example, if positions, normals, and texture coordinates are stored noninterleaved in a single vertex buffer, the `OffsetInBytes` property can be used to select the starting point for pulling from a vertex buffer:

```
Vector3F [] positions = // ...
Vector3F [] normals = // ...
Vector2H [] textureCoordinates = // ...

int normalsOffset = ArraySizeInBytes.Size(positions);
int textureCoordinatesOffset =
    normalsOffset + ArraySizeInBytes.Size(normals);

// ...

va.Attributes [0] = new VertexBufferAttribute(positionBuffer,
    ComponentDatatype.Float, 3);
va.Attributes [1] = new VertexBufferAttribute(normalBuffer,
    ComponentDatatype.Float, 3, false, normalsOffset, 0);
va.Attributes [2] = new VertexBufferAttribute(
    textureCoordinatesBuffer, ComponentDatatype.HalfFloat, 2,
    false, textureCoordinatesOffset, 0);
```

Finally, if attributes are interleaved in the same vertex buffer, the `StrideInBytes` property is also used to set the stride between each attribute since attributes are no longer adjacent to each other:

```
int normalsOffset = SizeInBytes<Vector3F >.Value;
int textureCoordinatesOffset =
    normalsOffset + SizeInBytes<Vector3F >.Value;

va.Attributes [0] = new VertexBufferAttribute(
    positionBuffer, ComponentDatatype.Float, 3,
    false, SizeInBytes<InterleavedVertex >.Value);
va.Attributes [1] = new VertexBufferAttribute(
    normalBuffer, ComponentDatatype.Float, 3,
    false, normalsOffset, SizeInBytes<InterleavedVertex >.Value);
va.Attributes [2] = new VertexBufferAttribute(
    textureCoordinatesBuffer, ComponentDatatype.HalfFloat, 2,
    false, textureCoordinatesOffset,
    SizeInBytes<InterleavedVertex >.Value);
```

As shown earlier in Listing 3.5 and Figure 3.4, DrawState has a Vertex Array member that provides vertices for rendering. The renderer's public interface has no global state for the "currently bound vertex array"; instead, client code provides a vertex array to every draw call.

3.5.4   GL Renderer Implementation

The GL calls that configure a vertex array are in VertexArrayGL3x and
VertexBufferAttributeGL3x. A GL name for the vertex array is created
with `glGenVertexArrays` and, of course, eventually deleted with `glDelete`
`VertexArrays`. Assigning a component or index buffer to a vertex array does not immediately result in any GL calls. Instead the calls are
delayed until the next draw call that uses the vertex array, to simplify
state management [32]. When a vertex array is modified, it is marked
as dirty. When a draw call is made, the GL vertex array is bound with
`glBindVertexArray`. If it is dirty, its dirty components are cleaned by calling
`glDisableVertexAttribArray` or calling `glEnableVertexAttribArray`, `gl`
`BindBuffer`, and `glVertexAttribPointer` to actually modify the GL vertex
array. Likewise, a call to `glBindBuffer` with `GL_ELEMENT_ARRAY_BUFFER` is
used to clean the vertex array's index buffer.

In ContextGL3x, the actual draw call is issued with `glDrawRange`
`Elements` if an index buffer is used or with `glDrawArrays` if no index buffer
is present.

3.5.5   Vertex Data in Direct3D

Our renderer abstractions for vertex buffers, index buffers, and vertex arrays also map nicely to D3D. In D3D, vertex and index buffers are created with ID3D11Device::`CreateBuffer`, similar to calling GL's `glBuffer`
`Data`. The function `CreateBuffer` has a `D3D11_BUFFER_DESC` argument that
describes the type of buffer to create, its size in bytes, and a usage hint,
among other things. The data for the buffer itself (e.g., the array passed
to our `CopyFromSystemMemory` methods) are passed to D3D's `CreateBuffer`
as part of the `D3D11_SUBRESOURCE_DATA` argument.

A buffer can be updated using ID3D11DeviceContext::`UpdateSub`
`resource`, similar to GL's `glBufferSubData`, or by mapping a pointer to
the buffer using ID3D11DeviceContext::`Map`, similar to GL's `glMapBuffer`.

Although D3D does not have an identical object to GL's vertex array
object, our VertexArray interface can still be implemented through D3D
methods for binding an input layout, vertex buffer(s), and an optional
index buffer to the input-assembler stage, similar to what `glBindVertex`
`Array` achieves in GL. An input-layout object is created using ID
3D11Device::`CreateInputLayout`, which describes the components in a vertex
buffer. This object is bound to the input-assembler stage before rendering
using ID3D11DeviceContext::`IASetInputLayout`. One or more vertex buffers
are bound using ID3D11DeviceContext::`IASetVertexBuffers`, and an optional
index buffer is bound using ID3D11DeviceContext::`IASetIndexBuffer`. The

index data type is specified as an argument, which can be either a 16-bit or 32-bit unsigned integer.

Draw calls are issued using methods on the ID3D11DeviceContext interface. In addition to binding the input layout, vertex buffer(s), and index buffer, the primitive type is also bound using IASetPrimitive Topology. Finally, a draw call is issued. A call to Draw would be used in place of GL's glDrawArrays, and a call to DrawIndexed would be used in place of glDrawRangeElements.

3.5.6 Meshes

Even with our renderer abstractions, creating vertex buffers, index buffers, and vertex arrays requires quite a bit of bookkeeping. In many cases, in particular those for rendering static meshes, client code shouldn't have to concern itself with how many bytes to allocate, how to organize attributes into one or more vertex buffers, or which vertex-array attributes correspond to which shader attribute locations. Luckily, the renderer allows us to raise the level of abstraction and simplify this process. Low-level vertex and index buffers are available to clients, but clients can also use higher-level types when ease of use is more important than fine-grained control.

OpenGlobe.Core contains a Mesh class shown in Listing 3.21 and Figure 3.12. A mesh describes geometry. It may represent the ellipsoidal surface of the globe, a tile of terrain, geometry for billboards, a model for a building, etc. The Mesh class is similar to VertexArray in that it contains vertex attributes and optional indices. A key difference is Mesh has strongly typed vertex attributes, as opposed to the renderer's VertexBuffer, which requires a VertexBufferAttribute to interpret its attributes.

The Mesh class is defined in OpenGlobe.Core because it is just a container. It is not used directly for rendering, it does not depend on any renderer types, and it does not create any GL objects under the hood, so it doesn't belong in OpenGlobe.Renderer. It simply contains data. An overload of Context.CreateVertexArray takes a mesh as input and creates a vertex array containing vertex and index buffers based on the mesh. This

```
public class Mesh
{
  public VertexAttributeCollection Attributes { get; }
  public IndicesBase Indices { get; set; }

  public PrimitiveType PrimitiveType { get; set; }
  public WindingOrder FrontFaceWindingOrder { get; set; }
}
```

Listing 3.21. Mesh interface.

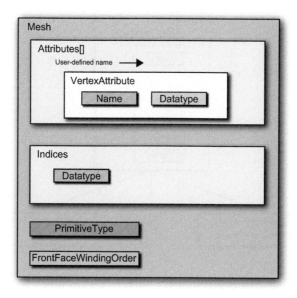

Figure 3.12. A Mesh contains a collection of vertex attributes, an optional collection of indices, and members describing the primitive type and winding order. These types are easier to work with than the lower-level renderer types. A Context.**CreateVertexArray** overload can create a vertex array containing vertex and index buffers, given a mesh, allowing the user to avoid the bookkeeping of dealing with renderer types directly.

allows client code to use the higher-level mesh class to describe geometry and let Context.**CreateVertexArray** do the work of laying out vertex buffers and other bookkeeping. This **CreateVertexArray** overload only uses publicly exposed renderer types, so it does not need to be rewritten for each supported rendering API.

Another benefit of Mesh is that algorithms that compute geometry can create a mesh object. This decouples geometric computation from rendering, which makes sense; for example, an algorithm to compute triangles for an ellipsoid should not depend on our renderer types. In Chapter 4, we cover several algorithms that create a mesh object that approximates the ellipsoidal surface of the globe. In fact, we use the Mesh type throughout the book when building geometry.

A mesh contains a collection of vertex attributes, similar to how a vertex array can have separate vertex buffers for each attribute. The VertexAttributeType enumeration in Listing 3.22 lists the supported data types. There is a concrete class for each type that inherits from Vertex Attribute<**T**> and VertexAttribute, as shown in Listing 3.22 and

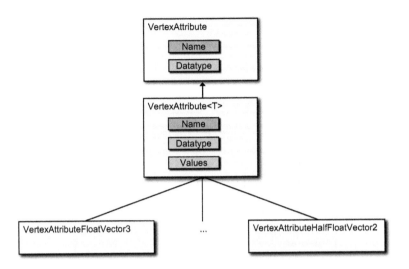

Figure 3.13. A Mesh contains a collection of VertexAttributes indexed by a user-defined name. Actual implementations of VertexAttribute contain a collection of fully typed attributes, as opposed to VertexBuffer, which is untyped and can therefore store multiple attribute types in the same collection.

```
public enum VertexAttributeType
{
    UnsignedByte,
    HalfFloat,
    HalfFloatVector2,
    HalfFloatVector3,
    HalfFloatVector4,
    Float,
    FloatVector2,
    FloatVector3,
    FloatVector4,
    EmulatedDoubleVector3
}

public abstract class VertexAttribute
{
    protected VertexAttribute(string name,
                              VertexAttributeType type);
    public string Name { get; }
    public VertexAttributeType Datatype { get; }
}

public class VertexAttribute<T> : VertexAttribute
{
    public IList<T> Values { get; }
}
```

Listing 3.22. VertexAttribute and VertexAttribute<T> interfaces.

```
public enum IndicesType
{
  UnsignedShort ,
  UnsignedInt
}

public abstract class IndicesBase
{
  protected IndicesBase(IndicesType type);
  public IndicesType Datatype { get; }
}

public class IndicesUnsignedShort : IndicesBase
{
  public IndicesUnsignedShort();
  public IndicesUnsignedShort(int capacity);
  public IList<short> Values { get; } // Add one index
  public void AddTriangle(TriangleIndicesUnsignedShort triangle);
     // Add three indices
}
```

Listing 3.23.   IndicesBase interface and an example concrete class, IndicesUnsignedShort. IndicesByte and IndicesUnsignedInt are similar.

Figure 3.13. Each attribute has a user-defined name (e.g., "position," "normal") and its data type. A concrete attribute class has a strongly typed **Values** collection that contains the actual attributes. This allows client code to create a strongly typed collection (e.g., VertexAttributeFloatVector3), populate it, and then add it to a mesh. Then, Context.**CreateVertexArray** can inspect the attribute's type using its **Datatype** member and cast to the appropriate concrete class. This is very similar to the class hierarchy design used for uniforms in Section 3.4.4.

A mesh's indices are handled similar to vertex attributes, except a mesh only has a single indices collection, not a collection of them. As shown in Listing 3.23 and Figure 3.14, two different index data types are supported: unsigned shorts and unsigned ints. Again, client code can create the specific type of indices desired (e.g., IndicesUnsignedShort), and Context .**CreateVertexArray** will use the base class's **Datatype** property to cast to the correct type and create the appropriate index buffer.

Listing 3.24 shows example client code using a Mesh and **context.Cre ateVertexArray** to create a vertex array containing a single triangle. First, a Mesh object is created, and its primitive type and winding order are set to triangles and counterclockwise, respectively. Next, memory is allocated for three single-precision 3D vector vertex attributes. These attributes are added to the mesh's attributes collection. Similarly, memory is allocated for three indices of type **unsigned** short, which are added to the mesh. Then, the actual vertex attributes and indices are assigned. Note the use of a helper class, TriangleIndicesUnsignedShort, to add indices for one triangle in a single line of code, as opposed to calling **indices.Values.Add** three times.

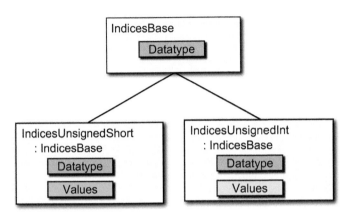

Figure 3.14. A Mesh contains an IndicesBase member. Each concrete class derived from IndicesBase contains a collection of indices of unsigned shorts or unsigned ints.

Finally, a vertex array is created from the mesh. A shader's list of attributes is passed to `CreateVertexArray` along with the mesh to match up the mesh's attribute names with the names of the shader's attributes.

It is helpful to think of Mesh and its related types as geometry in system memory for general application use and to think of VertexArray and related

```
Mesh mesh = new Mesh();
mesh.PrimitiveType = PrimitiveType.Triangles;
mesh.FrontFaceWindingOrder = WindingOrder.Counterclockwise;

VertexAttributeFloatVector3 positionsAttribute =
    new VertexAttributeFloatVector3("position", 3);
mesh.Attributes.Add(positionsAttribute);

IndicesUnsignedShort indices = new IndicesUnsignedShort(3);
mesh.Indices = indices;

IList<Vector3F> positions = positionsAttribute.Values;
positions.Add(new Vector3F(0, 0, 0));
positions.Add(new Vector3F(1, 0, 0));
positions.Add(new Vector3F(0, 0, 1));

indices.AddTriangle(new TriangleIndicesUnsignedShort(0, 1, 2));

// ...

ShaderProgram sp = Device.CreateShaderProgram(vs, fs);
VertexArray va = context.CreateVertexArray(
    mesh, sp.VertexAttributes, BufferHint.StaticDraw);
```

Listing 3.24. Creating a VertexArray for a Mesh containing a single triangle.

types as geometry in driver-controlled memory for rendering use. Strictly speaking, though, a vertex array does not contain geometry; it references and interprets vertex and index buffers.

Although we often favor the ease of use of Mesh and Context.**Create VertexArray**, there are times when the flexibility of vertex and index buffers makes it worthwhile to create them directly. In particular, if several vertex arrays should share a vertex or index buffer, the renderer types should be used directly.

Our implementation of Context.**CreateVertexArray** creates a separate vertex buffer for each vertex attribute. Implement two variations: one that stores attributes in a single noninterleaved vertex buffer and another that stores attributes in a single interleaved buffer. What are the performance differences?

○○○○ **Try This**

3.6   Textures

Textures represent image data in driver-controlled memory. In virtual globes, games, and many graphics applications, textures can account for more memory usage than vertex data. Textures are used to render the high-resolution imagery that has become a commodity in virtual globes. The texels in textures do not have to represent pixels though; as we shall see in Chapter 11, textures are also useful for terrain heights and other data.

Our renderer exposes a handful of classes for using 2D textures, including the texture itself, pixel buffers for transferring data to and from textures, a sampler describing filtering and wrap modes, and texture units for assigning textures and samplers for rendering.

3.6.1   Creating Textures

In order to create a texture, client code must first create a description of the texture, Texture2DDescription, shown in Listing 3.25 and Figure 3.15. A description defines the width and height of the texture, its internal format, and if mipmapping should be used. It also has three derived properties based on the format: `ColorRenderable`, `DepthRenderable`, and `DepthStencilRenderable`. These describe where a texture with the given description can be attached when using framebuffers (see Section 3.7). For

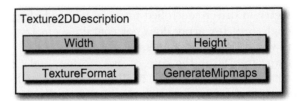

Figure 3.15. A Texture2DDescription describes the immutable state of a texture. A Texture2DDescription is passed to Device.**CreateTexture2D** to describe the resolution, internal format, and mipmaps of the texture to create.

```
public enum TextureFormat
{
  RedGreenBlue8 ,
  RedGreenBlueAlpha8 ,
  Red8 ,
  Red32f ,
  Depth24 ,
  Depth32f ,
  Depth24Stencil8 ,
  // ... See TextureFormat.cs for the full list .
}

public struct Texture2DDescription :
    IEquatable<Texture2DDescription>
{
  public Texture2DDescription(int width , int height ,
                              TextureFormat format );
  // ... Constructor overload with generateMipmaps .

  public int Width { get; }
  public int Height { get; }
  public TextureFormat TextureFormat { get; }
  public bool GenerateMipmaps { get; }

  public bool ColorRenderable { get; }
  public bool DepthRenderable { get; }
  public bool DepthStencilRenderable { get; }
}
```

Listing 3.25. Texture2DDescription interface.

example, if ColorRenderable is false, the texture cannot be attached to a framebuffer's color attachments.

Once a Texture2DDescription is created, it is passed to Device.**Create Texture2D**, to create an actual texture of type Texture2D:

```
Texture2DDescription description = new Texture2DDescription (
    256, 256, TextureFormat.RedGreenBlueAlpha8 );
Texture2D texture = Device.CreateTexture2D(description );
```

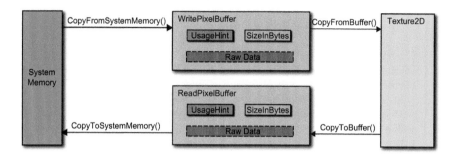

Figure 3.16. Pixel buffers are used to transfer data between system memory and a texture. The method WritePixelBuffer is used to transfer from system memory to a texture, while ReadPixelBuffer goes in the opposite direction. Pixel buffers are untyped, similar to vertex buffers.

The obvious next step is to provide the texture with data. This is done using pixel buffers, as shown in Figure 3.16. There are two types of pixel buffers: WritePixelBuffer, which is used to transfer data from system memory to a texture, and ReadPixelBuffer, for transferring from a texture to system memory. In other words, write pixel buffers write to textures, and read pixel buffers read from textures. Pixel buffers look and feel very much like vertex buffers. They are untyped, meaning they only contain raw bytes, but they provide `CopyFromSystemMemory` and `CopyToSystemMemory` overloads with a generic parameter `T` so client code doesn't have to cast.

Pixel buffers are so similar to vertex buffers that both write and read pixel buffers have almost an identical interface to vertex buffers, shown previously in Listing 3.18. The main difference is that pixel buffers also support methods to copy to and from .NET's Bitmap instead of just arrays. These methods are appropriately named `CopyFromBitmap` and `CopyTo Bitmap`.

One could argue that vertex buffers and pixel buffers should use the same abstract class, or at least write and read pixel buffers should use the same class. We avoid doing so because we prefer the strong typing provided by separate classes. A method that takes a write pixel buffer, like Texture2D.`CopyFromBuffer`, can only take a write pixel buffer. This is checked at compile time. If both pixel-buffer types used the same class, the check would have to be done at runtime, making the method harder to use and less efficient. Although we're not too concerned about the cost of an extra if statement, we are concerned about designing methods that are easy to use correctly and hard to use incorrectly.

Question ○○○○

> If vertex and pixel buffers need to be treated polymorphically, it makes sense to introduce a Buffer abstract base class, with derived abstract classes VertexBuffer, WritePixelBuffer, and ReadPixelBuffer. Methods that want a specific type use the derived class, and methods that work with any buffer use the base class. One use case is to render to a pixel buffer, then interpret this buffer as vertex data. Can you think of another?

Given that pixel buffers work so similarly to vertex buffers, it is no wonder that client code to copy data to pixel buffers looks similar. The example in Listing 3.26 copies two *red, green, blue, alpha* (RGBA) pixels to a pixel buffer.

Client code can also use `CopyFromBitmap` to copy data from a Bitmap to a pixel buffer:

```
Bitmap bitmap = new Bitmap(filename);
WritePixelBuffer pixelBuffer = Device.CreateWritePixelBuffer(
    PixelBufferHint.Stream,
    BitmapAlgorithms.SizeOfPixelsInBytes(bitmap));
pixelBuffer.CopyFromBitmap(bitmap);
```

Given the interface for Texture2D shown in Listing 3.27, it is easy to see that a `CopyFromBuffer` method should be used to copy data from a pixel buffer to a texture. Two arguments to this method are used to interpret the raw bytes stored in the pixel buffer, similar to how a VertexBufferAttribute is used to interpret the raw bytes in a vertex buffer. This requires specifying the format (e.g., RGBA) and data type (e.g., un-

```
BlittableRGBA[] pixels = new BlittableRGBA[]
{
  new BlittableRGBA(Color.Red),
  new BlittableRGBA(Color.Green)
};

int sizeInBytes = ArraySizeInBytes.Size(pixels);
WritePixelBuffer pixelBuffer = Device.CreateWritePixelBuffer(
    PixelBufferHint.Stream, sizeInBytes);
pixelBuffer.CopyFromSystemMemory(pixels);
```

Listing 3.26. Copying data to a WritePixelBuffer.

signed byte) of the data in the pixel buffer. The necessary conversions will take place to convert from this format and data type to the internal texture format that was specified in the description when the texture was created. To copy the entire pixel buffer created above using an array of BlittableRGBAs, the following could be used:

```
texture.CopyFromBuffer(writePixelBuffer,
    ImageFormat.RedGreenBlueAlpha, ImageDatatype.UnsignedByte);
```

Other overloads to CopyFromBuffer allow client code to modify only a subsection of the texture and to specify the row alignment.

```
public enum ImageFormat
{
  DepthComponent,
  Red,
  RedGreenBlue,
  RedGreenBlueAlpha,
  // ... See ImageFormat.cs for the full list.
}

public enum ImageDatatype
{
  UnsignedByte,
  UnsignedInt,
  Float,
  // ... See ImageDatatype.cs for the full list.
}

public abstract class Texture2D : Disposable
{
  public virtual void CopyFromBuffer(
      WritePixelBuffer pixelBuffer,
      ImageFormat format,
      ImageDatatype dataType);
  public abstract void CopyFromBuffer(
      WritePixelBuffer pixelBuffer,
      int xOffset,
      int yOffset,
      int width,
      int height,
      ImageFormat format,
      ImageDatatype dataType,
      int rowAlignment);
  // ... Other CopyFromBuffer overloads

  public virtual ReadPixelBuffer CopyToBuffer(
      ImageFormat format, ImageDatatype dataType)
  // ... Other CopyToBuffer overloads

  public abstract Texture2DDescription Description { get; }
  public virtual void Save(string filename);
}
```

Listing 3.27. Texture2D interface.

Texture2D also has a `Description` property that returns the description used to create the texture. This object is immutable; the resolution, format, and mipmapping behavior of a texture cannot be changed once it is created. Texture2D also contains a method, `Save`, to save the texture to disk, which is useful for debugging.

To simplify texture creation, an overload to Device.`CreateText ure2D` takes a Bitmap. This can be used to create a texture from a file on disk in a single line of code, such as the following:

```
Texture2D texture = Device.CreateTexture2D(new Bitmap(filename),
    TextureFormat.RedGreenBlue8, generateMipmaps);
```

Similar to the `CreateVertexArray` overload that takes a Mesh, this favors ease of use over flexibility, and we use it in many examples throughout this book.

Texture rectangles. In addition to regular 2D textures that are accessed in a shader using normalized texture coordinates (e.g., $(0, width)$ and $(0, height)$ are each mapped to the normalized range $(0, 1)$), our renderer also supports 2D texture rectangles that are accessed using unnormalized texture coordinates in the range $(0, width)$ and $(0, height)$. Addressing a texture in this manner makes some algorithms cleaner, like ray-casting height fields in Section 11.2.3. A texture rectangle still uses Texture2D but is created with Device.`CreateTexture2DRectangle`. Texture rectangles do not support mipmapping and samplers with any kind of repeat wrapping.

3.6.2   Samplers

When a texture is used for rendering, client code must also specify the parameters to use for sampling. This includes the type of filtering that should occur for minification and magnification, how to handle wrapping texture coordinates outside of the $[0, 1]$ range, and the degree of anisotropic filtering. Filtering can affect quality and performance. In particular, anisotropic filtering is useful for improving the visual quality for horizon views of textured terrain in virtual globes. Texture wrapping has several uses, including tiling detail textures for terrain shading, as discussed in Section 11.4.

Like Direct3D and recent versions of OpenGL, our renderer decouples textures and samplers. Textures are represented using Texture2D, and samplers are represented using TextureSampler, shown in Listing 3.28 and Figure 3.17. Client code can explicitly create a sampler using Device.`Create Texture2DSampler`:

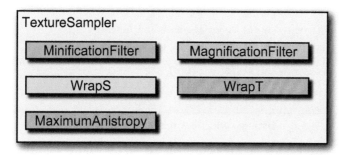

Figure 3.17. A TextureSampler describes sampling state, including filtering and wrap modes.

```
TextureSampler sampler = Device.CreateTexture2DSampler(
    TextureMinificationFilter.Linear,
    TextureMagnificationFilter.Linear,
    TextureWrap.Repeat,
    TextureWrap.Repeat);
```

The Device also contains a collection of common samplers, so the above call to **CreateTexture2DSampler** could be replaced with just the following:

```
TextureSampler sampler = Device.Samplers.LinearRepeat;
```

The latter approach has the benefit of not creating an additional renderer object, and thus, another GL object.

```
public enum TextureMinificationFilter
{
    Nearest,
    Linear,
    // .. Mipmapping filters
}

public enum TextureMagnificationFilter
{
    Nearest,
    Linear
}

public enum TextureWrap
{
    Clamp,
    Repeat,
    MirroredRepeat
}
```

```
public abstract class TextureSampler : Disposable
{
  public TextureMinificationFilter  MinificationFilter { get; }
  public TextureMagnificationFilter MagnificationFilter { get; }
  public TextureWrap WrapS { get; }
  public TextureWrap WrapT { get; }
  public float MaximumAnistropy { get; }
}
```

Listing 3.28. TextureSampler interface.

Try This ○○○○

> The property Device.**Samplers** contains four premade samplers: **NearestClamp**, **LinearClamp**, **NearestRepeat**, and **LinearRepeat**. Would a sampler cache similar to the shader cache in Section 3.4.6 be more useful? If so, design and implement it. If not, why?

3.6.3   Rendering with Textures

Given a Texture2D and a TextureSampler, it is a simple matter to tell the Context we want to use them for rendering. In fact, we can tell the context we want to use multiple textures, perhaps each with a different sampler, for the same draw call. This ability for shaders to read from multiple textures is called multitexturing. It is widely used in general, and throughout this book; for example, Section 4.2.5 uses multitexturing to shade the side of the globe in sunlight with a day texture and the other side with a night texture, and Section 11.4 uses multitexturing in a variety of terrain-shading techniques. In addition to reading from multiple textures, a shader can also read multiple times from the same texture.

The number of texture units, and thus, the maximum number of unique texture/sampler combinations that can be used at once, is defined by Device.**NumberOfTextureUnits**. Context contains a collection of Text ureUnits, as shown in Figure 3.18. Each texture unit is accessed with an index between 0 and Device.**NumberOfTextureUnits** − 1. Before calling Context.Draw, client code assigns a texture and sampler to each texture unit it needs. For example, if day and night textures are used, the client code may look like the following:

```
context.TextureUnits[0].Texture = dayTexture;
context.TextureUnits[0].TextureSampler =
   Device.TextureSamplers.LinearClamp;
context.TextureUnits[1].Texture = nightTexture;
```

```
context.TextureUnits[1].TextureSampler =
    Device.TextureSamplers.LinearClamp;
context.Draw(/* ... */);
```

The GLSL shader can define two sampler2D uniforms to access the texture units using the renderer's automatic uniforms:

```
uniform sampler2D og_texture0;   // day   -  texture  unit  0
uniform sampler2D og_texture1;   // night  -  texture  unit  1
```

Alternatively, custom-named uniforms can be used and explicitly set to the appropriate texture unit in client code using Uniform<int>.

Why is the collection of texture units a part of the context and not part of the draw state? What use cases does this make easier? What are the downsides?

○○○○ Question

It is often useful to compute mipmaps offline when more time can be spent on high-quality filtering, instead of at runtime with an API call like **glGenerateMipmap**. Modify Texture2D and related classes to support precomputed mipmaps.

○○○○ Try This

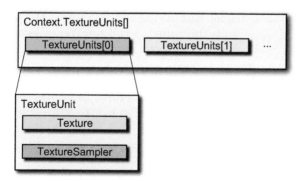

Figure 3.18. Each Context has Device.**NumberOfTextureUnits** texture units. A texture and a sampler are assigned to a texture unit for rendering.

Try This ○○○○

> Our renderer only supports 2D textures because, believe it or not, 2D textures were the only texture type needed to write the example code for this entire book! There are many other useful texture types: 1D textures, 3D textures, cube maps, compressed textures, and texture arrays. Adding renderer support for these types is a straightforward extension of the existing design. Have at it.

3.6.4   GL Renderer Implementation

The GL implementation for textures is spread across several classes. WritePixelBufferGL3x and ReadPixelBufferGL3x contain the implementations for write and read pixel buffers, respectively. These classes use the same GL buffer object patterns used by vertex buffers, described in Section 3.5.1. Namely, `glBufferData` is used to allocate memory in the constructor, and `CopyFromSystemMemory` is implemented with `glBufferSubData`. The pixel- and vertex-buffer implementations actually share code under the hood. One notable difference is that write pixel buffers use a target of `GL_PIXEL_UNPACK_BUFFER`, and read pixel buffers use a target of `GL_PIXEL_PACK_BUFFER`.

Texture2DGL3x is the the implementation for an actual 2D texture. Its constructor creates the GL texture and allocates memory for it. First, a GL name is created using `glGenTexture`. Next, a call to `glBindBuffer` is made, with a target of `GL_PIXEL_UNPACK_BUFFER` and a buffer of zero. This ensures that no buffer object is bound that could provide the texture with data. Then, `glActiveTexture` is used to activate the last texture unit; for example, if Device.`NumberOfTextureUnits` is 32, it will activate texture unit 31. This somewhat nonintuitive step is required so that we know what texture unit is affected and can account for it before the next draw call if client code wants to render with a different texture on the same texture unit. Once the GL state is configured, a call to `glTexImage2D` with a null data argument finally allocates memory for the texture.

The implementation for `CopyFromBuffer` has some similarities to the constructor. First, a call to `glBindBuffer` is made to bind the write pixel buffer (unpack pixel-buffer object in GL terms), which has the image data we wish to copy to the texture. Next, a call to `glPixelStore` sets the row alignment. Then, `glTexSubImage2D` is used to transfer the data. If the texture is to be mipmapped, a final call to `glGenerateMipmap` generates mipmaps for the texture.

TextureSamplerGL3x is the implementation for texture samplers. Given that this class is immutable, and given the clean GL API for sampler objects, TextureSamplerGL3x has one of the cleanest implementations in our renderer. A GL name for the sampler is created in the constructor with **glGenSamplers** and deleted later with **glDeleteSamplers** when the object is disposed. The constructor calls **glSamplerParameter** several times to define the sampler's filtering and wrapping parameters.

Finally, texture units are implemented in TextureUnitsGL3x and TextureUnitGL3x. It is the texture unit's responsibility to bind textures and samplers for rendering. A delayed approach is used, similar to the approach used for uniforms and vertex arrays. When client code assigns a texture or sampler to a texture unit, the texture unit is not activated with **glActiveTexture**, nor are the texture or sampler GL objects bound immediately. Instead, the texture unit is marked as dirty. When a call to Context.**Draw** is made, it iterates over the dirty texture units and cleans them. That is, it calls **glActiveTexture** to activate the texture unit and **glBindTexture** and **glBindSampler** to bind the texture and sampler, respectively. The last texture unit, which may have been modified in Texture2D. **CopyFromBuffer**, is treated as a special case and explicitly binds its texture and sampler if they are not null.

3.6.5 Textures in Direct3D

Textures in D3D have some similarities to our renderer interfaces. In D3D, a description, **D3D11_TEXTURE2D_DESC**, is used to create a texture with ID3D11Device::**CreateTexture2D**. A difference is that the usage hint (e.g., static, dynamic, or stream) is specified in the D3D description; in our renderer, it is only specified as part of the write or read pixel buffer. Our texture's **CopyFromBuffer** and **CopyToBuffer** can be implemented with ID3D11DeviceContext::**CopyResource**. It is also possible to implement a write pixel buffer using just an array in system memory and then implement the texture's **CopyFromBuffer** using ID3D11DeviceContext::**UpdateSubresource** or ID3D11DeviceContext::**Map**. Our pixel buffers don't map directly to D3D, so it may be better to do some redesign rather than to complicate the D3D implementation.

Since our renderer's sampler is immutable, D3D samplers map quite nicely to ours. In D3D, sampler parameters are described using a **D3D11_SAMPLER_DESC**, which is passed to ID3D11Device::**CreateSamplerState** to create an immutable sampler. A nice feature of D3D is that if a sampler already exists with the same description that is passed to **CreateSamplerState**, the existing sampler is returned, making a renderer or application-level cache less important.

To access data in a texture for rendering, D3D requires a shader-resource view to be created using ID3D11Device::**CreateShaderResourceView**. The texture's resource view and sampler can be bound to different stages of the pipeline on slots (texture units) using ID3D

`llDeviceContext` methods like `PSSetShaderResources` and `PSSetSamp`
`lers`. Similar to D3D shaders themselves, resource views and samplers
are bound to individual pipeline stages, whereas in GL and our renderer,
they are bound to the entire pipeline.

In D3D, the origin for texture coordinates is the upper left corner, and
rows go from the the top of the image to the bottom. In GL, the origin is
the lower left corner, and rows go from the bottom to the top. This doesn't
cause a problem when reading textures in a shader; the same texture data
can be used with the same texture coordinates with both APIs. However, this origin difference does cause a problem when writing to a texture
attached to a framebuffer. In this case, rendering needs to be "flipped"
by reconfiguring the viewport transform, depth range, culling state, and
projection matrix and utilizing ARB_fragment_coord_conventions [84].

3.7 Framebuffers

The last major component in our renderer is the framebuffer. Framebuffers
are containers of textures used for render to texture techniques. Some of
these techniques are quite simple. For example, the scene might be rendered
to a high-resolution texture attached to a framebuffer that is then saved
to disk, to implement taking screen captures at a resolution much higher
than the monitor's resolution.[22] Other techniques using framebuffers are
more involved. For example, deferred shading writes to multiple textures
in the first rendering pass, outputting information such as depth, normals,
and material properties. In the second pass, a full screen quad is rendered
that reads from these textures to shade the scene. Many techniques using
framebuffers will write to textures in the first pass, then read from them
in a later pass.

```
public abstract class ColorAttachments
{
    public abstract Texture2D this [int index] { get; set; }
    public abstract int Count { get; }
    // ...
}

public abstract class Framebuffer : Disposable
{
    public abstract ColorAttachments ColorAttachments { get; }
    public abstract Texture2D DepthAttachment { get; set; }
    public abstract Texture2D DepthStencilAttachment { get; set; }
}
```

Listing 3.29. ColorAttachments and Framebuffer interfaces.

---

[22]This is exactly how we created many figures in this book.

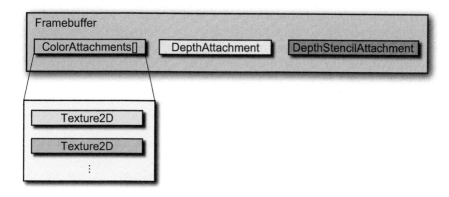

Figure 3.19. A Framebuffer can have up to Device.`MaximumNumberOfColor Attachments` color attachments and a depth or depth/stencil attachment. All attachments are Texture2D objects.

In our renderer, framebuffers are represented by Framebuffer, shown in Listing 3.29 and Figure 3.19. Clients can create framebuffers with Context.`CreateFramebuffer`. Like vertex arrays, framebuffers are lightweight container objects and are not sharable among contexts, which is why `CreateFramebuffer` is a method on Context, not Device. Once a framebuffer is created, we can attach multiple textures as color attachments and another texture as a depth or depth/stencil attachment.

It is important that the texture's format is compatible with the framebuffer attachment type, which is why Texture2DDescription contains `ColorRenderable`, `DepthRenderable`, and `DepthStencilRenderable` properties mentioned in Section 3.6.1. The example on the next page creates a framebuffer with a color and depth attachment.

We render to a framebuffer by assigning it to Context.`Framebuffer`. If the depth-testing renderer state is enabled, the framebuffer must have a depth or depth/stencil attachment. Forgetting a depth attachment is a common OpenGL beginner mistake. In this case, our renderer throws an exception in Context.`Draw`. If only OpenGL was so informative!

```
Framebuffer framebuffer = context.CreateFramebuffer();
framebuffer.ColorAttachments[0] = Device.CreateTexture2D(
    new Texture2DDescription(640, 480,
        TextureFormat.RedGreenBlue8, false));
framebuffer.DepthAttachment = Device.CreateTexture2D(
    new Texture2DDescription(640, 480,
        TextureFormat.Depth32f, false));
```

When multiple color attachments are used, it is important to match up the fragment shader's output variable with the framebuffer's color attachment's index, similar to how vertex-shader attribute inputs are matched up with vertex-array attributes. For example, a fragment shader may have two output variables:

```
out vec4 dayColor;
out vec4 nightColor;
```

As described in Section 3.4.3, the color attachment indices for the fragment-shader output variables can be queried with ShaderProgram.**FragmentOutputs**. This index can then be used to assign a texture to the appropriate color attachment, associating the shader's output variable with the texture:

```
ShaderProgram sp = // ...
framebuffer.ColorAttachments[sp.FragmentOutputs("dayColor")] =
    dayTexture;
framebuffer.ColorAttachments[sp.FragmentOutputs("nightColor")] =
    nightTexture;
```

After a texture attached to a framebuffer is written to, it can be assigned to a texture unit and read in a later rendering pass, but it is not possible to both write to and read from the same texture during a single draw call.

We could move the **Framebuffer** property from Context to Draw State to reduce the amount of context-level global state. It is convenient to keep the framebuffer as part of the context because it allows objects to render without having to know which framebuffer they are rendering to. Although this design relies on context-level state, a framebuffer is much easier to manage than the entire renderer state.

3.7.1 GL Renderer Implementation

The GL implementation for framebuffers is in FramebufferGL3x and ColorAttachmentsGL3x. A name for the GL framebuffer object is created and deleted with **glGenFramebuffers** and **glDeleteFramebuffers**. As we've seen throughout this chapter, a delayed technique is used so no other GL calls are made until client code calls Context.**Draw** or Context.**Clear**. Many objects like shaders and textures do not affect clears, so it is important to remember that clears depend on the framebuffer.

As usual, when client code assigns textures to framebuffer attachments, the attachment is marked as dirty. Before drawing or clearing, the framebuffer is bound with `glBindFramebuffer`, and the attachments are cleaned by calling `glFramebufferTexture`. If no texture is attached to a dirty attachment (e.g., because it was assigned null), then `glFramebuffer Texture` is called with a texture argument of zero. Likewise, if Context. `Framebuffer` was assigned null, `glBindFramebuffer` is called with a framebuffer argument of zero, which means the default framebuffer is used.

3.7.2   Framebuffers in Direct3D

In D3D, render-target and depth/stencil views can be used to implement our renderer's Framebuffer. A render-target view can be created for a color attachment using ID3D11Device::`CreateRenderTargetView`. Likewise, a depth/stencil view can be created for the depth/stencil attachment using ID3D11Device::`CreateDepthStencilView`. For rendering, these views can be bound to the output-merger pipeline stage using ID3D11Device Context::`OMSetRenderTargets`.

Given that D3D treats color output (render targets) and depth/stencil output separately, implementing our Context.`Clear` requires calling both ID3D11DeviceContext::`ClearRenderTargetView` and ID3D11DeviceContext :: `ClearDepthStencilView` if both color and depth/stencil buffers are to be cleared.

3.8   Putting It All Together: Rendering a Triangle

Chapter03Triangle is an example that uses the renderer to render a single red triangle. It pulls together many of the concepts and code snippets in this chapter into an example you can run, step through in the debugger, and most importantly, build your own code upon. The Chapter03Triangle project references `OpenGlobe.Core` and `OpenGlobe.Renderer`. It has no need to call OpenGL directly; all rendering is done through publicly exposed renderer types.

Chapter03Triangle contains just a single class. Its private members are shown in Listing 3.30. All that is needed is a window to render to and the state types used for issuing a draw call. Of course, the draw state includes

```
private readonly GraphicsWindow _window;
private readonly SceneState _sceneState;
private readonly ClearState _clearState;
private readonly DrawState _drawState;
```

Listing 3.30. Chapter03Triangle private members.

a shader program, vertex array, and render state, but these are not directly held as members in the example.

The bulk of the work is done in the constructor shown in Listing 3.31. It creates a window and hooks up events for window resizing and rendering. A window needs to be created before other renderer types because a window creates the GL context under the hood.

Question ○○○○

> How would you change the GL renderer implementation to remove this restriction that a window needs to be created before other renderer types? Is it worth the effort?

A default scene state and clear state are created. Recall that the scene state contains state used to set automatic uniforms, and the clear state contains state for clearing the framebuffer. For this example, the defaults are acceptable. In the scene state, the camera's default position is $(0, -1, 0)$, and it is looking at the origin with an up vector of $(0, 0, 1)$, so it is looking at the xz-plane with x to the right and z being up.

A shader program consisting of simple vertex and fragment shaders is created next. The shader sources are hard coded in strings. If they were more complicated, we would store them in separate files as embedded resources, as described in Section 3.4. The vertex shader transforms the input position using the automatic uniform `og_modelViewPerspectiveMatrix`, so the example doesn't have to explicitly set the transformation matrix. The fragment shader outputs a solid color based on the `u_color` uniform, which is set to red.

Next, a vertex array for an isosceles right triangle in the xz-plane, with legs of unit length is created by first creating a Mesh, then calling Context.`CreateVertexArray`, similar to Listing 3.24.

Then, a render state is created with disabled facet culling and depth testing. Even though this is the default state in OpenGL, these members default to true in our renderer since that is the common use case. This example disables them for demonstration purposes; they could also be left alone without affecting the output.

Finally, a draw state is created that contains the shader program, vertex array, and render state previously created. A camera helper method, `ZoomToTarget`, adjusts the camera's position so the triangle is fully visible.

Once all the renderer objects are created, rendering is simple, as shown in Listing 3.32. The framebuffer is cleared, and a draw call draws the triangle, instructing the renderer to interpret the vertices as triangles.

```
public Triangle()
{
  _window = Device.CreateWindow(800, 600,
                                "Chapter 3:   Triangle");
  _window.Resize += OnResize;
  _window.RenderFrame += OnRenderFrame;
  _sceneState = new SceneState();
  _clearState = new ClearState();

  string vs =
      @"in vec4 position;
        uniform mat4 og_modelViewPerspectiveMatrix;
        void main() { gl_Position = og_modelViewPerspectiveMatrix←
            *
                                    position; }";
  string fs =
      @"out vec3 fragmentColor;
        uniform vec3 u_color;
        void main() { fragmentColor = u_color; }";
  ShaderProgram sp = Device.CreateShaderProgram(vs, fs);
  ((Uniform<Vector3F>)sp.Uniforms["u_color"]).Value =
      new Vector3F(1, 0, 0);

  Mesh mesh = new Mesh();
  VertexAttributeFloatVector3 positionsAttribute =
      new VertexAttributeFloatVector3("position", 3);
  mesh.Attributes.Add(positionsAttribute);
  IndicesUnsignedShort indices = new IndicesUnsignedShort(3);
  mesh.Indices = indices;
  IList<Vector3F> positions = positionsAttribute.Values;
  positions.Add(new Vector3F(0, 0, 0));
  positions.Add(new Vector3F(1, 0, 0));
  positions.Add(new Vector3F(0, 0, 1));
  indices.AddTriangle(new TriangleIndicesUnsignedShort(0, 1, 2));
  VertexArray va = _window.Context.CreateVertexArray(
      mesh, sp.VertexAttributes, BufferHint.StaticDraw);

  RenderState renderState = new RenderState();
  renderState.FacetCulling.Enabled = false;
  renderState.DepthTest.Enabled = false;

  _drawState = new DrawState(renderState, sp, va);

  _sceneState.Camera.ZoomToTarget(1);
}
```

Listing 3.31. Chapter03Triangle's constructor creates renderer objects required to render the triangle.

```
private void OnRenderFrame()
{
  Context context = _window.Context;
  context.Clear(_clearState);
  context.Draw(PrimitiveType.Triangles, _drawState, _sceneState);
}
```

Listing 3.32. Chapter03Triangle's `OnRenderFrame` issues the draw call that renders the triangle.

Try This ○○○○

> Change the PrimitiveType.**Triangles** argument to PrimitiveType. **LineLoop**. What is the result?

When issuing the draw call, we could also explicitly state the index offset and count, like the following, which uses an offset of zero and a count of three:

```
context.Draw(PrimitiveType.Triangles, 0, 3,
    _drawState, _sceneState);
```

Of course, for this index buffer containing three indices, this has the same result as not providing an offset and count; it draws using the entire index buffer. Although Chapter03Triangle is a simple example, the rendering code for many examples is nearly this easy: issue Context.**Draw** and let the renderer bind the shader program, set automatic uniforms, bind the vertex array, set the render states, and finally draw! The only difference is that more advanced examples may also assign the framebuffer, textures, and samplers and set shader uniforms. Rendering is usually concise; the bulk of the work is in creating and updating renderer objects.

Try This ○○○○

> Modify Chapter03Triangle to apply a texture to the triangle. Use the Device.**CreateTexture2D** overload that takes a Bitmap to create a texture from an image on disk. Add texture coordinates to the mesh using VertexAttributeHalfFloatVector2. Assign the texture and a sampler to a texture unit, and modify the vertex and fragment shader accordingly.

Try This ○○○○

> Modify Chapter03Triangle to render to a framebuffer, then render the framebuffer's color attachment to the screen using a textured full screen quad. Create the framebuffer using Context.**CreateFramebuffer**, and assign a texture that is the same size as the window to one of the framebuffer's color attachments. Consider using ViewportQuad in **Open Globe.Scene** to render the quad that reads the texture and renders to the screen.

3.9 Resources

A quick Google search will reveal a plethora of information on OpenGL and Direct3D. The OpenGL SDK[23] and the DirectX Developer Center[24] are among the best references.

It is instructive to look at different renderer designs. Eric Bruneton's Ork[25] (OpenGL rendering kernel) is a C++ API built on top of OpenGL. It is fairly similar to our renderer but also includes a resource framework, scene graph, and task graph, making it closer to a full 3D engine.

Wojciech Sterna's *GPU Pro* article discusses the differences between D3D 9 and OpenGL 2 [157]. The source code for the Blossom Engine, available on the *GPU Pro* website,[26] includes a renderer that supports both D3D 9 and GL 2 using preprocessor defines, avoiding the virtual-call overhead of using abstract classes like our design. Although this couples the implementations, it is instructive to see the differences between D3D and GL implementations right next to each other.

David Eberly's Wild Magic Engine[27] contains an abstract renderer with both D3D 9 and GL implementations. The engine has been used and described in many of his books, including *3D Game Engine Design* [43].

OGRE[28] is a popular open source 3D engine that supports both OpenGL and Direct3D via an abstract RenderSystem class.

The web has good articles on sorting by state before rendering [47, 52].

Two articles in *Game Engine Gems 2* go into detail on the delayed GL techniques used throughout this chapter and the approach for automatic uniforms used in Section 3.4.5 [31, 32].

Later in this book, Section 10.4 describes multithreading with OpenGL and related abstractions in our renderer.

> Perhaps the ultimate exercise for this chapter is to write a D3D implementation of the OpenGlobe renderer. SlimDX, a DirectX .NET wrapper, can be used to make Direct3D calls from C#.[29] A good first step is to make the GL implementation work more like D3D by checking out the following extensions, which are core features since OpenGL 3.2: ARB_vertex_array_bgra [86], ARB_fragment_coord_conventions, and ARB_provoking _vertex [85]. How will you handle shaders?

○○○○ **Try This**

---

[23] http://www.opengl.org/sdk/
[24] http://msdn.microsoft.com/en-us/directx/
[25] http://ork.gforge.inria.fr/
[26] http://www.akpeters.com/gpupro/
[27] http://www.geometrictools.com/
[28] http://www.ogre3d.org/

Try This ○○○○

> Write an OpenGL ES 2.0 implementation of the OpenGlobe renderer. How much code from the GL 3.x implementation can be reused? How do you handle lack of geometry shaders?

Patrick Says ○○○○

> If you embark on either of the above adventures, we would love to hear about it.

Globe Rendering

This chapter covers a central topic in virtual globes: rendering a globe, with a focus on ellipsoid tessellation and shading algorithms.

We cover three tessellation algorithms used to produce triangles approximating a globe's surface. First, we discuss a simple subdivision-surfaces algorithm for a unit sphere that is typical of an introductory computer graphics course. Then, we cover ellipsoid tessellations based on a cube map and, finally, the geodetic grid.

Our shading discussion begins by reviewing simple per-fragment lighting and texture mapping with a low-resolution satellite-derived image of Earth. We compare per-fragment procedural generation of normals and texture coordinates to precomputed normals and texture coordinates. Specific virtual globe techniques for rendering a latitude-longitude grid and night lights using fragment shaders finish the shading discussion.

The chapter concludes with a technique for rendering a globe without tessellation by using GPU ray casting.

4.1 Tessellation

GPUs primarily render triangles, so to render an ellipsoid-based globe, triangles approximating the ellipsoid's surface are computed in a process called *tessellation* and then fed to the GPU for rendering. This section covers three popular ellipsoid tessellation algorithms and the trade-offs between them.

4.1.1 Subdivision Surfaces

Our first algorithm tessellates a unit sphere centered at the origin. This easily extends to ellipsoids by scaling each point by the ellipsoid's radii,

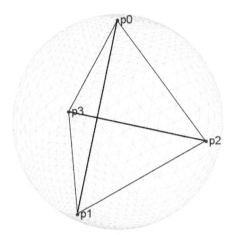

Figure 4.1. A tetrahedron with endpoints on the unit sphere. The faces of the tetrahedron are recursively subdivided and projected onto the unit sphere to approximate the sphere's surface.

(a, b, c), and then computing geodetic surface normals (see Section 2.2.3) for shading.

The subdivision-surfaces algorithm is concise and easy to implement and doesn't even require any trigonometry. Start with a *regular tetrahedron*, that is, a pyramid composed of four equilateral triangles with endpoints that lie on the unit sphere, as shown in Figure 4.1.

The tetrahedron endpoints, $p0$, $p1$, $p2$, and $p3$, are defined as

$$p0 = (0, 0, 1),$$
$$p1 = \frac{(0, 2\sqrt{2}, -1)}{3},$$
$$p2 = \frac{(-\sqrt{6}, -\sqrt{2}, -1)}{3}, \tag{4.1}$$
$$p3 = \frac{(\sqrt{6}, -\sqrt{2}, -1)}{3}.$$

The tetrahedron is a very coarse polygonal approximation to a unit sphere. In order to get a better approximation, each triangle is first subdivided into four new equilateral triangles, as shown in Figures 4.2(a) and 4.2(b). For a given triangle, introducing four new triangles creates three new points, each of which is a midpoint of the original triangle's edge:

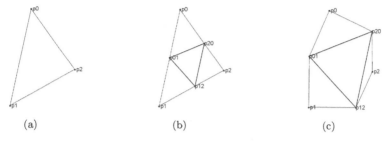

(a)   (b)   (c)

Figure 4.2. Subdividing a triangle and projecting new points onto the unit sphere. This process is repeated recursively until the tessellation sufficiently approximates the sphere's surface. (a) A triangle with its endpoints on the unit sphere. (b) A triangle is subdivided into four new triangles. (c) New points are normalized so they lie on the unit sphere.

$$p01 = \frac{(p0 + p1)}{2},$$
$$p12 = \frac{(p1 + p2)}{2}, \quad\quad\quad (4.2)$$
$$p20 = \frac{(p2 + p0)}{2}.$$

Splitting a triangle into four new triangles alone does not improve the approximation; the midpoints are in the same plane as the original triangle and, therefore, not on the unit sphere. To project the midpoints on the unit sphere and improve our approximation, simply normalize each point:

$$p01 = \frac{p01}{\|\mathbf{p01}\|},$$
$$p12 = \frac{p12}{\|\mathbf{p12}\|}, \quad\quad\quad (4.3)$$
$$p20 = \frac{p20}{\|\mathbf{p20}\|}.$$

The geometric result is shown in Figure 4.2(c).

This choice of subdivision produces nearly equilateral triangles.[1] Other subdivision choices, such as creating three new triangles using the original triangle's centroid, create long, thin triangles that can adversely affect shading due to interpolation over a long distance.

Subdivision continues recursively until a stopping condition is satisfied. A simple stopping condition is the number of subdivisions, n, to perform.

---

[1] After normalization, the triangles are not all equilateral.

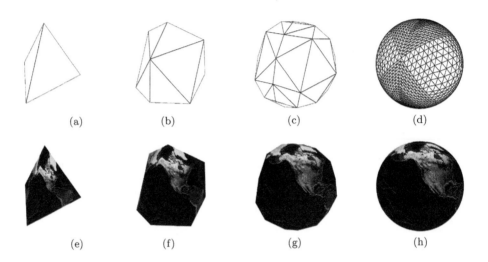

Figure 4.3. (a) and (e) $n = 0$ yields four triangles. (b) and (f) $n = 1$ yields 16 triangles. (c) and (g) $n = 2$ yields 64 triangles. (d) and (h) $n = 5$ yields 4,096 triangles.

For $n = 0$, four (4^1) triangles representing the original tetrahedron are produced; for $n = 1$, 16 (4^2) triangles are produced; for $n = 2$, 64 (4^3) triangles are produced. In general, n subdivisions yield 4^{n+1} triangles. Figure 4.3 shows the wireframe and textured globe for various values of n.

Larger values of n result in a closer approximation to a sphere at the cost of extra computation time and more memory. When n is too small, the visual quality is affected, because the underlying polygonal approximation of the sphere is seen. When n is too large, performance suffers because of extra computation and memory usage. Fortunately, n does not need to be the same for all viewer locations. When the viewer is near to the sphere, a higher value of n can be used. Likewise, when the viewer is far from the sphere, and thus the sphere covers fewer pixels, a lower value of n can be used. This general technique, called *level of detail*, is discussed in Section 12.1.

The tetrahedron is one of five *platonic solids*. Platonic solids are convex objects composed of equivalent faces, where each face is a congruent convex regular polygon. We use a tetrahedron because it is the simplest, having only four faces, but any platonic solid can be subdivided into a sphere using the same algorithm. Each has different trade-offs: platonic solids with more faces require less subdivision. Example code and discussion on subdividing an octahedron (eight faces) and an icosahedron (20 faces) is provided by Whitacre [179] and Laugerotte [95].

4.1.2   Subdivision-Surfaces Implementation

OpenGlobe includes example implementations of the globe-tessellation algorithms described in this chapter. Subdivision-surface tessellation for a unit sphere is implemented in SubdivisionSphereTessellatorSimple. The public interface, shown in Listing 4.1, is a static class with one method named Compute that takes the number of subdivisions and returns an indexed triangle list representing the sphere's approximation.

The first order of business for Compute is to initialize a Mesh object used to return the triangle list. This mesh, initialized in Listing 4.2, simply contains a collection of position attributes and a collection of indices. The tessellation is performed such that the winding order for front-facing triangles is counterclockwise.

For efficiency, the expected number of vertices and triangles are computed and used to set the capacity for the positions and indices collections, respectively. Doing so avoids unnecessary copying and memory allocations.

Listing 4.3 shows the real work of Compute, which is to create the initial tetrahedron using the four endpoints from Equation (4.1). Once these points are added to the position attribute collection, a call is made to Subdivide for each of the four triangles that make up the initial tetrahedron. The method Subdivide is a private method that recursively subdivides a triangle. The indices passed to each Subdivide call define a counterclockwise winding order (see Figure 4.1).

```
public  static  class  SubdivisionSphereTessellatorSimple
{
    public  static  Mesh  Compute(int  numberOfSubdivisions)
    {  /* ... */  }
}
```

Listing 4.1. SubdivisionSphereTessellatorSimple public interface.

```
Mesh  mesh  =  new  Mesh();
mesh.PrimitiveType  =  PrimitiveType.Triangles;
mesh.FrontFaceWindingOrder  =  WindingOrder.Counterclockwise;

VertexAttributeDoubleVector3  positionsAttribute  =
    new  VertexAttributeDoubleVector3(
        "position",  SubdivisionUtility.NumberOfVertices(
            numberOfSubdivisions)
mesh.Attributes.Add(positionsAttribute);

IndicesInt  indices  =  new  IndicesInt(3  *
    SubdivisionUtility.NumberOfTriangles(numberOfSubdivisions));
mesh.Indices  =  indices;
```

Listing 4.2. SubdivisionSphereTessellatorSimple.Compute mesh initialization.

```
double negativeRootTwoOverThree = -Math.Sqrt(2.0) / 3.0;
const double negativeOneThird = -1.0 / 3.0;
double rootSixOverThree = Math.Sqrt(6.0) / 3.0;

IList<Vector3d> positions = positionsAttribute.Values;
positions.Add(new Vector3d(0, 0, 1));
positions.Add(new Vector3d(
    0, 2.0 * Math.Sqrt(2.0) / 3.0, negativeOneThird));
positions.Add(new Vector3d(
    -rootSixOverThree, negativeRootTwoOverThree, negativeOneThird↵
    ));
positions.Add(new Vector3d(
    rootSixOverThree, negativeRootTwoOverThree, negativeOneThird)↵
    );

Subdivide(positions, indices,
    new TriangleIndices<int>(0, 1, 2),numberOfSubdivisions);
Subdivide(positions, indices,
    new TriangleIndices<int>(0, 2, 3), numberOfSubdivisions);
Subdivide(positions, indices,
    new TriangleIndices<int>(0, 3, 1), numberOfSubdivisions);
Subdivide(positions, indices,
    new TriangleIndices<int>(1, 3, 2), numberOfSubdivisions);

return mesh;
```

Listing 4.3. SubdivisionSphereTessellatorSimple.Compute initial tetrahedron.

```
private static void Subdivide(IList<Vector3d> positions,
    IndicesInt indices, TriangleIndices<int> triangle, int level)
{
  if (level > 0)
  {
    positions.Add(Vector3d.Normalize((positions[triangle.I0]
        + positions[triangle.I1]) * 0.5));
    positions.Add(Vector3d.Normalize((positions[triangle.I1]
        + positions[triangle.I2]) * 0.5));
    positions.Add(Vector3d.Normalize((positions[triangle.I2]
        + positions[triangle.I0]) * 0.5));
    int i01 = positions.Count - 3;
    int i12 = positions.Count - 2;
    int i20 = positions.Count - 1;
    --level;
    Subdivide(positions, indices, new TriangleIndices<int>(
        triangle.I0, i01, i20), level);
    Subdivide(positions, indices, new TriangleIndices<int>(
        i01, triangle.I1, i12), level);
    Subdivide(positions, indices, new TriangleIndices<int>(
        i01, i12, i20), level);
    Subdivide(positions, indices, new TriangleIndices<int>(
        i20, i12, triangle.I2), level);
  }
  else
  {
    indices.Values.Add(triangle.I0);
    indices.Values.Add(triangle.I1);
    indices.Values.Add(triangle.I2);
  }
}
```

Listing 4.4. SubdivisionSphereTessellatorSimple.Subdivide.

The recursive call to `Subdivide` (see Listing 4.4) is the key step in this algorithm. If no more subdivision is required (i.e., `level = 0`), then the method simply adds three indices for the input triangle to the mesh and returns. Therefore, in the simplest case when the user calls `Subdivision SphereTessellatorSimple.Compute(0)`, `Subdivide` is called once per tetrahedron triangle, resulting in the mesh shown in Figure 4.3(a). If more subdivision is required (i.e., `level > 0`), three new points on the unit sphere are computed by normalizing the midpoint of each triangle edge using Equations (4.2) and (4.3). The current level is decreased, and `Subdivide` is called four more times, each with indices for a new triangle. The four new triangles better approximate the sphere than does the input triangle. Indices for the input triangle are never added to the mesh's indices collection until recursion terminates. Doing so would add unnecessary triangles underneath the mesh.

Run Chapter04SubdivisionSphere1 to see SubdivisionSphereTessellatorSimple in action. Experiment with the number of subdivisions performed, and verify that your output matches Figure 4.3.

○○○○ **Try This**

This implementation introduces duplicate vertices because triangles with shared edges create the same vertex at the edge's midpoint. Improve the implementation to avoid duplicate vertices.

○○○○ **Try This**

4.1.3   Cube-Map Tessellation

An interesting approach to tessellating an ellipsoid is to project a cube with tessellated planar faces onto the ellipsoid's surface. This is called *cube-map tessellation*. Triangles crossing the IDL can be avoided by using an axis-aligned cube centered around the origin in WGS84 coordinates and rotated 45° around z, such that the x and y components for the corners are $(-1, 0)$, $(0, -1)$, $(1, 0)$, and $(0, 1)$.

Cube-map tessellation starts with such a cube. Each face is then tessellated into a regular grid such that the face remains planar but is made up of additional triangles.[2] Cubes with two, four, and eight grid partitions

---

[2]Tessellations like this were also used to improve the quality of specular highlights and spotlights in fixed-function OpenGL when only per-vertex lighting was possible [88].

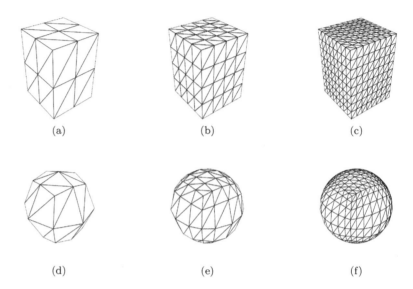

Figure 4.4. Cube-map tessellation tessellates the planar faces of a cube, then projects the points onto an ellipsoid. (a) Two partitions. (b) Four partitions. (c) Eight partitions. (d) Two partitions. (e) Four partitions. (f) Eight partitions.

are shown in Figures 4.4(a), (b), and (c), respectively. More partitions ultimately lead to a better ellipsoid approximation. Each point on the tessellated cube is then normalized to create the unit sphere. An ellipsoid is created by multiplying the normalized point by the ellipsoid's radii. Figures 4.4(d), (e), and (f) show ellipsoids created from two, four, and eight grid partitions.

Normalizing and scaling the original points act as a perspective projection. This results in straight lines on the cube becoming geodesic curves on the ellipsoid, not necessarily lines of constant latitude or longitude. As can be seen in Figure 4.4(f), some distortion occurs, particularly with triangles at a cube face's corner. Grid squares in the middle of the face are not badly distorted though. The distortion is not a showstopper; in fact, Spore used cube-map tessellation for procedurally generated spherical planets, favoring it over the pole distortion of other tessellation algorithms [28].

Like the subdivision-surfaces algorithm, this approach avoids oversampling at the poles but may create triangles that cross over the poles. Conceptually, cube-map tessellation is as simple as other subdivision techniques, and its implementation is only slightly more involved.

An implementation is provided in CubeMapEllipsoidTessellator. Given an ellipsoid shape, the number of grid partitions, and the desired vertex attributes, CubeMapEllipsoidTessellator.`Compute` returns a mesh representing a polygonal approximation to the ellipsoid. First, it creates the eight corner points for the cube. Then it computes positions along each of the cube's 12 edges at a granularity determined by the number of partitions; for example, two partitions would result in one position added to each edge at the edge's midpoint, as in Figure 4.4(a). Next, each face is tessellated row by row, starting with the bottom row. A final pass is made over each point to project the point onto the ellipsoid and optionally compute a normal and texture coordinates.

> Experiment with cube-map tessellation by replacing the call to SubdivisionSphereTessellatorSimple.`Compute` in Chapter04Subdivision Sphere1 with a call to CubeMapEllipsoidTessellator.`Compute`.

○○○○ **Try This**

An alternative implementation for tessellating cube faces is to recursively divide each face into four new faces similar to a quadtree. Each face can serve as a separate tree using chunked LOD (see Chapter 14), as done by Wittens [182].

An alternative to tessellating each face separately is to tessellate a single face, then rotate it into place, and project it onto the ellipsoid. This can be done on the CPU or GPU; for the latter, memory is only required for a single cube's face.

Miller and Gaskins describe an experimental globe-tessellation algorithm for NASA World Wind that is similar to cube-map tessellation [116]. The faces of a cube are mapped onto the ellipsoid with one edge of the cube aligned with the IDL. The four nonpolar faces are subdivided into quadrilaterals representing a rectangular grid along lines of constant latitude and longitude. Note that this is different than cube-map tessellation, which creates geodesic curves.

The polar faces are tessellated differently to minimize issues at the poles. First, a polar face is divided into four regions, similar to placing an X through the face (see Figure 4.5(a)). Afterwards, polar regions are subdivided into four regions at their edge's midpoints (see Figure 4.5(b)). Note that edges in the polar regions are not lines of constant latitude or longitude but instead geodesic curves. Special care is taken to avoid cracks when triangulating quadrilaterals that share an edge with a polar and nonpolar region.

Miller and Gaskins suggest two criteria for selecting a latitude range for polar regions. If the goal is to maximize the number of lines of constant

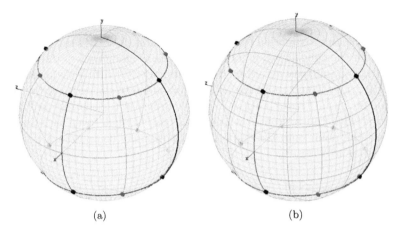

(a)   (b)

Figure 4.5. NASA World Wind experimental globe-tessellation algorithm. (Images courtesy of James Miller, University of Kansas, and Tom Gaskins, NASA [116].)

latitude or longitude, only regions above a high latitude (e.g., 60° or 70°) or below a low latitude should be polar regions. Ultimately, they use a range of above/below +/- 40° for polar regions so all quadrilaterals are approximately the same size.

4.1.4 Geographic-Grid Tessellation

Perhaps the most intuitive ellipsoid-tessellation approach is to use a *geographic grid*. This algorithm has two steps.

First, a set of points distributed over the ellipsoid are computed. This can be done with a nested for loop that iterates over the spherical coordinates $\phi \in [0, \pi]$ and $\theta \in [0, 2\pi]$ at a chosen granularity. Recall that ϕ is similar to latitude and θ is similar to longitude. Most applications use a different granularity for ϕ and θ since their ranges are different.[3] Each grid point is converted from spherical coordinates to Cartesian coordinates using Equation (4.4):

$$
\begin{aligned}
x &= a \cos\theta \sin\phi, \\
y &= b \sin\theta \sin\phi, \\
z &= c \cos\phi.
\end{aligned}
\tag{4.4}
$$

---

[3]For example, GLUT's gluSphere function takes the number of stacks along the z-axis and number of slices around the z-axis. These are used to determine the granularity for ϕ and θ, respectively.

```
IList<Vector3d> positions = // ...
positions.Add(new Vector3d(0, 0, ellipsoid.Radii.Z));

for (int i = 1; i < numberOfStackPartitions; ++i)
{
  double phi = Math.PI * (((double)i) /
                  numberOfStackPartitions);
  double cosPhi = Math.Cos(phi);
  double sinPhi = Math.Sin(phi);

  for (int j = 0; j < numberOfSlicePartitions; ++j)
  {
    double theta = (2.0 * Math.PI) *
                  (((double)j) / numberOfSlicePartitions);
    double cosTheta = Math.Cos(theta);
    double sinTheta = Math.Sin(theta);

    positions.Add(new Vector3d(
        ellipsoid.Radii.X * cosTheta * sinPhi,
        ellipsoid.Radii.Y * sinTheta * sinPhi,
        ellipsoid.Radii.Z * cosPhi));
  }
}
positions.Add(new Vector3d(0, 0, -ellipsoid.Radii.Z));
```

Listing 4.5. Computing points for geographic-grid tessellation.

Here, (a, b, c) is the radii of the ellipsoid, which is $(1, 1, 1)$ in the trivial case of the unit sphere. Since $\sin(0) = \sin \pi = 0$, the north pole ($\phi = 0$) and south pole ($\phi = \pi$) are treated as special cases outside of the for loop. A single point, $(0, 0, \pm c)$, is used for each. This step of the algorithm is shown in Listing 4.5. Note that Math.Cos(theta) and Math.Sin(theta) could be computed once and stored in a lookup table.

Once points for the tessellation are computed, the second step of the algorithm is to compute triangle indices. Indices for triangle strips are generated in between every row, except the rows adjacent to the poles. To avoid

(a)

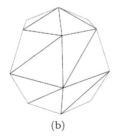
(b)

(c)

Figure 4.6. Geographic-grid tessellation connects a distribution of points on the ellipsoid with triangles. (a) $\Delta\theta = 2\pi/3$, $\Delta\phi = \pi/2$. (b) $\Delta\theta = \pi/3$, $\Delta\phi = \pi/4$. (c) $\Delta\theta = \pi/8$, $\Delta\phi = \pi/8$.

Figure 4.7. Standard geographic-grid tessellation results in too many triangles at the poles.

degenerate triangles, these rows are connected to their closest pole using triangle fans. For a complete implementation, see `GeographicGridEllipsoid Tessellator`.

Figure 4.6(a) shows a very coarse approximation to an ellipsoid with just two rows of triangle fans. In Figure 4.6(b), the nonpolar rows are triangle strips. Figure 4.6(c) begins to look like an ellipsoid.

This algorithm is conceptually straightforward and widely used. It was used in the initial release of NASA World Wind [116] and in STK and Insight3D. It avoids triangles that cross the IDL and produces triangles whose edges are lines of constant latitude and longitude, except, of course, the bisecting edges.

The major weakness of this approach is the singularity created at the poles, shown in Figure 4.7. The thin triangles near the poles can lead to lighting and texturing artifacts. Since fragments on shared edges are redundantly shaded, this can even add fragment-shading costs [139]. The overcrowding of triangles at the poles leads to additional performance issues because view frustum culling is now less effective near the poles (e.g., a view at a pole contains many more triangles than a similar view at the equator).

Over-tessellation at the poles can be dealt with by adjusting the tessellation based on latitude. The closer a row is to a pole, the fewer points, and thus fewer triangles, in between rows are needed. Gerstner suggests scaling by $\sin \phi$ to relate the area of a triangle to the area of the spherical section it projects onto [61].

4.1.5 Tessellation-Algorithm Comparisons

It is hard to pick a clear winner between the three ellipsoid-tessellation algorithms discussed. They are all fairly straightforward to implement and used successfully in a wide array of applications. Table 4.1 lists some

Algorithm	Oversampling at Poles	Triangles Avoid Poles	Triangles Avoid IDL	Similar Shape Triangles	Aligned with Constant Latitude/Longitude Lines
Subdivision Surfaces	No	No	No	Yes	No
Cube Map	No	No	Yes	No	No
Geographic Grid	Yes	Yes	Yes	No	Yes

Table 4.1. Properties of ellipsoid-tessellation algorithms.

properties to consider when choosing among them. This isn't a hard and fast list; algorithm variations can change their properties. For example, geographic-grid tessellation can minimize oversampling at the poles by adjusting the tessellation based on latitude. Likewise, any tessellation can avoid triangles that cross the IDL by splitting triangles that cross it.

4.2 Shading

Tessellation is the first step in globe rendering. The next step is *shading*, that is, simulating the interaction of light and material to produce a pixel's color. First basic lighting and texturing are reviewed, then specific virtual globe techniques for rendering a latitude-longitude grid and night lights are covered.

4.2.1 Lighting

This section briefly examines lighting computations in a vertex and fragment shader used to shade a globe. Starting with the pass-through shaders of Listing 4.6, diffuse and specular terms are incrementally added.

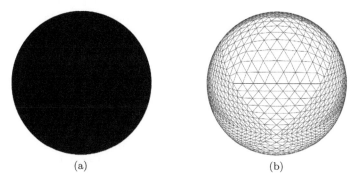

(a) (b)

Figure 4.8. The pass-through shader produces a solid color without texturing or lighting. (a) Solid globe. (b) Wireframe globe.

```
// Vertex shader
in vec4 position;
uniform mat4 og_modelViewPerspectiveMatrix;

void main()
{
    gl_Position = og_modelViewPerspectiveMatrix * position;
}

// Fragment shader
out vec3 fragmentColor;
void main() { fragmentColor = vec3(0.0, 0.0, 0.0); }
```

Listing 4.6. Pass-through shaders.

The vertex shader transforms the input `position` to clip coordinates by multiplying by the automatic uniform, `og_modelViewPerspective`. Like all OpenGlobe automatic uniforms, simply declaring this uniform in a shader is enough to use it. There is no need to explicitly assign it in C# code; it is automatically set in Context.`Draw` based on the values in SceneState.

As you can imagine, these shaders produce a solid black globe, as shown in Figure 4.8(a).

Try This ○○○○

> While reading this section, replace the **vs** and **fs** strings in Sub divisionSphere1's constructor with the vertex and fragment shaders in this section, and see if you can reproduce the figures.

Without lighting, the curvature of the globe is not apparent. To light the globe, we will use a single point light source that is located at the eye position, as if the viewer were holding a lantern next to his or her head. To keep the shaders concise, lighting is computed in world coordinates. In many applications, lighting is computed in eye coordinates. Although this makes some lighting computations simpler since the eye is located at $(0, 0, 0)$, using eye coordinates also requires transforming the world position and normal into eye coordinates. For applications with multiple lights and objects in different model coordinates, this is an excellent approach. Since our example only has a single coordinate space, light is computed in world coordinates for conciseness.

For visual quality, lighting is computed per fragment. This prevents the underlying tessellation of the globe from showing through and eliminates artifacts with specular highlights that are common with per-vertex lighting. Per-fragment lighting also integrates well with texture-based effects discussed in Section 4.2.5.

```
in vec4 position;
out vec3 worldPosition;
out vec3 positionToLight;

uniform mat4 og_modelViewPerspectiveMatrix;
uniform vec3 og_cameraLightPosition;

void main()
{
    gl_Position = og_modelViewPerspectiveMatrix * position;
    worldPosition = position.xyz;
    positionToLight = og_cameraLightPosition - worldPosition;
}
```

Listing 4.7. Diffuse-lighting vertex shader.

```
in vec3 worldPosition;
in vec3 positionToLight;
out vec3 fragmentColor;

void main()
{
    vec3 toLight = normalize(positionToLight);
    vec3 normal = normalize(worldPosition);
    float diffuse = max(dot(toLight, normal), 0.0);
    fragmentColor = vec3(diffuse, diffuse, diffuse);
}
```

Listing 4.8. Diffuse-lighting fragment shader.

In order to compute per-fragment *Phong* lighting, a linear combination of *diffuse*, *specular*, and *ambient* terms determines the intensity of the light in the range $[0, 1]$. First, consider just the diffuse term. A characteristic of rough surfaces, diffuse light scatters in all directions and generally defines the shape of an object. Diffuse light is dependent on the light position and surface orientation but not the view parameters. To compute the diffuse term, the pass-through vertex shader needs to pass the world position and position-to-light vector to the fragment shader, as shown in Listing 4.7.

This vertex shader utilizes og_cameraLightPosition, an automatic uniform representing the position of the light in world coordinates. The fragment shader in Listing 4.8 uses the diffuse term to compute the fragment's intensity.

The dot product is used to approximate the diffuse term, since diffuse light is most intense when the vector from the fragment to the light is the same as the surface normal at the fragment $(\cos(0) = 1)$ and intensity drops off as the angle between the vectors increases (cos approaches zero). The vector positionToLight is normalized in the fragment shader since it may not be unit length due to interpolation between the vertex and fragment shader, even if it were normalized in the vertex shader. Since the globe is

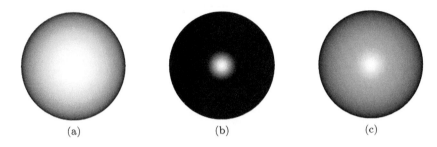

（a）   （b）   （c）

Figure 4.9. Phong lighting with one light located at the camera. (a) Diffuse light scatters in all directions. (b) Specular light highlights the reflection of shiny surfaces. (c) 70% diffuse and 30% specular light is used.

approximated by a sphere in this example, the fragment's surface normal is computed analytically by simply normalizing the world position. The general case of computing the geodetic surface normal for a point on an ellipsoid requires using `GeodeticSurfaceNormal`, introduced in Section 2.2.2. Figure 4.9(a) shows the globe shaded with just diffuse light.

Specular light shows highlights capturing the reflection of shiny surfaces, such as bodies of water. Figure 4.9(b) shows the globe shaded with just specular light, and Figure 4.9(c) shows the globe shaded with both diffuse and specular light. Specular light is dependent on the eye location, so the vertex shader in Listing 4.7 needs to output a vector from the vertex to the eye. Listing 4.9 shows the final vertex shader for Phong lighting. This shader forms the foundation for many shaders throughout this book.

```
in vec4 position;
out vec3 worldPosition;
out vec3 positionToLight;
out vec3 positionToEye;

uniform mat4 og_modelViewPerspectiveMatrix;
uniform vec3 og_cameraEye;
uniform vec3 og_cameraLightPosition;

void main()
{
    gl_Position = og_modelViewPerspectiveMatrix * position;
    worldPosition = position.xyz;
    positionToLight = og_cameraLightPosition - worldPosition;
    positionToEye = og_cameraEye - worldPosition;
}
```

Listing 4.9. Vertex shader for diffuse and specular lighting.

```
in vec3 worldPosition;
in vec3 positionToLight;
in vec3 positionToEye;
out vec3 fragmentColor;

uniform vec4 og_diffuseSpecularAmbientShininess;

float LightIntensity(vec3 normal, vec3 toLight, vec3 toEye,
                     vec4 diffuseSpecularAmbientShininess)
{
  vec3 toReflectedLight = reflect(-toLight, normal);

  float diffuse = max(dot(toLight, normal), 0.0);
  float specular = max(dot(toReflectedLight, toEye), 0.0);
  specular = pow(specular, diffuseSpecularAmbientShininess.w);

  return (diffuseSpecularAmbientShininess.x * diffuse) +
         (diffuseSpecularAmbientShininess.y * specular) +
          diffuseSpecularAmbientShininess.z;
}

void main()
{
  vec3 normal = normalize(worldPosition);
  float intensity = LightIntensity(normal,
      normalize(positionToLight), normalize(positionToEye),
      og_diffuseSpecularAmbientShininess);
  fragmentColor = vec3(intensity, intensity, intensity);
}
```

Listing 4.10. Final Phong-lighting fragment shader.

The specular term is approximated in the fragment shader by reflecting the vector to the light around the surface normal and computing the dot product of this vector with the vector to the eye. The specular term is most intense when the angle between the vector to the eye and the normal is the same as the angle between the vector to the light and the normal. Intensity drops off as the difference in angles increases. Listing 4.10 shows the final Phong-lighting fragment shader.

The bulk of the work is done in a new function `LightIntensity`, which is used throughout the rest of this book. The diffuse term is computed as before, and the specular term is computed as described above. The specular term is then raised to a power to determine its sharpness. Small exponents produce a large, dull specular highlight, and high values create a tight highlight. Finally, the diffuse and specular terms are multiplied by user-defined coefficients, and an ambient term, representing light throughout the entire scene, is added to create a final light intensity.

4.2.2 Texturing

Lighting the globe makes its curvature apparent, but the real fun begins with texture mapping. Texture mapping helps achieve one of the primary

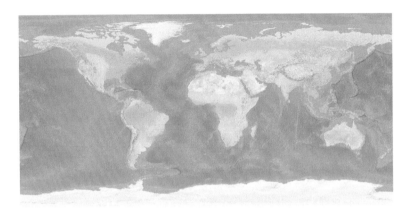

Figure 4.10. Satellite-derived land imagery with shaded relief and water from Natural Earth.

tasks of virtual globes: displaying high-resolution imagery. Our discussion focuses on computing texture coordinates per fragment, similar to how normals were computed per fragment in the previous section. For now, it is assumed the texture fits into memory, and there is enough precision in GLSL's float for texture coordinates spanning the globe.

World rasters, like the one in Figure 4.10, typically have a 2 : 1 aspect ratio and WGS84 data. To map such a texture onto the globe, texture coordinates need to be computed. Given a fragment on the globe with a geodetic surface normal, n, with components in the range $[-1, 1]$, the goal is to compute texture coordinates, s and t, in the range $[0, 1]$. This mapping is shown in Equation (4.5):

$$s = \frac{\text{atan2}(n_y, n_x)}{2\pi} + 0.5,$$
$$t = \frac{\arcsin n_z}{\pi} + 0.5. \tag{4.5}$$

First, consider t since it is the simpler of the two expressions. Since n_z relates to latitude, it can be used to compute t, the vertical texture coordinate. Intuitively, t should be 1 when n_z is 1 (north pole), 0.5 when n_z is 0 (equator), and 0 when n_z is -1 (south pole). The mapping cannot be linear because of the curvature of the Earth, so arcsin is used to capture the curvature. Given a value, x, in the range $[-1, 1]$, arcsin returns the angle, ϕ, in the range $[-\pi/2, \pi/2]$ that satisfies $x = \sin(\theta)$. Above, $\frac{\arcsin n_z}{\pi}$ is in the range $[-0.5, 0.5]$. Adding 0.5 puts t in the desired $[0, 1]$ range.

```
in vec3 worldPosition;
out vec3 fragmentColor;

uniform sampler2D og_texture0;
uniform vec3 u_globeOneOverRadiiSquared;

vec3 GeodeticSurfaceNormal(vec3 positionOnEllipsoid,
    vec3 oneOverEllipsoidRadiiSquared)
{
  return normalize(positionOnEllipsoid *
                   oneOverEllipsoidRadiiSquared);
}

vec2 ComputeTextureCoordinates(vec3 normal)
{
  return vec2(
      atan(normal.y, normal.x) * og_oneOverTwoPi + 0.5,
      asin(normal.z) * og_oneOverPi + 0.5);
}

void main()
{
  vec3 normal = GeodeticSurfaceNormal(
      worldPosition, u_globeOneOverRadiiSquared);
  vec2 textureCoordinate = ComputeTextureCoordinates(normal);
  fragmentColor = texture(og_texture0, textureCoordinate).rgb;
}
```

Listing 4.11. Globe-texturing fragment shader.

Likewise, n_x and n_y relate to longitude and can be used to compute s, the horizontal texture coordinate. Starting with $n_x = -1$ and $n_y = 0$ (International Date Line), sweep the vector $[n_x, n_y]$ around the positive z-axis. As θ, the angle between the vector and the negative x-axis in the xy-plane, increases, $\mathrm{atan2}(n_y, n_x)$ increases from $-\pi$ to π. Dividing this by 2π puts it in the range $[-0.5, 0.5]$. Finally, adding 0.5 puts s in the desired $[0, 1]$ range.

A fragment shader for texture mapping the globe based on Equation (4.5) is shown in Listing 4.11. The function that computes texture coordinates given the fragment's normal, ComputeTextureCoordinates, is used throughout this book. Note that the GLSL function atan is equivalent to atan2.

The result of this fragment shader is shown in Figure 4.11(a). Although textured nicely, the globe lacks curvature because the shader does not include lighting. A simple way to combine lighting and texture mapping is to modulate the intensity of the light with the color of the texture:

```
fragmentColor = intensity *
                texture(og_texture0, textureCoordinate).rgb;
```

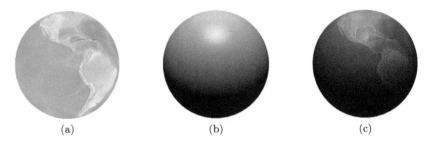

(a)   (b)   (c)

Figure 4.11. Combining texturing and lighting. (a) Texture mapping without lighting provides color but not shape. (b) Lighting without texture mapping provides shape but not color. (c) Combining texture mapping and lighting yields both shape and color.

The result is shown in Figure 4.11(c). The final fragment color is the product of the color from Figures 4.11(a) and 4.11(b).

Mapping a rectangular texture to a globe results in problems at the poles because there is a high ratio of texels to pixels. Texture filtering doesn't help and can even make things worse. EVE Online solves this by interpolating between two texture coordinates: a planar projection for the poles and a spherical projection for the "belly" [109]. Another approach to avoiding distortion at the poles is to store the globe texture as a cube map [57]. This has the additional benefit of spreading texel detail more evenly throughout the globe, but the cube map introduces slight distortion at the boundary of each cube face.

4.2.3   CPU/GPU Trade-offs

Now that we have successfully computed geometry for a globe and shaded it, let's think critically about the algorithms with respect to processor and memory usage. Only positions for the ellipsoid are computed on the CPU and stored in vertex attributes. Normals and texture coordinates, required for shading, are computed per fragment on the GPU. Alternatively, normals and texture coordinates could be computed and stored once, per vertex, on the CPU. The per-fragment approach has several advantages:

- Reduced memory usage since per-vertex normals and texture coordinates do not need to be stored. Less vertex data results in less system bus traffic, lower memory bandwidth requirements, and lower vertex assembly costs [123].

- Improved visual quality since the normal is analytically computed per fragment, not interpolated across a triangle. This quality improvement is similar to normal and bump mapping, which modify normals

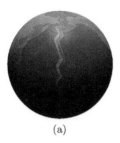 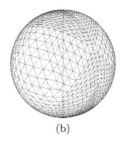

(a) (b)

Figure 4.12. Per-vertex texture-coordinate interpolation across the IDL causes artifacts unless special care is taken for triangles crossing the IDL. (a) Artifacts along the IDL. (b) Wireframe showing triangles crossing the IDL.

per fragment based on texture lookups, making a flat triangle appear to have more detail due to lighting changes.

• More concise code since the CPU tessellation is simplified without adding significant complexity to the shaders.

To better understand the trade-offs between per-vertex and per-fragment approaches, experiment with the per-vertex example, Chapter04SubdivisionSphere2, and see how it differs from the per-fragment example, Chapter04SubdivisionSphere1. In particular, SubdivisionSphereTessellator, the tessellator that computes positions, normals, and texture coordinates, is nearly twice the length of SubdivisionSphereTessellatorSimple, the tessellator that just computes positions.

○○○○ Try This

The last item is particularly important because, when it comes to graphics code, it is easy to get caught up in performance and not realize how important it is to produce simple, clean, concise code. The per-fragment approach simplifies CPU tessellation because the tessellator does not need to compute and store normals and texture coordinates per vertex. Since per-vertex texture coordinates are interpolated across triangles, special care needs to be taken for triangles that cross the IDL and poles [132]. Simply using Equation (4.5) results in a triangle's vertex on one side of the IDL having an s texture coordinate near one and a vertex on the other side of the IDL having s near zero, so almost the entire texture is interpolated across the triangle. The resulting artifacts are shown in Figure 4.12. Of course, this can be detected, and the texture coordinate can be adjusted for repeat filtering, but the per-fragment approach eliminates the need for special case code.

There are downsides to the per-fragment approach. On current GPUs, inverse trigonometry functions are not well optimized or highly precise.

4.2.4   Latitude-Longitude Grid

Almost all virtual globes have the ability to show a latitude-longitude grid, also called a lat/lon grid, on the globe, as shown in Figure 4.13. The grid makes it easy for users to quickly identify the approximate geographic position of points on the globe.

The grid is composed of lines of constant latitude, often highlighting the equator, Arctic Circle, Tropic of Cancer, Tropic of Capricorn, and Antarctic Circle, and lines of constant longitude, often highlighting the prime meridian and the IDL. As the viewer zooms in, the resolution of the grid usually increases, similar to how a level-of-detail algorithm refines a mesh as the viewer zooms in (see Figures 4.14(a)–4.14(c)).

Most grid-rendering approaches compute lines of latitude and lines of longitude at a resolution appropriate for the current view and render the grid using line or line-strip primitives. Using indexed lines can have an advantage over line strips since lines of latitude and lines of longitude typically share the same vertices. If the viewer doesn't move drastically, the same grid tessellation can be reused for many frames.

This approach is tried and true, and in fact, we use it in STK, with the exception that the resolution is defined by the user, not the viewer's current zoom level. Even so, we present a shader-based approach here that

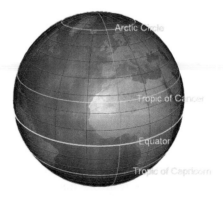

Figure 4.13. Latitude-longitude grid with annotations.

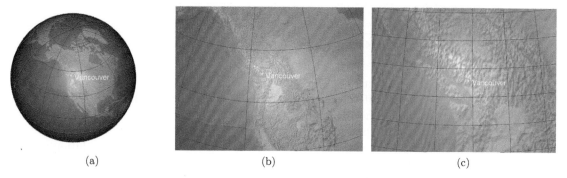

(a) (b) (c)

Figure 4.14. The latitude-longitude grid increases resolution as the viewer zooms in. A view-dependent grid resolution ensures the grid doesn't clutter the globe at distant views and has enough detail at close views.

procedurally generates the grid in the fragment shader during the same pass the globe is rendered. This has several advantages:

- CPU time is not required to compute the grid.

- An extra pass is not needed to render the grid.

- No additional memory or bus traffic is required.

- Z-fighting between the grid and globe does not occur.

There are, of course, disadvantages:

- The time required to render the globe and grid in one pass is higher than the time required to just render the globe.

- The accuracy is limited by the GPU's 32-bit floating-point precision.

- Text annotations still need to be handled separately.

Even with the disadvantages, procedurally generating the grid in the fragment shader is an attractive approach.

Our typical lighting and texturing fragment shader needs the additional uniforms shown in Listing 4.12. The x component of each uniform corresponds to lines of longitude and the y component corresponds to lines of latitude. The uniform u_gridLineWidth defines the line width and u_gridResolution defines the grid resolution, that is, the spacing between lines, in the range $[0, 1]$. The number of partitions is determined by $\frac{1}{resolution}$. For example, a latitude line spacing of 0.05 results in 20 partitions made by lines of constant latitude.

```
uniform  vec2  u_gridLineWidth;
uniform  vec2  u_gridResolution;
```

Listing 4.12. Uniforms for the latitude-longitude grid fragment shader.

The texture coordinate is used to determine if the fragment is on a grid line. If so, the fragment is shaded as such, otherwise, the fragment is shaded normally using the globe's texture. The s and t texture coordinates span the globe from west to east and south to north, respectively. The mod function is used to determine if the line resolution evenly divides the texture coordinate. If so, the fragment is on the grid line. The simplest approach detects a grid line if (any(equal(mod(`textureCoord inate`, `u_gridResolution`), vec2(0.0, 0.0))))). In practice, this results in no fragments on the grid because explicitly comparing floating-point numbers against zero is risky business.

The obvious next attempt is to test for values around zero, perhaps any(lessThan(mod(`textureCoordinate`, `u_gridResolution`), vec2 (0.001, 0.001))). This does in fact detect fragments on the grid, but the width of the grid lines, in pixels, varies with the viewer's zoom because of the hard-coded epsilon. When the viewer is far away from the globe, texture coordinates are changing rapidly from fragment to fragment in screen space, so 0.001 creates thin lines. When the viewer is close to the surface, texture coordinates are changing slowly from fragment to fragment, so 0.001 creates thick lines.

To shade lines with a view-independent constant pixel width, the epsilon should be related to the rate of change in texture coordinates from fragment to fragment. The key is to determine how much a texture coordinate is changing between adjacent fragments and use this, or some multiple of it, as the epsilon. Thankfully, GLSL provides functions that return the rate of change (i.e., the derivative) of a value from fragment to fragment. The function dFdx returns the rate of change of an expression in the x screen direction and dFdy returns the rate of change in the y direction. For example, dFdx(texture `Coordinate`) returns a vec2 describing how fast the s and t texture coordinates are changing in the x screen direction.

The epsilon to determine if a fragment is on a grid line should be a vec2, with its s component representing the maximum tolerance for the s texture coordinate in either x or y screen space direction. Likewise, its t component represents the maximum tolerance for the t texture coordinate in either x or y screen space direction. This is shown in Listing 4.13, along with the entire fragment shader main function. Run Chapter04LatitudeLongitudeGrid for a complete example.

Fragments on the grid are shaded solid red. A variety of shading options exist. The color could be a uniform or two uniforms, one for lines of latitude

```
void main()
{
  vec3 normal = GeodeticSurfaceNormal(
      worldPosition, u_globeOneOverRadiiSquared);
  vec2 textureCoordinate = ComputeTextureCoordinates(normal);
  vec2 distanceToLine = mod(textureCoordinate, u_gridResolution);
  vec2 dx = abs(dFdx(textureCoordinate));
  vec2 dy = abs(dFdy(textureCoordinate));
  vec2 dF = vec2(max(dx.s, dy.s),
                 max(dx.t, dy.t)) * u_gridLineWidth;

  if (any(lessThan(distanceToLine, dF)))
  {
    fragmentColor = vec3(1.0, 0.0, 0.0);
  }
  else
  {
    float intensity =
        LightIntensity(normal,
                       normalize(positionToLight),
                       normalize(positionToEye),
                       og_diffuseSpecularAmbientShininess);
    fragmentColor = intensity *
                    texture(og_texture0, textureCoordinate).rgb;
  }
}
```

Listing 4.13. Lat/lon fragment grid shader.

and one for lines of longitude. Certain lines, like the equator and prime meridian, can be detected based on the texture coordinate and shaded a different color, as done in Figure 4.13. The line color can be blended with the globe texture and made to fade in/out based on the viewer's zoom. Finally, antialiasing and line patterns can be computed.

> Add antialiasing to the grid lines in Chapter04LatitudeLongitudeGrid by utilizing the prefiltering technique in the **OpenGlobe.Scene.Wireframe** fragment shader based on Brentzen [10]. Add a pattern to the grid lines based on the example described by Gateau [56].

○○○○ Try This

The grid resolution can be made view-dependent by modifying u_gridResolution on the CPU based on the viewer's height. Any function that maps the viewer's height to a grid resolution can be used. A flexible approach is to define a series of nonoverlapping height intervals (e.g., $[0, 100)$, $[100, 1,000)$, $[1,000, 10,000)$) that each correspond to a grid resolution. When the globe is rendered, or when the viewer moves, the interval containing the viewer's height is found and the corresponding grid resolution is used.

For a large number of intervals, a tree data structure can be used to quickly find the interval containing the viewer's height, but this is usually unnecessary. A simple sorted list of intervals can be used efficiently by exploiting temporal coherence; instead of testing against the entire list of intervals in order, first test the viewer's height against the interval it was in last time. If it is outside this interval, the adjacent intervals are tested, and so on. In the vast majority of cases, the lookup is constant time.

4.2.5   Night Lights

A common feature in virtual globes and planet-centric games is to show nighttime city lights on the side of the globe not illuminated by the sun. This is a classic use of multitexturing: a daytime texture is used where the sun illuminates the surface and a night-lights texture is used where it does not. The texture in Figure 4.15 shows Earth's city lights at nighttime.

Most of our night-lights implementation is in the fragment shader. Our normal pass-through vertex shader, such as the one from Chapter04Sub-divisionSphere1, can be used with the exception that the light position uniform should be the sun position, `og_sunPosition`, instead of the light attached to the camera, `og_cameraLightPosition`.

Our normal fragment shader needs more modifications, including the four new uniforms shown in Listing 4.14. A day texture, `u_day Texture`, and night texture, `u_nightTexture`, are required. Instead of having a sharp transition between day and night for areas of the globe experiencing

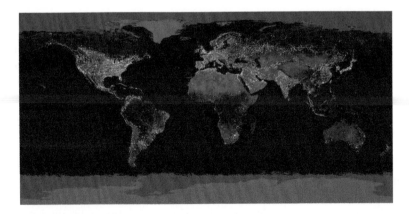

Figure 4.15. This city-lights texture, from NASA Visible Earth, is applied to the surface of the globe not illuminated by the sun.

```
uniform sampler2D u_dayTexture;
uniform sampler2D u_nightTexture;
uniform float u_blendDuration;
uniform float u_blendDurationScale;
```

Listing 4.14. Uniforms for the night-lights shader.

dusk and dawn, the day and night textures are blended based on `u_blend`
`Duration`, the duration of the transition period in the range $[0, 1]$. The
function `u_blendDurationScale` is simply precomputed as $\frac{1}{(2*u\_blendDuration)}$
to avoid redundant computation in the fragment shader.

The bulk of the fragment shader is shown in Listing 4.15. The diffuse-
lighting component is used to determine if the fragment should be shaded as
day, night, or dusk/dawn. If diffuse is greater than the blend duration, then
the fragment is shaded using the day texture and Phong lighting. Think
of it as if there were no transition period ($u\_blendDuration = 0$); if the dot
product between the fragment's surface normal and the vector to the light
is positive, then the fragment is illuminated by the sun. Likewise, if diffuse
is less than `-u_blendDuration`, then the fragment should be shaded using
just the night texture and no lighting. Dusk/dawn is handled by the final
condition when diffuse is in the range [`-u_blendDuration`, `u_blendDuration`]
by blending the day color and night color using mix.

```
vec3 NightColor(vec3 normal)
{
  return texture(u_nightTexture,
              ComputeTextureCoordinates(normal)).rgb;
}

vec3 DayColor(vec3 normal, vec3 toLight, vec3 toEye,
    float diffuseDot, vec4 diffuseSpecularAmbientShininess)
{
  float intensity = LightIntensity(normal, toLight, toEye,
      diffuseDot, diffuseSpecularAmbientShininess);
  return intensity * texture(u_dayTexture,
      ComputeTextureCoordinates(normal)).rgb;
}

void main()
{
  vec3 normal = normalize(worldPosition);
  vec3 toLight = normalize(positionToLight);
  float diffuse = dot(toLight, normal);

  if (diffuse > u_blendDuration)
  {
    fragmentColor = DayColor(normal, toLight,
                          normalize(positionToEye), diffuse,
                          og_diffuseSpecularAmbientShininess);
  }
  else if (diffuse < -u_blendDuration)
  {
```

```
        fragmentColor = NightColor ( normal ) ;
    }
    else
    {
        vec3 night = NightColor ( normal ) ;
        vec3 day = DayColor ( normal , toLight ,
                             normalize ( positionToEye ) , diffuse ,
                             og_diffuseSpecularAmbientShininess ) ;
        fragmentColor = mix ( night , day ,
            ( diffuse + u_blendDuration ) * u_blendDurationScale ) ;
    }
}
```

Listing 4.15. Night-lights fragment shader.

Example images are shown in Figure 4.16, and a complete example is provided in Chapter04NightLights. The shader can be written more concisely if both the day and night color are always computed. For fragments not in dusk/dawn, this results in a wasted texture read and possibly a wasted call to **LightIntensity**.

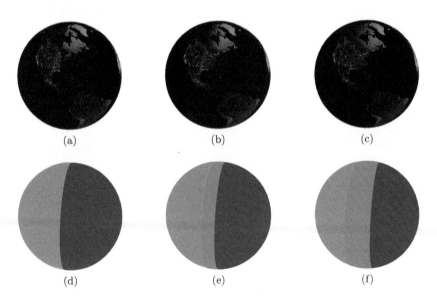

(a)   (b)   (c)

(d)   (e)   (f)

Figure 4.16. Night lights with various blend durations. In the bottom row, night is gray, day is blue, and the transition is orange. The top row shows the shaded result. (a) **u_blendDuration** = 0. (b) **u_blendDuration** = 0.1f. (c) **u_blendDuration** = 0.2f. (d) **u_blendDuration** = 0. (e) **u_blendDuration** = 0.1f. (f) **u_blendDuration** = 0.2f.

> Run the night-lights example with a frame-rate utility. Note the frame
> rate when viewing just the daytime side of the globe and just the night-
> time side. Why is the frame rate higher for the nighttime side? Because
> the night-lights texture is a lower resolution than the daytime texture and
> does not require any lighting computations. This shows using dynamic
> branching to improve performance.

○○○○ **Try This**

Virtual globe applications use real-world data like this night-light texture derived from satellite imagery. On the other hand, video games generally focus on creating a wide array of convincing artificial data using very
little memory. For example, EVE Online takes an interesting approach to
rendering night lights for their planets [109]. Instead of relying on a night-
light texture whose texels are directly looked up, a texture atlas of night
lights is used. The spherically mapped texture coordinates are used to look
up surrogate texture coordinates, which map into the texture atlas. This
allows a lot of variation from a single texture atlas because sections can be
rotated and mirrored.

Rendering night lights is one of many uses for multitexturing in globe
rendering. Other uses include cloud textures and gloss maps to show specular highlights on bodies of waters [57,147]. Before the multitexturing hardware was available, effects like these required multiple rendering passes.
STK, being one of the first products to implement night lights, uses a
multiple-pass approach.

4.3   GPU Ray Casting

GPUs are built to rasterize triangles at very rapid rates. The purpose of
ellipsoid-tessellation algorithms is to create triangles that approximate the
shape of a globe. These triangles are fed to the GPU, which rapidly ras-
terizes them into shaded pixels, creating an interactive visualization of the
globe. This process is very fast because it is embarrassingly parallel; individual triangles and fragments are processed independently, in a massively
parallel fashion. Since tessellation is required, rendering a globe this way
is not without its flaws:

- No single tessellation is perfect; each has different strengths and weaknesses.

- Under-tessellation leads to a coarse triangle mesh that does not approximate the surface well, and over-tessellation creates too many

triangles, negatively affecting performance and memory usage. View-dependent level-of-detail algorithms are required for most applications to strike a balance.

- Although GPUs exploit the parallelism of rasterization, memories are not keeping pace with the increasing computation power, so a large number of triangles can negatively impact performance. This is especially true of some level-of-detail algorithms where new meshes are frequently sent over the system bus.

Ray tracing is an alternative to rasterization. Rasterization starts with triangles and ends with pixels. Ray tracing takes the opposite approach: it starts with pixels and asks what triangle(s), or objects in general, contribute to the color of this pixel. For perspective views, a ray is cast from the eye through each pixel into the scene. In the simplest case, called *ray casting*, the first object intersecting each ray is found, and lighting computations are performed to produce the final image.

A strength of ray casting is that objects do not need to be tessellated into triangles for rendering. If we can figure out how to intersect a ray with an object, then we can render it. Therefore, no tessellation is required to render a globe represented by an ellipsoid because there is a well-known equation for intersecting a ray with an ellipsoid's implicit surface. The benefits of ray casting a globe include the following:

- The ellipsoid is automatically rendered with an infinite level of detail. For example, as the viewer zooms in, the underlying triangle mesh does not become apparent because there is no triangle mesh; intersecting a ray with an ellipsoid produces an infinitely smooth surface.

- Since there are no triangles, there is no concern about creating thin triangles, triangles crossing the poles, or triangles crossing the IDL. Many of the weaknesses of tessellation algorithms go away.

- Significantly less memory is required since a triangle mesh is not stored or sent across the system bus. This is particularly important in a world where size is speed.

Since current GPUs are built for rasterization, you may wonder how to efficiently ray cast a globe. In a naïve CPU implementation, a nested for loop iterates over each pixel in the scene and performs a ray/ellipsoid intersection. Like rasterization, ray casting is embarrassingly parallel. Therefore, a wide array of optimizations are possible on today's CPUs, including casting each ray in a separate thread and utilizing single instruction multiple data (SIMD) instructions. Even with these optimizations, CPUs

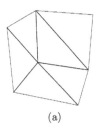

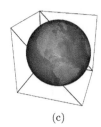

(a)   (b)   (c)

Figure 4.17. In GPU ray casting, (a) a box is rendered to (b) invoke a ray-casting fragment shader that finds the ellipsoid's visible surface. When an intersection is found, (c) the geodetic surface normal is used for shading.

do not support the massive parallelism of GPUs. Since GPUs are built for rasterization, the question is how do we use them for efficient ray casting?

Fragment shaders provide the perfect vehicle for ray casting on the GPU. Instead of tessellating an ellipsoid, create geometry for a bounding box around the ellipsoid. Then, render this box using normal rasterization and cast a ray from the eye to each fragment created by the box. If the ray intersects the inscribed ellipsoid, shade the fragment; otherwise, discard it.

The box is rendered with front-face culling, as shown in Figure 4.17(a). Front-facing culling is used instead of back-face culling so the globe still appears when the viewer is inside the box.

This is the only geometry that needs to be processed to render the ellipsoid, a constant vertex load of 12 triangles. With front-face culling, fragments for six of the triangles are processed for most views. The result is that a fragment shader is run for each fragment we want to cast a ray through. Since the fragment shader can access the camera's world-space position through a uniform, and the vertex shader can pass the vertex's interpolated world-space position to the fragment shader, a ray can be constructed from the eye through each fragment's position.[4] The ray simply has an origin of `og_cameraEye` and a direction of normalize(`world Position − og_cameraEye`).

The fragment shader also needs access to the ellipsoid's center and radii. Since it is assumed that the ellipsoid is centered at the origin, the fragment shader just needs a uniform for the ellipsoid's radii. In practice, intersecting a ray with an ellipsoid requires $\frac{1}{radii^2}$, so that should be precomputed once on the CPU and passed to the fragment shader as a uniform. Given the

---

[4]In this case, a ray is cast in world coordinates with the ellipsoid's center at the origin. It is also common to perform ray casting in eye coordinates, where the ray's origin is the coordinate system's origin. What really matters is that the ray and object are in the same coordinate system.

ray and ellipsoid information, Listing 4.16 shows a fragment shader that colors fragments green if a ray through the fragment intersects the ellipsoid or red if the ray does not intersect, as shown in Figure 4.17(b).

This shader has two shortcomings. First, it does not do any actual shading. Fortunately, given the position and surface normal of the ray intersection, shading can utilize the same techniques used throughout this chapter, namely LightIntensity() and ComputeTextureCoordinates(). Listing 4.17 adds shading by computing the position of the intersection along the ray using i.Time and shading as usual. If the ray does not intersect the ellipsoid, the fragment is discarded. Unfortunately, using discard has the adverse effect of disabling GPU depth buffer optimizations, including fine-grained early-z and coarse-grained z-cull, as discussed in Section 12.4.5.

```
in vec3 worldPosition;
out vec3 fragmentColor;
uniform vec3 og_cameraEye;
uniform vec3 u_globeOneOverRadiiSquared;

struct Intersection
{
  bool Intersects;
  float Time;        // Time of intersection along ray
};

Intersection RayIntersectEllipsoid(vec3 rayOrigin,
    vec3 rayDirection, vec3 oneOverEllipsoidRadiiSquared)
{ // ... }

void main()
{
  vec3 rayDirection = normalize(worldPosition - og_cameraEye);
  Intersection i = RayIntersectEllipsoid(og_cameraEye,
      rayDirection, u_globeOneOverRadiiSquared);
  fragmentColor = vec3(i.Intersects, !i.Intersects, 0.0);
}
```

Listing 4.16. Base GLSL fragment shader for ray casting.

```
// ...
vec3 GeodeticSurfaceNormal(vec3 positionOnEllipsoid,
                           vec3 oneOverEllipsoidRadiiSquared)
{
  return normalize(positionOnEllipsoid *
                   oneOverEllipsoidRadiiSquared);
}

void main()
{
  vec3 rayDirection = normalize(worldPosition - og_cameraEye);
  Intersection i = RayIntersectEllipsoid(og_cameraEye,
      rayDirection, u_globeOneOverRadiiSquared);
  if (i.Intersects)
  {
```

```
    vec3 position = og_cameraEye + (i.Time * rayDirection);
    vec3 normal = GeodeticSurfaceNormal(position,
        u_globeOneOverRadiiSquared);

    vec3 toLight = normalize(og_cameraLightPosition - position);
    vec3 toEye = normalize(og_cameraEye - position);
    float intensity = LightIntensity(normal, toLight, toEye,
        og_diffuseSpecularAmbientShininess);

    fragmentColor = intensity * texture(og_texture0,
        ComputeTextureCoordinates(normal)).rgb;
}
else
{
    discard;
}
}
```

Listing 4.17. Shading or discarding a fragment based on a ray cast.

```
float ComputeWorldPositionDepth(vec3 position)
{
    vec4 v = og_modelViewPerspectiveMatrix * vec4(position, 1);
    v.z /= v.w;
    v.z = (v.z + 1.0) * 0.5;
    return v.z;
}
```

Listing 4.18. Computing depth for a world-space position.

The remaining shortcoming, which may not be obvious until other objects are rendered in the scene, is that incorrect depth values are written. When an intersection occurs, the box's depth is written instead of the ellipsoid's depth. This can be corrected by computing the ellipsoid's depth, as shown in Listing 4.18, and writing it to gl_FragDepth. Depth is computed by transforming the world-space positions of the intersection into clip coordinates, then transforming this z-value into normalized device coordinates and, finally, into window coordinates. The final result of GPU ray casting, with shading and correct depth, is shown in Figure 4.17(c).

Since this algorithm doesn't have any overdraw, all the red pixels in Figure 4.17(b) are wasted fragment shading. A tessellated ellipsoid rendered with back-face culling does not have wasted fragments. On most GPUs, this is not as bad as it seems since the dynamic branch will avoid the shading computations [135, 144, 168], including the expensive inverse trigonometry for texture-coordinate generation. Furthermore, since the branches are coherent, that is, adjacent fragments in screen space are likely to take the same branch, except around the ellipsoid's silhouette, the GPU's parallelism is used well [168].

To reduce the number of rays that miss the ellipsoid, a viewport-aligned convex polygon bounding the ellipsoid from the viewer's perspective can be used instead of a bounding box [30]. The number of points in the bounding polygon determine how tight the fit is and, thus, how many rays miss the ellipsoid. This creates a trade-off between vertex and fragment processing.

GPU ray casting an ellipsoid fits seamlessly into the rasterization pipeline, making it an attractive alternative to rendering a tessellated approximation. In the general case, GPU ray casting, and full ray tracing in particular, is difficult. Not all objects have an efficient ray intersection test like an ellipsoid, and large scenes require hierarchical spatial data structures for quickly finding which objects a ray may intersect. These types of linked data structures are difficult to implement on today's GPUs, especially for dynamic scenes. Furthermore, in ray tracing, the number of rays quickly explodes with effects like soft shadows and antialiasing. Nonetheless, GPU ray tracing is a promising, active area of research [134, 178].

4.4   Resources

A detailed description of computing a polygonal approximation to a sphere using subdivision surfaces, aimed towards introductory graphics students, is provided by Angel [7]. The book is an excellent introduction to computer graphics in general. A survey of subdivision-surface algorithms is presented in *Real-Time Rendering* [3]. The book itself is an indispensable survey of real-time rendering. See "The Orange Book" for more information on using multitexturing in fragment shaders to render the Earth [147]. The book is generally useful as it thoroughly covers GLSL and provides a wide range of example shaders.

An ellipsoid tessellation based on the honeycomb [39], a figure derived from a soccer ball, may prove advantageous over subdividing platonic solids, which leads to a nonuniform tessellation. Another alternative to the tessellation algorithms discussed in this chapter is the HEALPix [65].

A series on procedurally generating 3D planets covers many relevant topics, including cube-map tessellation, level of detail, and shading [182]. An experimental globe-tessellation algorithm for NASA World Wind is described by Miller and Gaskins [116].

The entire field of real-time ray tracing is discussed by Wald [178], including GPU approaches. A high-level discussion on ray tracing virtual globes, with a focus on improving visual quality, is presented by Christen [26].

Part II

○○○○

Precision

5

Vertex Transform Precision

Consider a simple scene containing a few primitives: a point at the origin and two triangles in the plane $x = 0$ on either side of the point, spaced 1 m apart. Such a scene is shown in Figure 5.1(a). When the viewer zooms in, we expect the objects to appear coplanar, as shown in Figure 5.1(b).

What happens if we change the object positions such that they are drawn in the plane $x = 6,378,137$, which is Earth's equatorial radius in meters? When viewed from a distance, the objects still appear as expected—ignoring z-fighting artifacts, the subject of the following chapter. When zoomed in, other artifacts become noticeable: the objects appear to jitter, that is, they literally bounce around in a jerky manner as the viewer moves! In our example, this means the point and plane no longer appear

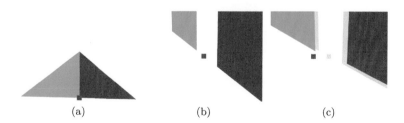

(a) (b) (c)

Figure 5.1. (a) A simple scene containing a point and two triangles. (b) Zoomed in on the scene, it is obvious that the point and triangles are in the same plane. (c) Since the scene is rendered with large world coordinates, jittering artifacts occur when zoomed in even closer than in (b). Here, the correct objects are shown in gray and the incorrectly positioned objects are shown in their original colors.

coplanar, as shown in Figure 5.1(c). Instead, they bounce in a view-dependent manner.

Jitter is caused by insufficient precision in a 32-bit floating-point value for a large value like $6,378,137$. Considering that typical positions in WGS84 coordinates are of similar magnitude, jitter is a significant challenge for virtual globes, where users expect to be able to zoom as close as possible to an object, without the object starting to bounce. This chapter explores why jittering occurs and approaches to combat it.

5.1 Jittering Explained

In a scene with large world coordinates, like virtual globes, jittering can occur when zoomed in and the viewer rotates or an object moves. Besides being visually disruptive, jitter makes it hard to determine exactly where

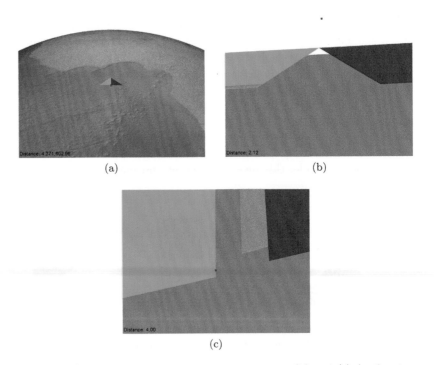

Figure 5.2. (a) From a distance, jitter is not apparent. (b) and (c) As the viewer zooms in, jittering becomes obvious. The correct point and triangle locations are shown in gray. The distance between the triangles is 1 m, and, in both images, the viewer is within a few meters of the point.

an object is located. Buildings bounce back and forth, vector data snap on and off of the underlying imagery, and terrain tiles overlap, then separate.

Jitter is shown in the series of images in Figure 5.2. The best way to get a feel for jitter artifacts is to run an application that exhibits it (not all virtual globes completely eliminate it) or to watch a video, such as the jittering space shuttle video in Ohlarik's article on the web [126].

Most of today's GPUs support only 32-bit floating-point values, which does not provide enough precision for manipulating large positions in WGS84 coordinates. CPUs use 64-bit double precision, which does provide enough precision, so many approaches to eliminating jitter utilize double-precision computation on the CPU or emulate it on the GPU. Although new Shader Model 5 hardware supports double precision, techniques described in this chapter are useful for supporting older hardware that is still in widespread use, and some techniques, namely RTC (see Section 5.2), are also useful for saving memory over using double precision.

5.1.1 Floating-Point Roundoff Error

Let's dive deeper into the behavior of floating-point numbers to understand why jittering occurs. It is not satisfactory to just say 32 bits is not enough precision—although that does summarize it. At the same time, we do not need to discuss every detail of the IEEE-754 specification [158]. Although GPUs sometimes deviate from the IEEE-754 for binary floating-point arithmetic [122], these deviations are not the culprit for jitter.

A rule of thumb is that 32-bit single-precision numbers have about seven accurate decimal digits, and 64-bit double-precision numbers have about 16. It is not possible to represent every rational number using 32, or even 64, bits. As such, many floating-point numbers are rounded to a nearby representable value. Listing 5.1 shows that the next representable 32-bit floating-point value after 6378137.0f is 6378137.5f. For example, 6378137.25f is not a representable value and is rounded down to 6378137.0f; likewise, 6378137.26f is not representable and is rounded up to 6378137.5f.

To make matters worse, gaps between representable floating-point values are dependent on the values themselves: larger values have larger gaps between them. In fact, if the numbers are large enough, whole numbers are skipped. Listing 5.2 shows that the next representable value after 17000000.0f is 17000002.0f. When modeling the world using meters, values in this 2-m gap are rounded up or down. Seventeen million meters may sound large considering that it is over ten million meters above Earth, but it is not impractical for all virtual globe applications. For example, in space applications, this is in the range for medium Earth orbit satellites.

```
float f  = 6378137.0f;      //  6378137.0
float f1 = 6378137.1f;      //  6378137.0
float f2 = 6378137.2f;      //  6378137.0
float f25 = 6378137.25f;    //  6378137.0
float f26 = 6378137.26f;    //  6378137.5
float f3 = 6378137.3f;      //  6378137.5
float f4 = 6378137.4f;      //  6378137.5
float f5 = 6378137.5f;      //  6378137.5
```

Listing 5.1. Roundoff error for 32-bit floating-point values.

```
float f  = 17000000.0f;     //  17000000.0
float f1 = 17000001.0f;     //  17000000.0
float f2 = 17000002.0f;     //  17000002.0
```

Listing 5.2. Roundoff error increases for large values to the point where whole numbers are skipped.

So far, we know there are a finite number of representable floating-point values, the gaps between these values increase as the values themselves increase, and nonrepresentable values are rounded to representable ones. There are other concerns too; in particular, addition and subtraction of values of different magnitudes drop least-significant bits, and thus precision, from the smaller values. The larger the variation between values, the more precision that is lost. To add icing to the cake, rounding errors accumulate from every floating-point operation.

5.1.2 Root Cause of Jittering

With the behavior of floating-point numbers in mind, let's revisit the jittery scene containing one point and two triangles, introduced at the beginning of the chapter. Consider the point located at $(6378137, 0, 0)$. When the viewer zooms in to within about 800 m, the point starts to jitter if the viewer rotates. Jitter becomes significant when zoomed in to just a few meters.

Why does it jitter? And why does jitter increase as the viewer gets closer?

The vertex shader used to transform this jittery point is shown in Listing 5.3. On the surface, everything looks OK: a model-view-perspective matrix transforms the input position from model coordinates (e.g., WGS84) to clip coordinates. This transformation matrix is computed on the CPU by multiplying individual model ($\mathbf{M}$), view ($\mathbf{V}$), and perspective ($\mathbf{P}$)

```
in vec4 position;
uniform mat4 og_modelViewPerspectiveMatrix;

void main()
{
    gl_Position = og_modelViewPerspectiveMatrix * position;
}
```

Listing 5.3. A seemingly innocent vertex shader can exhibit jitter.

matrices using double precision. Example values for these matrices, when the viewer is 800 m from the point, are shown below:

$$\mathbf{M} = \begin{pmatrix} 1 & 0 & 0 & 0 \\ 0 & 1 & 0 & 0 \\ 0 & 0 & 1 & 0 \\ 0 & 0 & 0 & 1 \end{pmatrix},$$

$$\mathbf{V} = \begin{pmatrix} 0.78 & 0.63 & 0.00 & -4,946,218.10 \\ 0.20 & -0.25 & 0.95 & -1,304,368.35 \\ 0.60 & -0.73 & -0.32 & -3,810,548.19 \\ 0.00 & 0.00 & 0.00 & 1.00 \end{pmatrix},$$

$$\mathbf{P} = \begin{pmatrix} 2.80 & 0.00 & 0.00 & 0.00 \\ 0.00 & 3.73 & 0.00 & 0.00 \\ 0.00 & 0.00 & -1.00 & -0.02 \\ 0.00 & 0.00 & -1.00 & 0.00 \end{pmatrix}. \tag{5.1}$$

The fourth column of $\mathbf{V}$, the translation, contains very large values because when the viewer is near the point, the viewer is far from the WGS84 origin. The model-view-perspective matrix, $\mathbf{MVP}$, is computed as $\mathbf{P} * \mathbf{V} * \mathbf{M}$. Values in its fourth column are even larger than in $\mathbf{V}$:

$$\mathbf{MVP} = \begin{pmatrix} 2.17 & 1.77 & 0.00 & -13,844,652.95 \\ 0.76 & -0.94 & 3.53 & -4,867,968.95 \\ -0.60 & 0.73 & 0.32 & 3,810,548.18 \\ -0.60 & 0.73 & 0.32 & 3,810,548.19 \end{pmatrix}. \tag{5.2}$$

When $\mathbf{MVP}$ is assigned to the vertex shader's uniform, it is converted from 64-bit to 32-bit values, which results in some roundoff error in the fourth column:

```
(float)-13,844,652.95 = -13,844,653.0f
(float)-4,867,968.95 = -4,867,969.0f
(float)3,810,548.18 = 3,810,548.25f
(float)3,810,548.19 = 3,810,548.25f
```

The vertex shader then multiplies the point's model position, p_{WGS84}, by **MVP** in 32-bit floating point:

$$p_{\text{clip}} = \mathbf{MVP} * p_{\text{WGS84}}$$

$$= \begin{pmatrix} 2.17f & 1.77f & 0.00f & -13,844,653.0f \\ 0.76f & -0.94f & 3.53f & -4,867,969.0f \\ -0.60f & 0.73f & 0.32f & 3,810,548.25f \\ -0.60f & 0.73f & 0.32f & 3,810,548.25f \end{pmatrix} \begin{pmatrix} 6,378,137.0f \\ 0.0f \\ 0.0f \\ 1.0f \end{pmatrix}$$

$$= \begin{pmatrix} (2.17f)(6,378,137.0f) + -13,844,653.0f \\ (0.76f)(6,378,137.0f) + -4,867,969.0f \\ (-0.60f)(6,378,137.0f) + 3,810,548.25f \\ (-0.60f)(6,378,137.0f) + 3,810,548.25f \end{pmatrix}.$$

This is where jittering artifacts are introduced: the matrix multiply is operating on large values with sparse 32-bit floating-point representations. A small change in the object or viewer position may not result in any change as roundoff errors produce the same results, but then suddenly roundoff error jumps to the next representable value and the object appears to jitter.

Jittering is noticeable when zoomed in (when considering Earth-scale scenes, 800 m is pretty close), but it is not noticeable when zoomed out. Why is there no jitter when the viewer is, say, 100,000 m from the point? Are the numbers smaller? The point didn't move, so p_{WGS84} is still the same, as are **M** and **P**, but **V**, and therefore **MVP**, changed because the viewer changed. For example,

$$\mathbf{V} = \begin{pmatrix} 0.78 & 0.63 & 0.00 & -4,946,218.10 \\ 0.20 & -0.25 & 0.95 & -1,304,368.35 \\ 0.60 & -0.73 & -0.32 & -3,909,748.19 \\ 0.00 & 0.00 & 0.00 & 1.00 \end{pmatrix},$$

$$\mathbf{MVP} = \begin{pmatrix} 2.17 & 1.77 & 0.00 & -13,844,652.95 \\ 0.76 & -0.94 & 3.53 & -4,867,968.95 \\ -0.60 & 0.73 & 0.32 & 3,909,748.18 \\ -0.60 & 0.73 & 0.32 & 3,909,748.19 \end{pmatrix}.$$

These matrices are different, but they look very similar to **V** and **MVP** when the viewer was 800 m away, in Equations (5.1) and (5.2), respectively. If the values are similar, shouldn't they exhibit similar floating-point precision problems?

Actually, the same behavior does occur—we just don't notice it because the amount of error introduced is less than a pixel. Jitter is view-dependent: when the viewer is very close to an object, a pixel may cover less than a meter, or even a centimeter, and therefore, jittering artifacts are prominent. When the viewer is zoomed out, a pixel may cover hundreds of meters or more, and precision errors do not introduce enough jitter to shift entire pixels; therefore, jitter is not noticeable. The field of view can also affect jitter since it affects pixel size; in narrower fields of view, each pixel covers less world space and is, therefore, more susceptible to jitter.

In summary, jitter is dependent on the magnitude of the components of an object's position and how a pixel covering the object relates to meters. Positions with larger components (e.g., positions far away from the origin) create more jitter. Pixels covering fewer meters, achieved by zooming close to the object or a narrow field of view or both, makes jitter noticeable. Scenes only need to be pixel perfect; jitter within a pixel is acceptable.

> Chapter05Jitter is the example project accompanying this chapter. It renders the simple scene composed of a point and two triangles and allows the user to change the algorithm used to eliminate jitter. In its default configuration, it doesn't eliminate jitter. Run the example, zoom in and out, and rotate to get a feel for when jitter starts to occur for this particular scene. How would changing the position of the object affect jitter?

○○○○ **Try This**

5.1.3   Scaling and Why It Doesn't Help

Since roundoff errors in large numbers create jitter, wouldn't eliminating large numbers also eliminate jitter? Instead of using meters as the unit, why not use a much larger unit—perhaps multiply each position in meters by $\frac{1}{6,378,137}$ so positions on Earth's surface have values near 1. There are many more representable 32-bit floating values near 1 than there are near 6,378,137.

Unfortunately, this doesn't actually fix jitter. Although there are many more representable floating-point values around 1 than there are around 6,378,137, there are still not enough because the units have scaled, requiring smaller gaps between representations. For example, there is a gap between 6378137.0f and 6378137.5f, which is 0.5 m when the world units are meters. If the world units are scaled by $\frac{1}{6,378,137}$, 0.5 m corresponds to $\frac{0.5}{6,378,137}$, or 0.000000078, so if there are representable values between 1.0f and 1.000000078f, then precision is gained. There are no such values though; in fact, there is a gap between 1.0f and 1.0000001f.

Scaling may make values themselves smaller, but it also requires smaller gaps between representable values, making it ineffective against jitter.

Try This ○○○○

> Run Chapter05Jitter and enable the option to scale world coordinates. This changes the units from meters to 6,378,137 m. Zoom in and rotate to see that this does not eliminate jitter. The point does not appear to jitter, but the triangles do. Why? What is unique about the point's position? What if we changed the point's position slightly? Does it start to jitter?

Since scaling doesn't eliminate jittering artifacts, let's look at some approaches that do, either by using the CPU's 64-bit double precision or emulating it on the GPU.

5.2   Rendering Relative to Center

One approach to eliminating jitter is to render an object's position relative to its center (RTC) in eye coordinates. For reasonably sized objects, this results in vertex positions and model-view-projection matrix values small enough for 32-bit floating-point precision, thus, eliminating jitter.

Consider vector data for an area on the globe, perhaps the boundary for a zip code or city. These data are typically provided in geographic coordinates and converted to WGS84 coordinates for rendering (see Section 2.3.1), but using such large positions results in noticeable jitter.

The first step in using RTC to eliminate jitter is to compute the center point of the vertex positions, $center_{WGS84}$, by determining the minimum and maximum x-, y-, and z-components of the positions and averaging them. See AxisAlignedBoundingBox and its Center property in OpenGlobe.

Next, subtract $center_{WGS84}$ from each position to translate the object from WGS84 to its own local coordinates with $center_{WGS84}$ as its origin. This subtraction is done using double precision on the CPU. The point $center_{WGS84}$ will contain large x-, y-, or z-components, but each position now has much smaller components than before because they are closer to $center_{WGS84}$ than the WGS84 origin.

If instead of using vector data, a 3D model, such as a building or bridge, were used, it would already be in this form: a center (or origin) in WGS84 coordinates with large components, and positions relative to the center

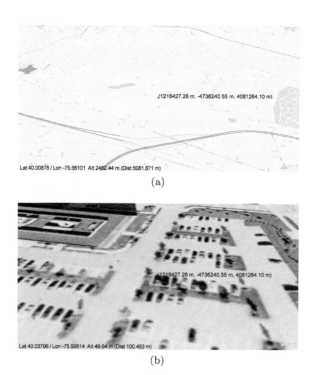

Figure 5.3. When the viewer zooms from (a) to (b), the WGS84 position of the object does not change, and its components remain too large for single precision, but the distance between the viewer and object becomes manageable for single precision. (Images taken using STK. Imagery © 2010 Microsoft Corporation and © 2010 NAVTEQ.)

manageable with 32-bit floating-point precision. The model would likely also have a different orientation than WGS84, but that's OK.

Having smaller vertex positions is half the solution; the other half is having smaller values in the translation (i.e., the fourth column) of **MVP**. This is achieved by transforming $\text{center}_{\text{WGS84}}$ to eye coordinates, $\text{center}_{\text{eye}}$, by multiplying $\text{center}_{\text{WGS84}}$ by **MV** on the CPU using double precision. As the viewer moves closer to the object's center, $\text{center}_{\text{eye}}$ becomes smaller because the distance between the viewer and the center decreases, as shown in Figure 5.3. Next, replace the x-, y-, and z-components in the fourth column of **MV** with $\text{center}_{\text{eye}}$, forming $\mathbf{MV_{RTC}}$:

$$\mathbf{MV_{RTC}} = \begin{pmatrix} MV_{00} & MV_{01} & MV_{02} & \text{center}_{\text{eye}}x \\ MV_{10} & MV_{11} & MV_{12} & \text{center}_{\text{eye}}y \\ MV_{20} & MV_{21} & MV_{22} & \text{center}_{\text{eye}}z \\ MV_{30} & MV_{31} & MV_{32} & MV_{33} \end{pmatrix}.$$

The final model-view-projection transform, $\mathbf{MVP_{RTC}}$, is formed by $\mathbf{P} * \mathbf{MV_{RTC}}$ on the CPU. This is then converted to 32-bit floating-point values, and used in the shader to transform positions relative to $center_{eye}$ to clip coordinates without jitter.

Example. Let's consider the point in our example scene, $p_{WGS84} = (6378137, 0, 0)$. We'll also pretend it is the only geometry, which makes $center_{WGS84} = (6378137, 0, 0)$. The point's position relative to center, p_{center}, is $p_{WGS84} - center_{WGS84}$, or $(6378137, 0, 0) - (6378137, 0, 0)$, which is $(0, 0, 0)$, a very manageable position for a 32-bit floating point number!

Using the same 800-m view distance as before, the model-view matrix, $\mathbf{MV}$, is

$$\mathbf{MV} = \mathbf{V} * \mathbf{M} = \begin{pmatrix} 0.78 & 0.63 & 0.00 & -4,946,218.10 \\ 0.20 & -0.25 & 0.95 & -1,304,368.35 \\ 0.60 & -0.73 & -0.32 & -3,810,548.19 \\ 0.00 & 0.00 & 0.00 & 1.00 \end{pmatrix}.$$

Before rendering the point, $\mathbf{MV}$ is used to compute $center_{eye}$:

$$center_{eye} = \mathbf{MV} * center_{WGS84}$$

$$= \begin{pmatrix} 0.78 & 0.63 & 0.00 & -4,946,218.10 \\ 0.20 & -0.25 & 0.95 & -1,304,368.35 \\ 0.60 & -0.73 & -0.32 & -3,810,548.19 \\ 0.00 & 0.00 & 0.00 & 1.00 \end{pmatrix} \begin{pmatrix} 6,378,137.0 \\ 0.0 \\ 0.0 \\ 1.0 \end{pmatrix} \quad (5.3)$$

$$= \begin{pmatrix} 0.0 \\ 0.0 \\ -800.0 \\ 1.0 \end{pmatrix}$$

Since the viewer is looking directly at the point in this case, $center_{eye}.xyz$ is $(0, 0, -800)$.[1] We can now create $\mathbf{MV_{RTC}}$ by replacing the large x-, y-, and z-components in the fourth column of $\mathbf{MV}$ with $center_{eye}$. The final model-view-perspective matrix is computed as $\mathbf{P} * \mathbf{MV_{RTC}}$. The vertex shader now deals with a manageable vertex position, $(0, 0, 0)$, and a manageable $\mathbf{MVP_{RTC}}$ fourth column, $(0, 0, -800, 1)^T$.

Implementation. The matrix $\mathbf{MVP_{RTC}}$ needs to be computed per object, or more precisely, per center if multiple objects share the same center. If the center changes or the viewer moves, $\mathbf{MVP_{RTC}}$ needs to be recomputed. This only requires a few lines of code, as shown in Listing 5.4. Recall that

---

[1]The multiplication in Equation (5.3) will not result in exactly this value because only two significant digits are shown here.

```
Matrix4D m = sceneState.ModelViewMatrix;
Vector4D centerEye = m * new Vector4D(_center, 1.0);
Matrix4D mv = new Matrix4D(
    m.Column0Row0, m.Column1Row0, m.Column2Row0, centerEye.X,
    m.Column0Row1, m.Column1Row1, m.Column2Row1, centerEye.Y,
    m.Column0Row2, m.Column1Row2, m.Column2Row2, centerEye.Z,
    m.Column0Row3, m.Column1Row3, m.Column2Row3, m.Column3Row3);

// Set shader uniform
_modelViewPerspectiveMatrixRelativeToCenter.Value =
    (sceneState.PerspectiveMatrix * mv).ToMatrix4F();
// .. draw call
```

Listing 5.4. Constructing the model-view-projection matrix for RTC.

the matrix multiply on the second line is using double precision. A complete RTC code example is in the RelativeToCenter class in Chapter05Jitter.

RTC eliminates jitter artifacts at the cost of requiring a custom model-view-perspective matrix per object (or per center). In addition, it requires a center be computed and positions defined relative to that center. This is not a problem for static data, but it can be extra CPU overhead for dynamic data that are changed every frame.

The real shortcoming of RTC is that it's not effective for large objects. RTC creates vertex positions small enough for 32-bit floating-point precision only when the positions are not too distant from the center. What happens if the object is large, and therefore, many positions are far from the center? Jittering will occur because vertex positions are too large for floating-point precision.

How large can an object be before jittering occurs? For 1 cm accuracy, Ohlarik recommends not exceeding a bounding radius of 131,071 m because 32-bit floating-point numbers generally have seven accurate decimal digits [126]. Many models fit into this size category, such as terrain tiles (as shown in Figure 5.4), buildings, and even cities, but many do not; consider vector data for states and countries, lines for satellite orbits, or in the extreme case, the plane in Figure 5.5.

Run Chapter05Jitter and zoom to 50 m or so, where rotating creates jitter. Switch to RTC and verify that jitter is eliminated. Zoom in to 2 m; the point and triangles start to jitter. Why? The point and triangles are rendered relative to a single center, but they are too large for RTC. The variable that determines the triangles' size, **triangleLength** in Jitter.**CreateAlgorithm**, is 200,000. Change this variable to a smaller value. Does jitter still occur using RTC?

○○○○ Try This

(a)   (b)

(c)

Figure 5.4. (a) A scene with jitter. (b) Wireframe overlaid to highlight jitter.
(c) The same scene with each terrain tile rendered using RTC. The "center" of a
tile is the center of its parent in the terrain's quadtree. The regular structure of a
quadtree makes it unnecessary to explicitly compute a center using each position.
(Images courtesy of Brano Kemen, Outerra.)

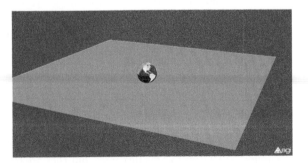

Figure 5.5. RTC does not eliminate jitter for large objects, such as this plane in
space. (Image taken using STK. The Blue Marble imagery is from NASA Visible
Earth.)

How then do we eliminate jitter for large objects? One approach is to split the object into multiple objects, each with a different center. Doing so hurts batching because each center requires a different $\mathbf{MVP_{RTC}}$ and, therefore, multiple draw calls. For some regular models, like terrain, this is natural, but for arbitrary models, it is usually not worth the trouble when there are other approaches such as rendering relative to eye.

5.3 Rendering Relative to Eye Using the CPU

To overcome the size limitations of RTC, objects can be rendered relative to eye (RTE). Both approaches utilize the CPU's double precision to ensure vertex positions and model-view-perspective matrix values are small enough for the GPU's single precision.

In RTE, each vertex is rendered relative to the eye. On the CPU, the model-view-perspective matrix is modified so the viewer is at the center of the scene, and each vertex position is translated so it is relative to the eye, not its original model coordinates. Similar to center$_{\mathrm{eye}}$ in RTC, in RTE, vertex position components get smaller as the viewer approaches because the distance between the position and the viewer decreases.

Implementation. To implement RTE, first set the translation in $\mathbf{MV}$ to zero, so the scene is centered at the viewer:

$$\mathbf{MV_{RTE}} = \begin{pmatrix} MV_{00} & MV_{01} & MV_{02} & 0 \\ MV_{10} & MV_{11} & MV_{12} & 0 \\ MV_{20} & MV_{21} & MV_{22} & 0 \\ MV_{30} & MV_{31} & MV_{32} & MV_{33} \end{pmatrix}.$$

Unlike RTC, this matrix can be used for all objects in the same model coordinate system. Each time the viewer moves or a position is updated, position(s) relative to eye are written to a dynamic vertex buffer. A position in model coordinates is made RTE by subtracting off the eye position:

$$p_{\mathrm{RTE}} = p_{\mathrm{WGS84}} - \mathrm{eye}_{\mathrm{WGS84}}$$

This subtraction is done on the 64-bit CPU using double precision, then p_{RTE} is casted to a 32-bit float and written to a dynamic vertex buffer before a draw call is issued. There is a distinction between being relative to eye and being in eye coordinates. The former involves only a translation, no rotation.

Try This ○○○○

> Run Chapter05Jitter and switch to CPU RTE. Does the scene ever jitter? Switch between CPU RTE and RTC. Given the large default size of the triangles (200,000 m), RTC results in jitter when zoomed very close, but CPU RTE does not.

Listing 5.5 shows the **MVP** and position updates used in CPU RTE. First, the model-view matrix is constructed with a zero translation so it is RTE. Then, the eye position, `eye`, is subtracted from each position and copied into a temporary array, which is then used to update the dynamic-position vertex buffer. The complete RTC code example is in the CPURelativeToEye class in Chapter05Jitter.

If just n positions change, only those n positions need to be written. If the viewer moves, all positions need to be written. Essentially, a previously static object, such as the border for a country, is now dynamic—it doesn't move in model coordinates, but it moves relative to the eye every time the eye moves!

Other vertex attributes for static objects can still be stored in a static vertex buffer; a dynamic vertex buffer is only required for the RTE positions. Both vertex buffers can be referenced from the same vertex-array object (see Section 3.5.3), and, therefore, used in the same draw call.

```
Matrix4D m = sceneState.ModelViewMatrix;
Matrix4D mv = new Matrix4D(
    m.Column0Row0, m.Column1Row0, m.Column2Row0, 0.0,
    m.Column0Row1, m.Column1Row1, m.Column2Row1, 0.0,
    m.Column0Row2, m.Column1Row2, m.Column2Row2, 0.0,
    m.Column0Row3, m.Column1Row3, m.Column2Row3, m.Column3Row3);

// Set shader uniform
_modelViewPerspectiveMatrixRelativeToEye.Value =
    (sceneState.PerspectiveMatrix * mv).ToMatrix4F();

for (int i = 0; i < _positions.Length; ++i)
{
    _positionsRelativeToEye[i] =
        (_positions[i] - eye).ToVector3F();
}

_positionBuffer.CopyFromSystemMemory(_positionsRelativeToEye);
// .. draw call

// ... elsewhere:
private readonly Vector3D[] _positions;
private readonly Vector3F[] _positionsRelativeToEye;
private readonly VertexBuffer _positionBuffer;
```

Listing 5.5. Setup for CPU RTE.

This RTE approach is called CPU RTE because the subtraction that transforms from model coordinates to RTE, $p_{\text{RTE}} = p_{\text{WGS84}} - \text{eye}_{\text{WGS84}}$, happens on the CPU. This eliminates jitter for all practical virtual globe purposes, resulting in excellent fidelity. The downside is performance: CPU RTE introduces CPU overhead, increases system bus traffic, and requires storing the original double-precision positions in system memory. This is especially poor for stationary objects that can no longer use static vertex buffers for positions.

CPU RTE takes advantage of the CPU's double precision at the cost of touching each position every time the viewer moves. Isn't the vertex shader a much better place to compute p_{RTE}? Using a vertex shader will allow us to take advantage of the incredible parallelism of the GPU (see Section 10.1.2), free up the CPU, reduce system bus traffic, and use static vertex buffers for static objects again. How can we utilize the vertex shader on 32-bit GPUs, given RTE relies on the 64-bit subtraction?

5.4 Rendering Relative to Eye Using the GPU

To overcome the performance shortcomings of CPU RTE, a double-precision position can be approximately encoded as two 32-bit floating-point positions and stored in two vertex attributes. This allows emulating the double-precision subtraction, $p_{\text{WGS84}} - \text{eye}_{\text{WGS84}}$, in a vertex shader. This approach is called GPU RTE because the subtraction happens on the GPU.

Implementation. Like CPU RTE, **MV** is first centered around the viewer by setting its translation to $(0, 0, 0)^T$. Two important implementation decisions are how to encode a double using two floats and how to emulate a double-precision subtraction using these values.

Using Ohlarik's implementation, near 1-cm accuracy can be achieved [126]. This is not as accurate as CPU RTE, which uses true double precision, but we have found it sufficient for virtual globes purposes. A double is encoded in a fixed-point representation using the fraction (mantissa) bits of two floats, `low` and `high`, using the function in Listing 5.6.

Ignoring the bit used to capture overflows, there are 23 fraction bits in an IEEE-754 single-precision floating-point value. In Ohlarik's scheme, when a double is encoded in two floats, `low` uses seven fraction bits to store the number's fractional value (i.e., the part after the decimal point). The remaining 16 bits represent an integer in the range $[0, 2^{16} - 1]$, or 0 to 65,535. All 23 fraction bits of `high` are used to represent numbers in the range $[2^{16}, (2^{23} - 1)(2^{16})]$, or 65,536 to 549,755,748,352, in increments

```
private static void DoubleToTwoFloats(
    double value, out float high, out float low)
{
  if (value >= 0.0)
  {
    double doubleHigh = Math.Floor(value / 65536.0) * 65536.0;
    high = (float)doubleHigh;
    low = (float)(value - doubleHigh);
  }
  else
  {
    double doubleHigh = Math.Floor(-value / 65536.0) * 65536.0;
    high = (float)-doubleHigh;
    low = (float)(value + doubleHigh);
  }
}
```

Listing 5.6. A method to approximately encode a double in two floats.

of 65,536. The eight exponent bits in each float remain unused to enable the use of floating-point addition and subtraction, as we will see shortly. This encoding is shown in Figure 5.6.

The largest representable whole number is 549,755,748,352 m, well above the Earth's equatorial radius of 6,378,137 m, which is a typical magnitude for positions near the surface. The seven fraction bits used in low store the fractional part in 2^{-7}, or 0.0078125, increments. Therefore, in one dimension, the error, due to roundoff, is less than 1 cm; in three dimensions, the maximum error is $\sqrt{0.0078125^2 + 0.0078125^2 + 0.0078125^2}$, or 1.353 cm, which is the accuracy for this GPU RTE approach.

Unlike traditional floating-point representations, where precision depends on the magnitude of the value, this fixed-point encoding has constant accuracy throughout. That is, it is accurate to 1.353 cm for values near zero, as well as for values near 500 billion.

The number of bits used to represent the fraction and integer parts can be adjusted to trade range for accuracy, or vice versa. For many virtual globe uses, such as GIS, 39 integer and seven fraction bits have significantly more range than required, but for other uses, such as space applications, the range is necessary.

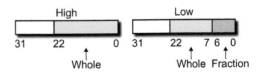

Figure 5.6. Encoding used for GPU RTE.

When a double-precision WGS84 position, p_{WGS84}, is encoded in two single precision values, p^{high} and p^{low}, and the WGS84 eye position, $\text{eye}_{\text{WGS84}}$, is encoded in eye^{high} and eye^{low}, $p_{\text{WGS84}} - \text{eye}_{\text{WGS84}}$ is computed as follows:

$$p_{\text{RTE}} = p_{\text{WGS84}} - \text{eye}_{\text{WGS84}}$$
$$\approx (p^{\text{high}} - \text{eye}^{\text{high}}) + (p^{\text{low}} - \text{eye}^{\text{low}}).$$

When the viewer is distant, $p^{\text{high}} - \text{eye}^{\text{high}}$ dominates p_{RTE}; when the viewer is near, $p^{\text{high}} - \text{eye}^{\text{high}}$ is zero, and p_{RTE} is defined by $p^{\text{low}} - \text{eye}^{\text{low}}$.

On the CPU, the x-, y-, and z-components of each position are encoded in two floating-point values and written to a vertex buffer, storing two attributes per position. Unlike CPU RTE, this vertex buffer does not need to be updated if the viewer moves because $p_{\text{WGS84}} - \text{eye}_{\text{WGS84}}$ occurs in the vertex shader, as shown in Listing 5.7. In addition to the encoded position, the shader also requires a uniform for the encoded eye position. A complete code example is in GPURelativeToEye in Chapter05Jitter.

Although the vertex shader requires only an additional two subtractions and one addition, GPU RTE doubles the amount of vertex buffer memory required for positions. This doesn't necessarily double the memory requirements unless only positions are stored. Unlike CPU RTE, this approach does not require keeping positions in system memory, nor does it require the significant CPU overhead and system bus traffic. With GPU RTE, stationary objects can store positions in static vertex buffers since the CPU is no longer involved in updating each position.

The main drawbacks to GPU RTE are that jittering can still occur for extremely close views, and for some applications, the extra vertex buffer memory may be significant.

```glsl
in vec3 positionHigh;
in vec3 positionLow;
uniform vec3 u_cameraEyeHigh;
uniform vec3 u_cameraEyeLow;
uniform mat4 u_modelViewPerspectiveMatrixRelativeToEye;

void main()
{
  vec3 highDifference = positionHigh - u_cameraEyeHigh;
  vec3 lowDifference = positionLow - u_cameraEyeLow;
  gl_Position = u_modelViewPerspectiveMatrixRelativeToEye *
                vec4(highDifference + lowDifference, 1.0);
}
```

Listing 5.7. The GPU RTE vertex shader.

Try This ○○○○

> Run Chapter05Jitter and compare CPU RTE to GPU RTE. You will notice jitter with GPU RTE when zoomed in very close given its 1.353-cm accuracy. Recall that the triangles are separated by only 1 m, so when zoomed in to the point where the triangles are outside of the viewport, a pixel can cover less than 1.353 cm, and jitter will occur. How would you change `DoubleToTwoFloats` in Listing 5.6 to decrease jitter? What are the trade-offs?

5.4.1   Improving Precision with DSFUN90

For use cases that require extremely close views, the accuracy of GPU RTE can be improved by using a different encoding scheme. The DSFUN90 Fortran library[2] contains routines for performing double-single arithmetic, which allows approximately 15 decimal digits, based on the work of Knuth [92]. Empirically, we've found that porting this library's double-single subtraction to GLSL results in a GPU RTE implementation that can be used in place of CPU RTE. We call this approach GPU RTE DSFUN90.

A double is encoded in two floats using the method in Listing 5.8. The `high` component is simply the double value cast to a float; `low` is the error introduced by this cast in single precision.

The vertex shader that uses this encoding to perform the double-single subtraction on the GPU is shown in Listing 5.9. With eight subtractions and four additions, this vertex shader is more expensive than the one for stock GPU RTE, which requires only two subtractions and one addition. Given that memory bandwidth and vertex attribute setup have the same cost for both approaches, the extra instructions are not prohibitively expensive. The complete DSFUN90 example is in GPURelativeToEyeDSFUN90 in Chapter05Jitter.

Other double-single arithmetic operations from DSFUN90 have been ported to the GPU using CUDA.[3] It would be straightforward to port these to GLSL.

```
private static void DoubleToTwoFloats (
    double value, out float high, out float low)
{
  high = (float)value;
  low = (float)(value - high);
}
```

Listing 5.8. Encoding a double in two floats for GPU RTE DSFUN90.

---

[2] http://crd.lbl.gov/~dhbailey/mpdist/
[3] http://sites.virtualglobebook.com/virtual-globe-book/files/dsmath.h

```
in vec3 positionHigh;
in vec3 positionLow;
uniform vec3 u_cameraEyeHigh;
uniform vec3 u_cameraEyeLow;
uniform mat4 u_modelViewPerspectiveMatrixRelativeToEye;

void main()
{
  vec3 t1 = positionLow - u_cameraEyeLow;
  vec3 e = t1 - positionLow;
  vec3 t2 = ((-u_cameraEyeLow - e) + (positionLow - (t1 - e))) +
              positionHigh - u_cameraEyeHigh;
  vec3 highDifference = t1 + t2;
  vec3 lowDifference = t2 - (highDifference - t1);

  gl_Position = u_modelViewPerspectiveMatrixRelativeToEye *
              vec4(highDifference + lowDifference, 1.0);
}
```

Listing 5.9. The GPU RTE DSFUN90 vertex shader.

> Run Chapter05Jitter and compare GPU RTE to GPU RTE DSFUN90.
> Zoom in until GPU RTE starts to jitter, then switch to DSFUN90. When
> would you use stock GPU RTE over DSFUN90? How does CPU RTE
> compare to DSFUN90? When would you use CPU RTE over DSFUN90?

○○○○ **Try This**

5.4.2 Precision LOD

In addition to increasing precision, GPU RTE DSFUN90 lends itself nicely
to precision level of detail (LOD). Take a second look at how a double-
precision value is encoded into two single-precision values:

```
double value = // ...
float high = (float)value;
float low = (float)(value - high);
```

The assignment to **high** casts the value to a float, which is the same
cast used if the object were naïvely rendered relative to world (RTW)
using an unmodified **MVP**, without any concern for jitter. This only
requires a single 32-bit vertex attribute but produces jitter when zoomed
in close, as explained in Section 5.1.2. A hybrid approach that uses RTW
when zoomed out and DSFUN90 when zoomed in can have the best of
both algorithms: high precision when zoomed in and low memory usage

when zoomed out—and no jittering anywhere. Similar to how geometric LOD changes geometry based on view parameters, precision LOD changes precision based on the view.

Both RTW and DSFUN90 share the vertex buffer used to store `high`, but only DSFUN90 requires the vertex buffer storing `low`. When the viewer is zoomed out, the high vertex buffer is used with the RTW shader, taking advantage of all the GPU memory savings and performance benefits of RTW. When the viewer zooms in, both the high and low vertex buffers are used with the DSFUN90 shader to render with full precision.

Both vertex buffers can be created at the same time, or the low vertex buffer can be created on demand to reduce the amount of GPU memory used, not just reduce the attribute setup cost and vertex-shader complexity. For most scenes, the viewer is only close to a small percentage of objects, so creating the vertex buffer on demand can be very beneficial. The downside is that the original double-precision positions, or at least the floating-point low part, need to be kept in system memory, or paged back in to create the low vertex buffer. Using separate vertex buffers, instead of interleaving high and low components in the same vertex buffer, can cause a slight performance penalty when both components are called for (see Section 3.5.1).

Precision LOD is a form of discrete LOD with just two discrete levels: low and high. As the viewer zooms in, the LOD level should switch to high before jitter becomes noticeable. This viewer distance should be determined by the position's maximum component, the viewer position's maximum component, and the field of view. This is similar to determining screen-space error.

Saving memory is always a good thing in a world where size is speed, but it is important to put into perspective how much memory can be saved with precision LOD. First, most scenes use significantly more memory for textures than for vertex data. Some virtual globe scenes, however, especially those heavy with vector data, can use a significant amount of memory for vertices. Next, for distant objects, precision LOD cuts the amount of memory required for positions in half, but this does not mean the amount of vertex data is also cut in half. Vertex data may also include attributes such as normals, texture coordinates, etc. If the shading is also LODed, some attributes may not be necessary for simplified shaders (e.g., a simplified lighting model may not require tangent vectors).

As hinted above, precision LOD can be combined with geometric and shader LOD. As the viewer zooms out, the number of vertices rendered, the size of each vertex, and the shading complexity may all be reduced. When shading complexity is reduced, it may also be beneficial to reduce the memory usage of other vertex attributes (e.g., replacing floating-point normals with half-floats).

The complete precision LOD example is in SceneGPURelativeTo EyeLOD in Chapter05Jitter.

Run Chapter05Jitter and compare GPU RTE DSFUN90 to precision LOD. At what distant interval does precision LOD jitter? Why? What needs to be adjusted to eliminate jitter in this example?

○○○○ **Try This**

5.5   Recommendations

Based on the trade-offs in Table 5.1, our recommendations are as follows:

- Use GPU RTE DSFUN90 for all objects with positions defined in WGS84 coordinates if you can afford the extra vertex memory and vertex-shader instructions. The increase in vertex-buffer memory usage may be insignificant compared to the memory used by textures. Likewise, the increase in vertex-shader complexity may not be significant compared to fragment shading, which is likely to be the case in games with complex fragment shaders. This results in the simplest software design and excellent precision.

- Use RTC for objects defined in their own coordinate system (e.g., a model for a building) with a bounding radius less than 131,071 m.

- For better performance and memory usage, consider a hybrid approach:

  - Use CPU RTE for highly dynamic geometry. If the geometry is changing frequently as the viewer moves, the additional CPU overhead is not significant. Using CPU RTE reduces the vertex memory requirements and reduces system bus traffic in this case.

  - Use RTC for all static geometry with a bounding radius less than 131,071 m. Although this requires a separate model matrix per object, each object only requires 32-bit positions, which do not need to be shadowed in system memory.

  - Use GPU RTE DSFUN90 for all static geometry with a bounding radius greater than 131,071 m. If using RTC results in many small batches due to the need to set the model matrix (or model-view matrix) uniform, use GPU RTE DSFUN90 to increase the batch size.

Algorithm	Strengths	Weaknesses
RTW (relative to world)	Simple, obvious, several objects can be rendered with the same model matrix, and only uses 32-bit positions.	Results in jitter on 32-bit GPUs when zoomed in, and even a few hundred meters away. Scaling the coordinates down can reduce, but not eliminate, jitter.
RTC (relative to center)	Natural for some geometry, like models, results in good precision for reasonably sized geometry, and only uses 32-bit positions.	Large geometry still jitters, but less so than with RTW. Each object requires a different model-view matrix, making it heavy on uniform updates and less large-batch friendly. Slightly unnatural for geometry typically defined in world coordinates like vector data.
CPU RTE (relative to eye)	Excellent precision, several objects can be rendered with the same model-view matrix, only uses 32-bit positions on the GPU, a handful of dynamic vertex buffers can be used instead of many static buffers, appropriate for dynamic geometry.	Significant CPU overhead hurts performance because the CPU needs to touch every position whenever the camera moves. Static geometry essentially becomes dynamic because of camera movement, disabling the use of static vertex buffers for positions. The only approach that requires keeping positions in system memory, along with scratch memory for translating them relative to the eye.
GPU RTE	Very good precision, even for large geometry, until zoomed in to a couple of meters; very little CPU overhead; several objects can be rendered with the same model-view matrix, appropriate for static geometry.	Jittering still occurs when zoomed in very close. A position requires two 32-bit components in a vertex buffer. Requires two subtractions and an addition in the vertex shader.
GPU RTE DS-FUN90	Similar to GPU RTE, only with no observed jitter; simpler CPU code than GPU RTE; works well with precision LOD.	Without LOD, a position requires two 32-bit components in a vertex buffer, and the vertex shader requires several instructions: eight subtractions and four additions.

Table 5.1. Summary of trade-offs between approaches to reducing jitter.

- It is reasonable to use GPU RTE instead of GPU RTE DSFUN90 if precision LOD will not be used, and the extra precision of the latter is not required.

When I joined AGI in 2004, most 3D code in STK used CPU RTE. Notable exceptions included models and terrain chunks, which were small enough to use RTC. Although CPU RTE doesn't sound ideal, a lot of geometry was dynamic, and CPU RTE actually made sense when the code was first written, before vertex processing was done on the GPU!

In 2008, while working on Insight3D, we reevaluated jitter. By this time, GPUs changed dramatically: they included fast programmable vertex shaders, and in many cases, the CPU couldn't feed the GPU quickly enough. We wanted to find a way to take advantage of shaders to stay off the CPU; then my colleague came up with GPU RTE. I couldn't have been happier, but we still thought we could do better.

For moderately-sized geometry, RTC can render with as much precision as GPU RTE, using half the amount of memory for positions. With the well-intentioned goal of saving memory, we developed a hybrid solution (simplified):

○○○○ Patrick Says

- If geometry is dynamic, use GPU RTE.

- If geometry is static with a boundary radius less than 131,071 m, use RTC.

- If geometry is static and greater than 131,071 m, divide the geometry in several meshes and render each using RTC. If some heuristic determines that this results in too many or too few batches, combine batches and render them with GPU RTE. Large triangles and lines were also rendered with GPU RTE, instead of subdividing them and creating multiple RTC batches.

We nicely abstracted this away and hooked up lots of code to use it. It turns out we had a lot of geometry that fit into the last category, and the divide step was problematic and slow. In addition, writing a vertex shader was a pain because the abstraction sometimes used GPU RTE, RTC, or both. Over time, we wound up replacing the abstraction with GPU RTE in many cases; we favored the simplicity over the memory savings of RTC.

Massive-world games also need to eliminate jitter but usually have the luxury of restricting the speed of the viewer, allowing additional techniques. For example, Microsoft Flight Simulator uses a technique similar to CPU

RTE [163]. The origin is not always at the eye; instead, it is only periodically updated to keep it within a few kilometers of the eye. Given the speed of the plane, the origin changes infrequently, not every time the viewer moves. When the origin changes, the updated vertices are double buffered because the update can take several frames. Double buffering has memory costs, so the full vertex memory savings of CPU RTE are not realized.

Flight Simulator's approach is effective because the viewer's speed is bound; in a virtual globe, the viewer could be somewhere completely different several frames later—the viewer can be zoomed out in space looking at the entire Earth one second, then zoomed in to a few meters, looking at high-resolution imagery the next second.

5.6   Resources

A solid understanding of floating point is useful for more than just transforming vertices. Calculations with large floating-point values are used everywhere in virtual globes. Thankfully, there are many resources on the topic, including Bush's example-filled article [23], Hecker's article in *Game Developer Magazine* [67], Ericson's slides on numerical robustness [46], and a chapter in Van Verth and Bishop's book on mathematics for games [176].

Jittering solutions have received a good bit of attention. Thorne's PhD thesis describes jittering and the continuous floating origin in depth [169]. Ohlarik's article introduces GPU RTE and compares it to other approaches [126].

Thall surveys strategies for double and quad precision on 32-bit GPUs and provides Cg implementations for arithmetic, exponential, and trigonometric functions [167]. The subtraction function can be used for implementing GPU RTE.

Forsyth recommends using RTC with fixed-point positions, instead of double-precision floating point, for better performance, constant precision, and a full 64 bits of precision [51]. The encoding scheme for GPU RTE introduced in Section 5.4 is a fixed-point representation embedded in the fractional bits of two single-precision floats, so a full 64 bits of precision is not achieved.

In addition to virtual globes, games with large worlds also have to combat jitter. Persson describes the technique used in "Just Cause 2" [140].

Depth Buffer Precision

At some point in their career, every graphics developer asks, "What should I set the near and far plane distances to?" Ambitious developers may even ask, "Why don't I just set the near plane to a really small value and the far plane to a really huge value?" When they see the dramatic impact of this, the surprised look on their faces is priceless!

We too were once baffled by visual artifacts, such as those shown in Figure 6.1(a). These artifacts are due to *z-fighting*—the depth buffer cannot tell which object is in front of which. Worst yet, the artifacts tend to flicker as the viewer moves around.

When we first experienced z-fighting, the near and far plane distances and their impact on depth buffer precision was the last place we thought to look.[1] Thankfully, there are good reasons for this behavior and several techniques for improving precision, thus reducing or completely eliminating artifacts. This chapter describes such techniques and their trade-offs, ranging from simple one liners to system-wide changes.

6.1 Causes of Depth Buffer Errors

Virtual globes are unique in that objects very close to the viewer and objects very far from the viewer are present in most scenes. Other 3D applications, such as first-person shooters in indoor environments, have a well-defined upper bound on view distance. In virtual globes and open-world games, the viewer can go anywhere and look in any direction—fun for them, not for us. The viewer may be zoomed in close to a model with

---

[1]Z-fighting can also occur when rendering coplanar geometry (e.g., two overlapping triangles lying on the same plane). This chapter does not consider this cause of z-fighting; instead, it focuses on z-fighting due to large worlds.

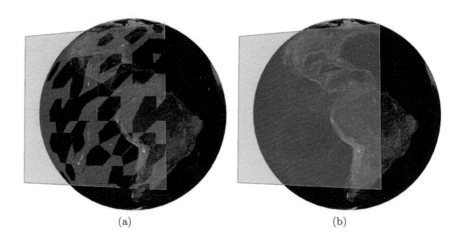

(a)   (b)

Figure 6.1. (a) A near-plane distance of 1 m creates artifacts due to limited depth precision for distant objects. (b) Moving the near plane out to 35 m eliminates the artifacts, at least for this view. In both cases, the gray plane is 1,000,000 m above Earth at its tangent point, and the far plane is set to a distant 27,000,000 m.

mountains in the distance, as shown in Figure 6.2(a), or the viewer may be zoomed into a satellite, with planets in the distance, as in Figure 6.2(c). Even a scene with just Earth requires a very large view distance, given Earth's semiminor axis of 6,356,752.3142 m (see Section 2.2.1).

Such views are difficult because they call for a close near plane and a very distant far plane. For orthographic projections, this usually isn't a concern because the relationship between eye space z, z_{eye}, and window space z, z_{window}, is linear. In perspective projections, this relationship is nonlinear [2,11]. Objects near the viewer have small z_{eye} values, with plenty of potential z_{window} values to map to, allowing the depth buffer to resolve visibility correctly. The farther away an object, the larger its z_{eye}. Due to the nonlinear relationship, these z_{eye} values have much less z_{window} values to map to, so the depth buffer has trouble resolving visibility of distant objects close to each other, which creates z-fighting artifacts like those in Figure 6.1(a).

Close objects (small z_{eye}) have lots of depth buffer precision (lots of potential z_{window}). The farther away an object, the less precision it has. This nonlinear relationship between z_{eye} and z_{window} is controlled by the ratio of the far plane, f, to the near plane, n. The greater the $\frac{f}{n}$ ratio, the greater the nonlinearity [2].

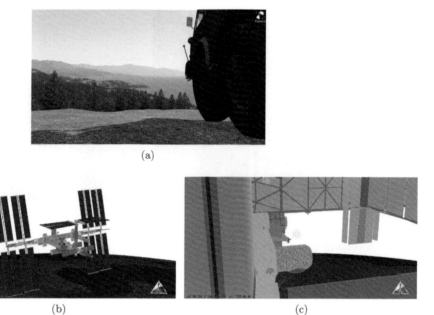

Figure 6.2. (a) It is challenging to view both the nearby model and distant terrain in the same scene without depth buffer artifacts. This figure was rendered using a logarithmic depth buffer, discussed in Section 6.4, to eliminate the artifacts. Image courtesy of Brano Kemen, Outerra. (b) The International Space Station (ISS) is shown orbiting Earth, with its solar panels pointed towards the sun. (c) As the viewer zooms towards the ISS, the lack of depth buffer precision in the distance needs to be accounted for. (Images in (b) and (c) were taken using STK, which uses multifrustum rendering to eliminate z-fighting, discussed in Section 6.5.)

The reason for this behavior requires a closer look at the perspective-projection matrix. The OpenGL version of this matrix is

$$\begin{pmatrix} \frac{2n}{r-l} & 0 & \frac{r+l}{r-l} & 0 \\ 0 & \frac{2n}{t-b} & \frac{t+b}{t-b} & 0 \\ 0 & 0 & -\frac{f+n}{f-n} & -\frac{2fn}{f-n} \\ 0 & 0 & -1 & 0 \end{pmatrix}.$$

The parameters (l, r, b, t, n, f) define a view frustum.[2] The variables l and r define left and right clipping planes, b and t define bottom and top clipping planes, and n and f define near and far clipping planes (see

---

[2]These are the same parameters passed to the now deprecated **glFrustum** function.

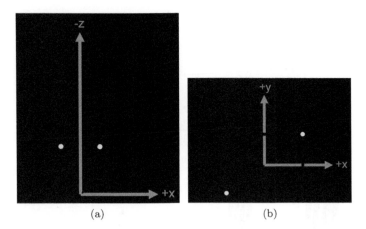

Figure 6.3. A frustum defined by parameters (l, r, b, t, n, f) used for perspective projection. (a) A frustum in the xz-plane. (b) A frustum in the xy-plane.

Figure 6.3). With the eye located at $(0, 0, 0)$, the corner points $(l, b, -n)$ and $(r, t, -n)$ are mapped to the lower left and upper right corners of the window.

Let's see how the perspective projection and subsequent transformations determine the final depth value, z_{window}, for a point in eye coordinates, $p_{\text{eye}} = (x_{\text{eye}}, y_{\text{eye}}, z_{\text{eye}})$. The perspective-projection matrix transforms p_{eye} to clip coordinates:

$$
p_{\text{clip}} = \begin{pmatrix} \frac{2n}{r-l} & 0 & \frac{r+l}{r-l} & 0 \\ 0 & \frac{2n}{t-b} & \frac{t+b}{t-b} & 0 \\ 0 & 0 & -\frac{f+n}{f-n} & -\frac{2fn}{f-n} \\ 0 & 0 & -1 & 0 \end{pmatrix} \begin{pmatrix} x_{\text{eye}} \\ y_{\text{eye}} \\ z_{\text{eye}} \\ 1 \end{pmatrix} = \begin{pmatrix} x_{\text{clip}} \\ y_{\text{clip}} \\ -\left(\frac{f+n}{f-n} z_{\text{eye}} + \frac{2fn}{f-n} \right) \\ -z_{\text{eye}} \end{pmatrix}.
$$

The next transform, perspective division, divides each p_{clip} component by w_{clip} yielding normalized device coordinates:

$$
p_{\text{ndc}} = \begin{pmatrix} \frac{x_{\text{clip}}}{-z_{\text{eye}}} \\ \frac{y_{\text{clip}}}{-z_{\text{eye}}} \\ \frac{-\left(\frac{f+n}{f-n} z_{\text{eye}} + \frac{2fn}{f-n} \right)}{-z_{\text{eye}}} \\ \frac{-z_{\text{eye}}}{-z_{\text{eye}}} \end{pmatrix} = \begin{pmatrix} x_{\text{ndc}} \\ y_{\text{ndc}} \\ \frac{\left(\frac{f+n}{f-n} z_{\text{eye}} + \frac{2fn}{f-n} \right)}{z_{\text{eye}}} \\ 1 \end{pmatrix}. \tag{6.1}
$$

Hence,

$$z_{ndc} = \frac{\left(\frac{f+n}{f-n} z_{eye} + \frac{2fn}{f-n}\right)}{z_{eye}},$$

$$= \frac{f+n}{f-n} + \frac{2fn}{z_{eye}(f-n)}.$$

In normalized device coordinates, the scene is in the axis-aligned cube with corners $(-1, -1, -1)$ and $(1, 1, 1)$; therefore, z_{ndc} is in the range $[-1, 1]$. The final transformation, the viewport transformation, maps z_{ndc} to z_{window}, where the range of z_{window} is defined by the application using $\texttt{glDepthRange}(n_{range}, f_{range})$:

$$z_{window} = z_{ndc} \left(\frac{f_{range} - n_{range}}{2}\right) + \frac{n_{range} + f_{range}}{2}$$

$$= \left(\frac{f+n}{f-n} + \frac{2fn}{z_{eye}(f-n)}\right) \left(\frac{f_{range} - n_{range}}{2}\right) + \frac{n_{range} + f_{range}}{2}.$$

In the common case of $n_{range} = 0$ and $f_{range} = 1$, z_{window} simplifies to

$$z_{window} = \left(\frac{f+n}{f-n} + \frac{2fn}{z_{eye}(f-n)}\right) \left(\frac{1}{2}\right) + \frac{1}{2}$$

$$= \frac{\left(\frac{f+n}{f-n} + \frac{2fn}{z_{eye}(f-n)}\right) + 1}{2}. \tag{6.2}$$

For a given perspective projection, everything on the right-hand side of Equation (6.2) is constant except for z_{eye}; therefore,

$$z_{window} \propto \frac{1}{z_{eye}}.$$

This relationship is the result of the perspective divide from Equation (6.1). This division causes perspective foreshortening, making objects in the distance appear smaller and parallel lines converge in the distance. Dividing by z_{eye} makes sense for x_{clip} and y_{clip}; when z_{eye} is big, objects in the distance are scaled down. At the same time, z_{clip} is divided, making it also inversely proportional to z_{eye}.

Consider Figure 6.4: when z_{eye} is small, $\frac{1}{z_{eye}}$ is a quickly moving function; hence, a small change in z_{eye} will yield a large enough change in z_{window} that the depth buffer can resolve depth correctly. As z_{eye} gets large, $\frac{1}{z_{eye}}$ becomes a slow-moving function; therefore, a small change in z_{eye} may yield the same z_{window}, creating rendering artifacts.

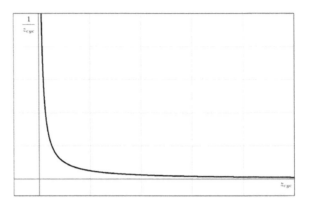

Figure 6.4. Given $z_{\text{window}} \propto \frac{1}{z_{\text{eye}}}$, a small change in z_{eye} when z_{eye} is small results in a large change to z_{window}. When z_{eye} is large, a small change in z_{eye} only very slightly changes $\frac{1}{z_{\text{eye}}}$, creating z-fighting artifacts.

The relationship between z_{eye} and z_{window} is also dependent on f and n. Figure 6.5 shows Equation (6.2) graphed for various values of n. Small values of n cram all the depth buffer precision at the near plane, larger values of n spread precision more evenly throughout. This is shown in the shape of each curve.

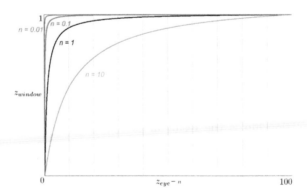

Figure 6.5. The relationship between z_{eye} and z_{window} for various values of near-plane distance, n, and a fixed-view distance, $f - n$, of 100. The x-axis shows the distance from the near plane. As n gets large, precision is spread more evenly throughout.

To see the effects of moving the near plane, run Chapter06Depth
BufferPrecision. Move the near plane out until z-fighting between the
gray plane and Earth is eliminated. Then move the gray plane closer
to Earth. The artifacts occur again because the world-space separation
between the objects is not large enough to create unique window space
z-values. Push the near plane out even farther to eliminate the artifacts.

○○○○ Try This

6.1.1 Minimum Triangle Separation

The minimum triangle separation, $S_{\min}$, for a distance from the eye, z_{eye},
is the minimum world-space separation between two triangles required for
correct depth occlusion. This is also referred to as the resolution of the
depth buffer at the given distance. Based on our discussion thus far, it is
easy to see that farther objects require a larger $S_{\min}$. For an x-bit fixed-
point depth buffer where the distance to the near plane, n, is much smaller
than the distance to the far plane, f, Baker [11] provides an approximation
to $S_{\min}$:

$$S_{\min} \approx \frac{z_{\text{eye}}^2}{2^x n - z_{\text{eye}}}.$$

Figure 6.6 shows $S_{\min}$ graphed for various values of n. As expected,
larger values of n result in a flatter curve. These lower values of $S_{\min}$ allow
objects to be closer together without creating rendering artifacts.

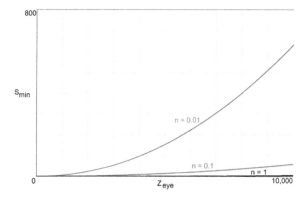

Figure 6.6. Approximate minimum triangle separation as a function of distance
for various near-plane values. Larger values of n result in a flatter curve and
allow proper depth occlusion for objects close together.

Although this nonlinear behavior is the most commonly cited reason for depth buffer errors, Akeley and Su [1] show that window coordinate precision, field of view, and error accumulated by single-precision projection, viewport, and rasterization arithmetic significantly contribute to effective depth buffer resolution.

Window coordinates, x_{window} and y_{window}, are commonly stored on the GPU in a fixed-point representation with 4 to 12 fractional bits. Prior to rasterization, a vertex is shifted into this representation, changing its position and thus its rasterized depth value. For a given z_{eye}, S_{min} depends on x_{eye} and y_{eye}; S_{min} increases as x_{eye} and y_{eye} move away from the origin. The wider the field of view, the greater the increase in S_{min}. The result is that depth buffer errors, and thus z-fighting, are more likely to occur along the window edges and in scenes with a wider field of view.

Now that we have an understanding of the root cause of z-fighting, it is time to turn our attention to the solutions. We start with some simple rules of thumb, then move on to more involved techniques.

6.2   Basic Solutions

Fortunately, some of the simplest approaches to eliminating artifacts due to depth buffer precision are also the most effective. Pushing out the near plane as far as possible ensures the best precision for distant objects. Also, the near plane doesn't need to be the same from frame to frame. For example, if the viewer zooms out to view Earth from space, the near plane may be pushed out dynamically. Of course, pushing out the near plane can create new rendering artifacts; objects are either entirely clipped by the near plane or the viewer can see "inside" partially clipped objects. This can be mitigated by using blending to fade out objects as they approach the near plane.

Likewise, pulling in the far plane as close as possible can help precision. Given that the far to near ratio, $\frac{f}{n}$, affects precision, pound for pound, moving the far plane in is less effective than moving the near plane out. For example, given an initial $n = 1$ and $f = 1,000$, the far plane would have to be moved in 500 m to get the same ratio as moving the near plane out 1 m. Nonetheless, minimizing the far-plane distance is still recommended. Objects can be gracefully faded out using blending or fog as they approach the far plane.

In addition to improving precision, setting the near plane and far plane as close as possible to each other can improve performance. More objects are likely to get culled on the CPU or at least clipped during the rendering pipeline, eliminating their rasterization and shading costs. A low $\frac{f}{n}$ ratio also helps GPU z-cull (also called hierarchical Z) optimizations. Z-cull tiles store a low-resolution depth value, so the lower the $\frac{f}{n}$, the more likely a tile's depth value will have enough precision to be safely culled when the depth difference is small [136].

Another effective technique commonly used in virtual globes is to aggressively remove or fade out objects in the distance. For example, in a scene with Earth and vector data for major highways, when the viewer reaches a high altitude, the highways will fade out. Removing distant objects can also declutter the scene, especially if the scene contains lots of text and billboards (see Chapter 9), whose pixel size does not decrease with distance. This is similar to LOD, which usually reduces the geometric or shader complexity of an object based on its pixel size to improve performance.

6.3   Complementary Depth Buffering

OpenGL supports three depth buffer formats: 16-bit fixed point, 24-bit fixed point, and 32-bit floating point. Most of today's applications use a 24-bit fixed-point depth buffer because it is supported by a wide array of video cards. A property of fixed-point representation is that values are uniformly distributed, unlike floating-point representation, where values are nonuniformly distributed. Recall that representable floating-point values are close together near zero and spaced farther and farther apart as values move away from zero.

A technique called complementary depth buffering [66,94,137] takes advantage of a 32-bit floating-point depth buffer to compensate for nonlinear depth mapping. For normal depth buffering with $n_{\text{range}} = 0$ and $f_{\text{range}} = 1$, the clear depth is typically one, the depth comparison function is "less," and the distance to the near plane is less than the distance to the far plane, $n < f$. For complementary depth buffering with the same n_{range} and f_{range}, clear depth to zero, use a depth compare of "greater," and swap n and f when constructing the perspective-projection matrix (compare Listings 6.1 and 6.2).

```
ClearState clearState = new ClearState();
clearState.Depth = 1;
context.Clear(clearState);
sceneState.Camera.PerspectiveNearPlaneDistance = nearDistance;
sceneState.Camera.PerspectiveFarPlaneDistance = farDistance;
renderState.DepthTest.Function = DepthTestFunction.Less;
```

Listing 6.1. Normal depth buffering. The depth buffer is cleared to one, the near and far planes are set as usual, and the "less" depth comparison function is used.

```
ClearState clearState = new ClearState();
clearState.Depth = 0;
context.Clear(clearState);
sceneState.Camera.PerspectiveNearPlaneDistance = farDistance;
sceneState.Camera.PerspectiveFarPlaneDistance = nearDistance;
renderState.DepthTest.Function = DepthTestFunction.Greater;
```

Listing 6.2. Complementary depth buffering improves precision for distant objects. The depth buffer is cleared to zero, the near and far planes are swapped, and the "greater" depth comparison function is used.

How does this improve depth buffer precision for distant objects? Floating-point representations are nonlinear, just like depth values. As floating-point values move far away from zero, there are less discrete rep-

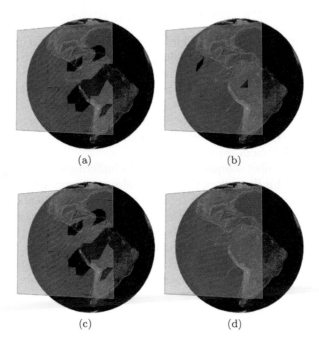

Figure 6.7. Complementary depth buffering with different depth buffer formats. The near plane is at 14 m. Similar to Figure 6.1, the gray plane is 1,000,000 m above Earth at its tangent point, and the far plane is set to a distant 27,000,000 m. In this scene, the difference between a (a) 24-bit fixed-point and (c) 32-bit floating-point buffer is unnoticeable. (a) and (b) Complementary depth buffering provides decent improvements when used with a 24-bit fixed-point buffer. (c) and (d) Complementary depth buffering is most effective when used with a 32-bit floating-point buffer.

resentations. By reversing the depth mapping, the highest floating-point precision is used at the far plane, where the lowest depth precision is found. Likewise, the lowest floating-point precision is used at the near plane, where the highest precision is found. These opposites balance each other out quite a bit. Persson recommends this technique even if a fixed-point buffer is used since it improves the precision during transformations [137]. Figure 6.7 shows complementary depth buffering with both fixed- and floating-point buffers.

To see this technique in action, run Chapter06DepthBufferPrecision. Swap between different depth buffer formats and toggle complementary depth buffering. The effectiveness of complementary depth buffering with a floating-point buffer can also be seen by running NVIDIA's simple depth float sample (see http://developer.download.nvidia.com/SDK/10. 5/opengl/samples.html).

○○○○ **Try This**

6.4   Logarithmic Depth Buffer

A logarithmic depth buffer improves depth buffer precision for distant objects by using a logarithmic distribution for z_{screen} [82]. It trades precision for close objects for precision for distant objects. It has a straightforward implementation, only requiring modification of z_{clip} in a shader using the following equation:[3]

$$z_{clip} = \frac{2 \ln(C z_{clip} + 1)}{\ln(Cf + 1)} - 1. \tag{6.3}$$

As in Section 6.1, z_{clip} is the z-coordinate in clip space, the space in the pipeline immediately after the perspective-projection transformation but before perspective divide. The variable f is the distance to the far plane, and C is a constant that determines the resolution near the viewer.

Decreasing C increases precision in the distance but decreases precision near the viewer. Given C and an x-bit fixed-point depth buffer, the approximate minimum triangle separation, S_{min}, at distance z_{eye} is

$$S_{min} = \frac{\ln(Cf + 1)}{(2^x - 1) \frac{C}{C z_{eye} + 1}}.$$

---

[3]This equation uses OpenGL's normalized device coordinates, where $z_{ndc} \in [-1, 1]$. For Direct3D, where $z_{ndc} \in [0, 1]$, the equation is $\frac{\ln(C z_{clip} + 1)}{\ln(Cf + 1)}$.

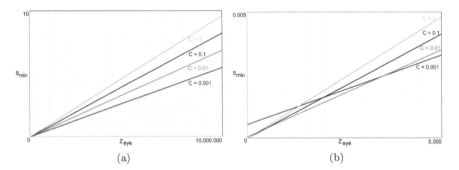

(a)   (b)

Figure 6.8. Approximate minimum triangle separation as a function of distance for various C values using a logarithmic depth buffer. Lower C values result in (a) more precision (lower S_{min}) in the distance but (b) less precision up close. Compare these graphs, including scale, to normal depth buffering shown in Figure 6.6.

Figure 6.8 graphs various C values for a 24-bit fixed-point depth buffer and a distant far plane of 10,000,000 m. Figure 6.8(a) shows that lower values of C provide increased precision in the distance. Figure 6.8(b) shows the adverse effect low values of C have on precision near the viewer. This precision is tolerable for many scenes, considering that $C = 0.001$ yields $S_{min} = 5.49$ at the far plane while still yielding $S_{min} = 0.0005$ at 1 m.

A logarithmic depth buffer can produce excellent results, as shown in Figure 6.9.

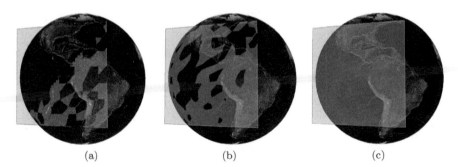

(a)   (b)   (c)

Figure 6.9. A comparison of (a) normal depth buffering, (b) complementary depth buffering, and (c) a logarithmic depth buffer. A logarithmic depth buffer produces excellent results for distant objects. In this scene, the near plane is at 1 m and the far plane is at 27,000,000 m. The gray plane is 20,000 m above Earth at its tangent point.

```
in vec4 position;

uniform mat4 og_modelViewPerspectiveMatrix;
uniform float og_perspectiveFarPlaneDistance;
uniform bool u_logarithmicDepth;
uniform float u_logarithmicDepthConstant;

vec4 ModelToClipCoordinates(
    vec4 position,
    mat4 modelViewPerspectiveMatrix,
    bool logarithmicDepth,
    float logarithmicDepthConstant,
    float perspectiveFarPlaneDistance)
{
  vec4 clip = modelViewPerspectiveMatrix * position;

  if (logarithmicDepth)
  {
    clip.z =
        ((2.0 * log(logarithmicDepthConstant * clip.z + 1.0) /
            log(logarithmicDepthConstant *
                perspectiveFarPlaneDistance + 1.0)) - 1.0) *
        clip.w;
  }

  return clip;
}

void main()
{
  gl_Position = ModelToClipCoordinates(position,
      og_modelViewPerspectiveMatrix,
      u_logarithmicDepth, u_logarithmicDepthConstant,
      og_perspectiveFarPlaneDistance);
}
```

Listing 6.3. Vertex shader for a logarithmic depth buffer.

Run Chapter06DepthBufferPrecision and enable the logarithmic depth buffer. Move the gray plane in the scene closer to Earth. Compare the results to that of complementary depth buffering or simple modification of the near plane.

○○○○ Try This

In a vertex-shader implementation, like the one shown in Listing 6.3, Equation (6.3) should be multiplied by w_{clip} to undo the perspective division that will follow. Using a vertex shader leads to an efficient implementation, especially considering that $\frac{2}{\ln(Cf+1)}$ is constant across all vertices and can, therefore, be made part of the projection matrix. But a vertex-shader implementation leads to visual artifacts for visible triangles with a vertex behind the viewer. The artifacts are caused because the fixed function computes perspective-correct depth by linearly interpolating $\frac{z}{w}$ and $\frac{1}{w}$. Multiplying z_{clip} by w_{clip} in the vertex shader means that $\frac{z}{w}$ used for interpolation is really just z. To eliminate this problem, Brebion writes depth in a fragment shader using Equation (6.3) and z_{clip} passed from the vertex shader [21].

6.5   Rendering with Multiple Frustums

Multifrustum rendering, also called depth partitioning, fixes the depth buffer precision problem by capitalizing on the fact that each object does not need to be rendered with the same near and far planes. Given that a reasonable $\frac{f}{n}$ ratio for a 24-bit fixed-point depth buffer is 1,000 [2], only a handful of unique frustums are required to cover a large distance with adequate depth precision.

For example, near- and far-plane distances for a virtual globe application might be 1 m and 100,000,000 m. Given Earth's semimajor axis of 6,378,137 m, these distances allow the viewer to zoom out quite a bit to view the entire planet and perhaps satellites orbiting it. Using multifrustum rendering with an $\frac{f}{n}$ ratio of 1,000, the frustum closest to the viewer would have a near plane of 1 m and a far plane of 1,000 m. The next closest frustum would have a near plane of 1,000 m and a far plane of 1,000,000 m. The farthest frustum would have a near plane of 1,000,000 m

Figure 6.10. Multifrustum rendering with a near plane of 1 m, a far plane of 100,000,000 m, and an $\frac{f}{n}$ ratio of 1,000 (not drawn to scale). When the scene is rendered, first the purple and green objects are drawn, then the cyan object, and finally the red object. To avoid artifacts, the frustums should overlap slightly.

and a far plane of 100,000,000 m (or even 1,000,000,000 m). This is shown in Figure 6.10. As explained in Section 6.2, once the near plane is pushed out far enough, a single frustum covers a large distance with adequate depth precision.

The scene is rendered starting with objects overlapping the frustum farthest from the viewer, then continuing with the next farthest frustum, and so on, until the closest frustum is rendered. Before rendering objects overlapping a frustum, the perspective-projection matrix is computed using the frustum's near and far planes, and the depth buffer is cleared. Clearing the depth buffer ensures that objects do not lose the depth test against objects in a farther frustum; objects in a closer frustum should always overwrite objects in a further frustum because the frustums are sorted far to near.

Rendering a scene like this requires determining which objects are in which frustums. When the viewer moves, the object/frustum relationship will change, even if the objects do not move. One approach to efficiently handling this is to first perform hierarchical-view frustum culling against the union of individual frustums (i.e., a frustum with the near plane of the closest frustum and a far plane of the farthest frustum). Next, sort the objects by distance to the viewer. Finally, use this sorted list to quickly group objects into the correct frustum(s) and render them in front-to-back order within each frustum to aid hardware z-buffer optimizations.

To avoid tearing artifacts, adjacent frustums should have slight overlap. In our example, the middle frustum has the same near plane, 1,000 m, as the first frustum's far plane. To avoid gaps in the rendered image, the middle frustum's near plane should overlap slightly (e.g., be set to 990 m).

> STK uses multifrustum rendering. It defaults to a near plane of 1 m, a far plane of 100,000,000 m, and an $\frac{f}{n}$ ratio of 1,000. This allows users to zoom in very close on a satellite and still accurately see planets in the distance. STK tries to minimize the total number of frustums by pushing the near plane out as far as possible based on the viewed object. There is an advanced panel that allows users to tweak the near plane, far plane, and $\frac{f}{n}$ ratio, but thankfully, this is almost never required.

○○○○ Patrick Says

Multifrustum rendering allows virtually infinite depth precision and even runs on old video cards. It allows a large enough view distance that fog and translucency tricks from Section 6.2 are usually unnecessary.

Multifrustum rendering is not without its drawbacks, however. It makes reconstructing position from depth, which is useful in deferred shading, difficult. It can also have a significant impact on performance if not used carefully.

6.5.1 Performance Implications

Care needs to be taken to avoid dramatic impacts to performance. For many scenes in STK and Insight3D, three or four frustums are required. This makes hierarchical-view frustum culling key to avoid wasting time for objects that are culled by the near or far plane in the majority of the frustums. Simply iterating over every object during each frustum leads to a large amount of wasted CPU time on frustum checks.

Other optimizations can be used in addition to, or instead of, hierarchical culling. In a first pass, objects can be culled by the left, bottom, right, and top planes. When each frustum is rendered, only the near and far planes need to be checked, improving the efficiency of the frustum check.

Another optimization exploits objects completely in a frustum. A data structure that allows efficient removal, such as a doubly linked list, is used to store objects to be rendered. The list is iterated over during each frustum. When an object is completely in a frustum, it is removed from the list since it will not be rendered in any other frustum. As each frustum is rendered, fewer and fewer objects are touched. Of course, the removed objects are added back before the next frame. This is particularly useful for dynamic objects, where maintaining a spatial data structure for hierarchical-view frustum culling can introduce more overhead than it saves.

The farther the near plane is from the viewer, the larger a frustum can be. Therefore, it is beneficial to move the near plane for the closest frustum as far out as possible to reduce the total number of frustums required.

Regardless of efficient culling, multifrustum rendering creates a performance and visual quality problem for objects straddling two or more frustums. In this case, the object must be rendered twice. This leads to a little overdraw at the fragment level, but it does cost extra CPU time, bus traffic, and vertex processing compared to objects only in a single frustum.

Multifrustum rendering also increases the tension between culling and batching. Batching many objects into a single draw call is important for performance [123,183]. Doing so typically leads to larger bounding volumes around objects. The larger a bounding volume, the more likely it is to overlap more than one frustum and, therefore, lead to objects rendered multiple times. This makes efficient use of batching harder to achieve. Hardware vendors started releasing extensions to reduce CPU overdraw for draw commands [124], which may help reduce the need for batching, but it obviously will not avoid application-specific per-object overhead.

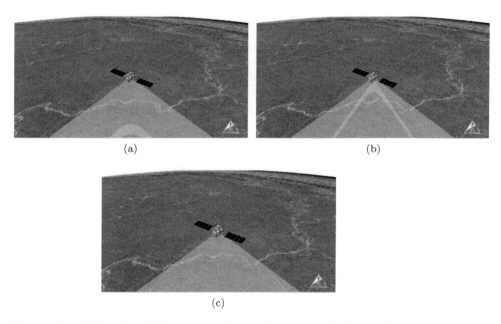

Figure 6.11. Although multifrustum rendering eliminates occlusion artifacts, it can introduce artifacts for translucent objects straddling two or more overlapping frustums. In the above image sequence the double translucency appears to slide across the yellow satellite sensor as the viewer zooms.

Besides causing performance issues, rendering objects multiple times can lead to artifacts shown in Figure 6.11. Given the slight overlap of adjacent frustums, redundantly rendered translucent objects appear darker or lighter in the overlapped region since the objects are blended on top of each other. As the viewer zooms in and out, this band can appear to slide along the object.

Multifrustum rendering also impacts performance by reducing the effectiveness of early-z (see Section 12.4.5), which benefits from rendering objects in near-to-far order from the viewer. In multifrustum rendering, frustums are rendered in the opposite order—far to near—although objects in each individual frustum can still be rendered near to far.

The final performance consideration for multifrustum rendering is to avoid redundant computations. Since an object may be rendered multiple times, it is important to avoid expensive computations in an object's render method, and instead, move such operations to a method that is called less frequently, like an update method.

Patrick Says ○○○○

> Although it solves an important problem, multifrustum rendering is a bit of a pain. Besides requiring careful attention to culling, it complicates the render loop and makes debugging harder since an object may be rendered more than once per frame. It even complicates less obvious code paths like picking.

6.6   W-Buffer

The w-buffer is an alternative to the z-buffer that can improve precision when resolving visibility for distant objects. The w-buffer uses w_{clip}, the original z_{eye}, to resolve visibility. This is linear in eye space and uniformly distributed between the near and far planes, within the constraints of a floating-point representation. Thus, compared to the z-buffer, the w-buffer has less precision close to the near plane but more precision moving towards the far plane. Table 6.1 compares the w-buffer to the z-buffer.

The problem is w is nonlinear in window space. This makes interpolating w for rasterization more expensive than interpolating z, which is linear in window space. Perspective-correct rasterization only has to calculate $\frac{1}{z}$ at each vertex and linearly interpolate across a triangle. Furthermore, z's linearity in window space improves hardware optimizations, including coarse-grained z-cull and z-compression, because the depth-value gradients are constant across a given triangle [137]. Today's hardware no longer implements w-buffers.

Buffer	Eye Space	Window Space
Z	Nonlinear	Linear
W	Linear	Nonlinear

Table 6.1. Relationship of the stored z- and w-buffer values to eye- and window-space values. The linear relationship of the w-buffer in eye space uniformly distributes precision through eye space at the cost of performance.

6.7   Algorithms Summary

Table 6.2 summarizes the major approaches for reducing or completely eliminating z-fighting caused by depth buffer precision errors. When de-

Algorithm	Summary
Adjust near/far plane	Easy to implement. Pushing out the near plane is more effective than pulling in the far plane. Use fog or blending to fade out objects before they are clipped.
Remove distant objects	Distant objects require a larger S_{min} to resolve visibility so removing distant objects that are near other objects can eliminate artifacts. This can also serve to declutter some scenes.
Complementary depth buffering	Easy to implement. Requires a floating-point depth buffer, which is available on most recent video cards, for maximum effectiveness. Doesn't account for truly massive view distances.
Logarithmic depth buffer	Highly effective for long view distances. Only requires a small vertex or fragment shader change, but all shaders need to be changed. Artifacts from using a vertex shader can be eliminated by using a fragment shader at the cost of losing hardware z-buffer optimizations.
Multifrustum rendering	Allows for arbitrarily long view distances and works on all video cards. Some objects may have to be rendered multiple times, which creates artifacts on translucent objects. A careful implementation is required for best performance.

Table 6.2. Summary of algorithms used to fix depth buffer precision errors.

ciding which approach to use, consider that many approaches can be combined. For example, complementary or logarithmic depth buffering can be used with multifrustum rendering to reduce the number of frustums required. Also, pushing out the near plane and pulling in the far plane will help in all cases.

Many of the approaches described in this chapter are implemented in Chapter06DepthBufferPrecision. Use this application to assist in selecting the approach that is best for you.

6.8 Resources

Solid coverage of transformations, including perspective projections, is found in "The Red Book" [155]. The "OpenGL Depth Buffer FAQ" contains excellent information on precision [110]. In addition, there are many worthwhile depth buffer precision articles on the web [11, 81, 82, 111, 137].

Another technique to overcome depth buffer precision errors is to use imposters, each rendered with their own projection matrix [130].

Reconstructing world- or view-space positions from depth is useful for deferred shading. The precision of the reconstructed position depends on the type of depth stored and the format it was stored in. A comparison of these, with sample code, is provided by Pettineo [141].

Part III

○○○○

Vector Data

7

$\bigcirc\bigcirc\bigcirc\bigcirc\circ$

Vector Data and Polylines

Virtual globes use two main types of data: raster and vector. Terrain and high-resolution imagery, which come in the form of raster data, create the breathtaking landscapes that attract so many of us to virtual globes. Vector data, on the other hand, may not have the same visual appeal but make virtual globes useful for "real work"—anything from urban planning to command and control applications.

Vector data come in the form of *polylines*, *polygons*, and *points*. Figure 7.1(b) shows examples of each; rivers are polylines, countries are polygons, major cities are points. When we compare Figure 7.1(a) to 7.1(b), it becomes clear how much richness vector data add to virtual globes.

This is the first of three chapters that discuss modern techniques for rendering polylines, polygons, and points on an ellipsoid globe. This discussion includes both rendering and preprocessing. For example, polygons need to be converted to triangles in a process called *triangulation*, and images for points should be combined into a single texture, called a *texture atlas*, in a process called *packing*. We suspect that you will be pleasantly surprised with all the interesting computer graphics techniques used to render vector data. On the surface, rendering polylines, polygons, and points sounds quite easy, but in reality, there are many things to consider.

7.1 Sources of Vector Data

Although our focus is on rendering algorithms, we should have an appreciation for the file formats used to store vector data. Perhaps the two most popular formats are Esri's shapefile [49] and the Open Geospatial Consortium (OGC) standard, KML [131].

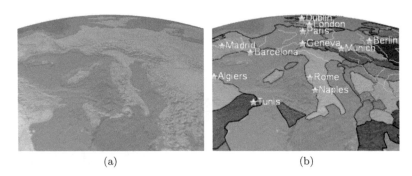

Figure 7.1. (a) A globe with just raster data. (b) The same globe also with vector data for rivers, countries, and major cities.

Shapefiles store nontopological geometry and attribute information for geospatial vector data, including line, area, and point features. Using our terminology, a shapefile's line feature is a polyline and an area feature is a polygon. The geometry for a feature is defined using a set of vector coordinates. A feature's attribute information is metadata, such as the city names shown in Figure 7.1(b).

A shapefile is actually composed of multiple files: the main `.shp` file containing the feature geometry, a `.shx` index file for efficient seeking, and a `.dbf` dBase IV file containing feature attributes. All features in the same shapefile must be of the same type (e.g., all polylines). The format is fully described in the *Esri Shapefile Technical Description* [49]. See `OpenGlobe.Core`.Shapefile for a class that can read point, polyline, and polygon shapefiles.

Another format, KML, formerly the Keyhole Markup Language, has gained widespread use in recent years due to the popularity of Google Earth. KML is an XML-based language for geographic visualization. It goes well beyond the geographic locations of polylines, polygons, and points to include such things as their graphical style, user navigation, and the ability to link to other KML files. The KML language is described in the KML 2.2 specification [131]. Google's open source library, *libkml*,[1] can be used to read and write KML files using C++, Java, or Python.

7.2 Combating Z-Fighting

Before getting to the fun stuff of actually rendering vector data, we need to consider the potential z-fighting artifacts discussed in Chapter 6. When

---

[1]http://code.google.com/p/libkml/

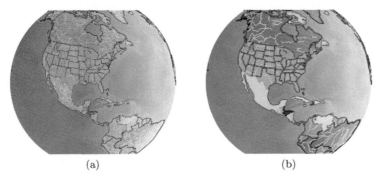

(a)   (b)

Figure 7.2. (a) Artifacts due to z-fighting and vector data failing the depth test. (b) Rendering the ellipsoid using the average depth value of its front and back faces eliminates the artifacts.

rendering vector data on a globe, care needs to be taken to avoid z-fighting between the vector data and the globe. The problem is somewhat different than the usual coplanar triangles fighting to win the depth test. Consider a tessellated ellipsoid and a polyline on the ellipsoid. Since both only approximate the actual curvature of the ellipsoid, in some cases, a line segment may be coplanar with one of the ellipsoid triangles; in other cases, the line segment may actually cut underneath triangles. In the former case, z-fighting occurs; in the latter case, either z-fighting occurs or the polyline loses the depth test. Even if the ellipsoid is ray casted, a la Section 4.3, artifacts still occur (see Figure 7.2(a)).

A simple technique can eliminate the artifacts. When rendering the ellipsoid, instead of writing the depth values of its front-facing triangles, use

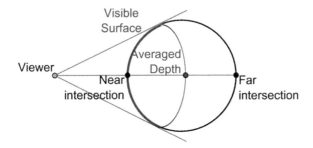

Figure 7.3. The averaged depth values of the front and back faces of the ellipsoid are shown in blue. Using these depth values, instead of the depth values of the front face, shown in red, can avoid z-fighting with vector data and still occlude objects on the other side of the ellipsoid.

the average of the depth values of its front-facing and back-facing triangles, as shown in Figure 7.3. This is easy to do when ray casting an ellipsoid in the fragment shader. The ray-ellipsoid intersection test needs to return both intersections:[2]

```
struct Intersection
{
  bool   Intersects;
  float  NearTime;  // Along ray
  float  FarTime;   // Along ray
};
```

The second and final step is to change the way gl_FragDepth is assigned. When writing depth using the front face, the code looks like this:

```
vec3 position = og_cameraEye + (i.NearTime * rayDirection);
gl_FragDepth = ComputeWorldPositionDepth(position,
    og_modelZToClipCoordinates);
```

It can simply be replaced by averaging i.NearTime and i.FarTime to get the distance along the ray:

```
vec3 averagePosition = og_cameraEye +
    mix(i.NearTime, i.FarTime, 0.5) * rayDirection;
gl_FragDepth = ComputeWorldPositionDepth(averagePosition,
    og_modelZToClipCoordinates);
```

The lighting computations should still use the original position, since that is the actual position being lit. See the fragment shader in OpenGlobe .Scene.RayCastedGlobe for a full implementation.

When averaged depth is used, objects on the back face of the ellipsoid still fail the depth test, but objects slightly under the front face pass. This is great for polylines whose line segments cut slightly underneath the ellipsoid, as can be seen in the southern boundaries of Georgia and Alabama in Figures 7.2(a) and 7.2(b). For other objects such as models, this sleight of hand may not be desirable. This approach can be extended to use two depth buffers: one with averaged depth and one with the the ellipsoid's front-facing triangle's depth values. Objects that may cause artifacts with the ellipsoid can be rendered with the averaged depth buffer, and other objects can be rendered with the regular depth values.

---

[2]When the ray intersects the ellipsoid at its tangent, both **NearTime** and **FarTime** will be the same value.

We use a variation of this approach, called the *depth cone*, in STK. First, the ellipsoid is rendered with its regular depth values, followed by any objects that should be clipped by the front faces of the ellipsoid. Then, a depth cone is computed on the CPU. The cone's base is formed from tangent points on the ellipsoid, and its apex is the center of the opposite side of the ellipsoid from the viewer's perspective. The cone is then rendered with only depth writes enabled, no depth test or color writes. Finally, objects that would have previously created artifacts are rendered "on the ellipsoid."

○○○○ Patrick Says

7.3   Polylines

It is amazing how much mileage virtual globes get out of the simple polyline. Polylines are used for boundaries around closed features, such as zip codes, states, countries, parks, school districts, and lakes. The list goes on and on. Polylines are also used to visualize open features, such as rivers, highways, and train tracks. Polylines are not limited to geographic information either; they can also represent the path for driving directions, the stages in the Tour de France, or even airplane flight paths and satellite orbits. Since not all polylines are on the ground, this section covers general polyline rendering. Section 7.3.5 discusses the special, but common, case of rendering polylines directly on the ground.

Most polyline datasets used by virtual globes come in the form of *line strips*, that is, a series of points where a line segment connects each point, except the first, to the previous point, as shown in Figure 7.4. To represent closed features, the first point can be duplicated and stored as the last

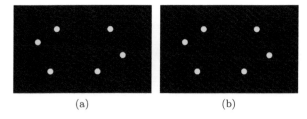

(a)   (b)

Figure 7.4. (a) An open line strip. A line segment implicitly connects each point, except the first, to the previous point. (b) A closed line strip is created by duplicating the first point at the end.

point. Another representation is a *line loop*, which is a line strip that always connects the last point to the first, avoiding the need to store an extra point for closed polylines.

7.3.1 Batching

It is easy enough to render a line strip using PrimitiveType.**Line Strip**—simply create a vertex buffer with a vertex per point in the polyline and issue a draw call using the line strip primitive. But how would you render multiple line strips? For a few dozen, or perhaps a few hundred line strips, it may be reasonable to create a vertex buffer per polyline and issue a draw call per polyline. The problem with this approach is that creating lots of small vertex buffers and issuing several draw calls can introduce significant application and driver CPU overhead. The solution is, of course, to combine line strips into fewer vertex buffers and render them with fewer draw calls. This is known as batching [123, 183].

There are three approaches to batching line strips:

1. *Lines.* The simplest approach is to use PrimitiveType.**Lines**, which treats every two consecutive points as an individual line segment, as in Figure 7.5(a). Since line segments do not need to be connected, this allows storing multiple line strips in one vertex buffer by duplicating each point in a line strip except the first and last points, as in Figure 7.5(b). This converts each line strip to a series of line segments. The line segments for all line strips can be stored in a single vertex buffer and rendered with a single draw call. The downside is that a line strip with n vertices, where $n \geq 2$, now requires $2 + 2(n - 2) = 2n - 2$ vertices—practically double the amount of memory! Nonetheless, in many cases, using a single draw call to render a vertex buffer of lines will outperform per-line strip vertex buffers and draw calls.

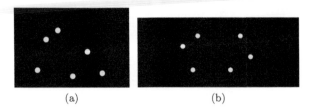

(a)   (b)

Figure 7.5. (a) Using the lines primitive, every two points define an individual line segment. (b) The lines primitive can be used to represent a line strip by duplicating interior points.

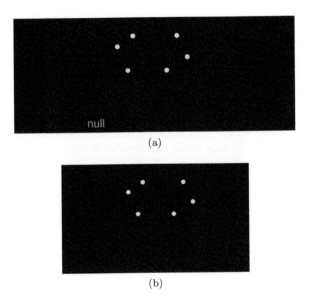

Figure 7.6. (a) Nonindexed lines duplicate points to store line strips. (b) Indexed lines only duplicate indices.

2. *Indexed lines.* The amount of memory used by PrimitiveType.**Lines** can be reduced by adding an index buffer (see Figure 7.6). In much the same way that indexed triangle lists use less memory than individual triangles for most meshes, indexed lines use less memory than nonindexed lines when most line segments are connected. In the case of batching line strips, all line segments, except the ones between strips, are connected. For $n \geq 2$, a line strip stored using indexed lines requires n vertices and $2n - 2$ indices. Since an index almost always requires less memory than a vertex, indexed lines generally consume less memory than nonindexed lines. For example, assuming 32-byte vertices and 2-byte indices, a line strip with n vertices requires the following:

- $32n$ bytes using a line strip,
- $32(2n - 2) = 64n - 64$ bytes using nonindexed lines,
- $32n + 2(2n - 2) = 40n - 4$ bytes using indexed lines.

In this example, for $n \geq 3$, indexed lines always use less memory than nonindexed lines and use only slightly more memory than line strips since vertex data dominate the memory usage. For example, a line strip with 1,000 points requires 32,000 bytes when stored as a line

```
RenderState renderState = new RenderState ();
renderState . PrimitiveRestart . Enabled  = true;
renderState . PrimitiveRestart . Index  = 0xFFFF;
```

Listing 7.1. Enabling primitive restart.

strip, 63,936 bytes as nonindexed lines, and 39,996 bytes as indexed lines. Of course, the gap between memory usage for line strips and indexed lines will vary based on the size of vertices and indices.

Indexed lines favor large vertices and small indices. When batching a large number of line strips, it is not uncommon for 4-byte indices to be required. If this becomes a concern, you can limit the number of vertices per batch such that vertices can be indexed using 2-byte indices.

Similar to indexed triangle lists, indexed lines can also take advantage of a GPU's vertex caches to avoid redundantly processing the same vertex.

3. *Indexed line strips and primitive restart.* Another memory-efficient way to batch together line strips is to use indexed line strips. Each line strip requires n vertices and only $n + 1$ indices, except the last line strip, which requires only n indices (see Figure 7.7). Interior line strips store an extra index at their end to signal a *primitive restart* [34], a special token that is interpreted as the end of a line strip and the start of a new one.

Primitive restart enables drawing multiple indexed line strips in a single draw call. It is part of the render state, which contains a flag to enable or disable it, as well as the index that signifies a restart, as in Listing 7.1.

In the previous example where indexed lines required 39,996 bytes, indexed line strips require 34,002 bytes, only 2,002 bytes more than

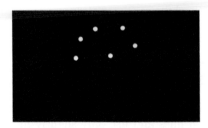

Figure 7.7. Multiple indexed line strips can be rendered in a single draw call by using primitive restart. In this figure, the restart index is $0xFFFF$.

line strips. When considering a single batch with at least one line strip with $n \geq 3$, indexed line strips always use less memory than indexed lines.

7.3.2 Static Buffers

In addition to batching, use static vertex and index buffers where appropriate to achieve the best polyline-rendering performance [120]. How often do country borders, park boundaries, or even highways change? Many polyline datasets are static. As such, they should be uploaded to the GPU once and stored in GPU memory. This makes draw calls more efficient by avoiding the traffic from system memory to video memory, resulting in dramatic performance improvements. A static buffer is requested by passing BufferHint.**StaticDraw** to Device.**CreateVertexBuffer**, Device.**CreateIndexBuffer**, or Context.**CreateVertexArray**.

It is even sometimes appropriate to use a static index buffer with a dynamic or stream vertex buffer. For example, consider a line strip for the path of an airplane currently in flight. When the application receives a position update, the new position is appended to the vertex buffer (which may need to grow every so often). If either indexed lines or indexed line strips are used, the new index or indices are known in advance. Therefore, a static index buffer can contain indices for a large number of vertices, and only the **count** argument passed to Context.**Draw** needs to change for a position update. The index buffer only needs to be reallocated and rewritten when it is no longer large enough.

7.3.3 Wide Lines

Let's turn our attention from performance to appearance. Simple techniques such as line width, solid color, and outline color go a long way in shading polylines. These simple visual cues can make a big difference. For example, a GIS user may want to make country borders wider than state borders. For another example, in a virtual globe used for battlespace visualization, threat zones may be outlined in red and safe zones outlined in green. If one thing is for sure, users want to be able to customize everything!

For line width, we'd like to be able to render lines with a width defined in pixels. In older versions of OpenGL, this was accomplished using the **glLineWidth** function. Since support for wide lines was deprecated in OpenGL 3, we turn to the geometry shader to render wide lines.[3]

---

[3]Interestingly enough, **glLineWidth** remains but width values greater than one generate an error.

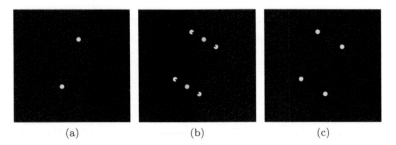

Figure 7.8. (a) A line segment in window coordinates showing normals pointing in opposite directions. (b) Each endpoint is offset along both normals to create four new points. (c) The new points are connected with a triangle strip to form a wide line.

The idea is simple: a geometry shader takes a line segment as input, transforms it into window coordinates, expands it into a two-triangle triangle strip of a given pixel width, then outputs the triangle strip (see Figure 7.8).

The vertex shader simply passes the vertex's model space position through, as in Listing 7.2.

The first order of business for the geometry shader is to project the line's endpoints into window coordinates. Using window coordinates makes it easy to create a triangle strip sized in pixels. In order to avoid artifacts, the line segment needs to be clipped to the near plane, just as the fixed function does against every frustum and clip plane after the geometry shader. Otherwise, one or both endpoints would be transformed to the wrong window coordinates.

Clipping is done by transforming the endpoints into clip coordinates and checking if they are on opposite sides of the near plane. If so, the line segment intersects the near plane and needs to be clipped by replacing the point behind the near plane with the point at the near plane. If both points are behind the near plane, the line segment should be culled. The function in Listing 7.3 transforms the endpoints from model to clip coordinates, clipped to the near plane.

```
in vec4 position;

void main()
{
    gl_Position = position;
}
```

Listing 7.2. Wide lines vertex shader.

```
void ClipLineSegmentToNearPlane(float nearPlaneDistance,
                                mat4 modelViewPerspectiveMatrix,
                                vec4 modelP0, vec4 modelP1,
                                out vec4 clipP0, out vec4 clipP1,
                                out bool culledByNearPlane)
{
  clipP0 = modelViewPerspectiveMatrix * modelP0;
  clipP1 = modelViewPerspectiveMatrix * modelP1;
  culledByNearPlane = false;

  float distanceToP0 = clipP0.z + nearPlaneDistance;
  float distanceToP1 = clipP1.z + nearPlaneDistance;

  if ((distanceToP0 * distanceToP1) < 0.0)
  {
    // Line segment intersects near plane
    float t = distanceToP0 / (distanceToP0 - distanceToP1);
    vec3 modelV = vec3(modelP0) +
                  t * (vec3(modelP1) - vec3(modelP0));
    vec4 clipV = modelViewPerspectiveMatrix * vec4(modelV, 1);

    // Replace the end point closest to the viewer
    // with the intersection point
    if (distanceToP0 < 0.0)
    {
      clipP0 = clipV;
    }
    else
    {
      clipP1 = clipV;
    }
  }
  else if (distanceToP0 < 0.0)
  {
    // Line segment is in front of near plane
    culledByNearPlane = true;
  }
}
```

Listing 7.3. Clipping a line segment to the near plane.

```
vec4 ClipToWindowCoordinates(vec4 v,
                             mat4 viewportTransformationMatrix)
{
  v.xyz /= v.w;
  v.xyz = (viewportTransformationMatrix * vec4(v.xyz, 1.0)).xyz;
  return v;
}
```

Listing 7.4. Transforming from clip to window coordinates.

If the line segment is not culled, the endpoints need to be transformed from clip to window coordinates. Perspective divide transforms clip into normalized device coordinates, then the viewport transformation is used to produce window coordinates. These transforms are done as part of the fixed-function stages after the geometry shader, but we also need to do them in the geometry shader to allow us to work in window coordinates. These transforms are encapsulated in a function, shown in Listing 7.4, that takes a point from clip to window coordinates.

Given the two endpoints in window coordinates, `windowP0` and `windowP1`, expanding the line into a triangle strip for a wide line is easy. Start by computing the line's normal. As shown in Figure 7.8(a), a 2D line has two normals: one pointing to the "left" and one to the "right." The normals are used to introduce two new points per endpoint; each new point is offset from the endpoint along each normal direction, as in Figure 7.8(b).

```
vec4 clipP0;
vec4 clipP1;
bool culledByNearPlane;
ClipLineSegmentToNearPlane(og_perspectiveNearPlaneDistance,
                           og_modelViewPerspectiveMatrix,
                           gl_in[0].gl_Position,
                           gl_in[1].gl_Position,
                           clipP0, clipP1, culledByNearPlane);
if (culledByNearPlane)
{
    return;
}

vec4 windowP0 = ClipToWindowCoordinates(clipP0,
    og_viewportTransformationMatrix);
vec4 windowP1 = ClipToWindowCoordinates(clipP1,
    og_viewportTransformationMatrix);

vec2 direction = windowP1.xy - windowP0.xy;
vec2 normal = normalize(vec2(direction.y, -direction.x));

vec4 v0 = vec4(windowP0.xy - (normal * u_fillDistance),
               -windowP0.z, 1.0);
vec4 v1 = vec4(windowP1.xy - (normal * u_fillDistance),
               -windowP1.z, 1.0);
vec4 v2 = vec4(windowP0.xy + (normal * u_fillDistance),
               -windowP0.z, 1.0);
vec4 v3 = vec4(windowP1.xy + (normal * u_fillDistance),
               -windowP1.z, 1.0);

gl_Position = og_viewportOrthographicMatrix * v0;
EmitVertex();

gl_Position = og_viewportOrthographicMatrix * v1;
EmitVertex();

gl_Position = og_viewportOrthographicMatrix * v2;
EmitVertex();

gl_Position = og_viewportOrthographicMatrix * v3;
EmitVertex();
```

Listing 7.5. Geometry shader `main()` function for wide lines.

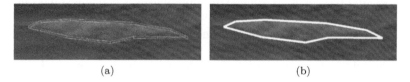

(a)   (b)

Figure 7.9. A ten-pixel-wide line rendered using a geometry shader. (a) Wireframe. (b) Solid.

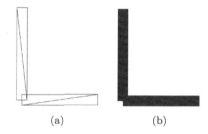

(a)   (b)

Figure 7.10. Wide line strips with sharp turns can create noticeable gaps.

The new points are offset using the line's pixel width. For an x-pixel-wide line, each point is offset by the normal scaled in both directions by $\frac{x}{2}$, which is u_fillDistance in Listing 7.5. The four new points are connected using a triangle strip and finally output from the geometry shader. The window-coordinate points are transformed by the orthographic projection matrix, just as we would when rendering anything in window coordinates, such as a heads-up display. Code for the geometry shader's main() function is shown in Listing 7.5.

The complete geometry shader for wide lines is in OpenGlobe.Scene. Polyline. Figure 7.9 shows both a wireframe and solid line rendered using this algorithm. Most polyline datasets used by virtual globes do not have sharp turns. But for those that do, this algorithm creates noticeable gaps between line segments if the lines are wide enough, as in Figure 7.10. A simple solution to round the turns is to render points the same size as the line width at each endpoint. Other solutions are described by McReynolds and Blythe [113].

7.3.4 Outlined Lines

In addition to line width, it is also useful to set a polyline's color. Of course, making each line strip, or even line segment, a different solid color is easy. What is more interesting is using two colors: one for the line's interior and one for its outline. Outlining makes a polyline stand out, as shown in Figure 7.11.

There are a variety of ways to render outlined lines. Perhaps the most straightforward is to render the line twice: first in the outline color with exterior width, then in the interior color with the interior width. Even if the second pass is rendered with a less-than-or-equal depth-comparison function, this naïve approach results in z-fighting as the outline and interior fight to win the depth test. Z-fighting can be avoided by using the stencil buffer to eliminate the depth test on the second pass, identical to decaling [113]. This is a good solution for supporting older GPUs and is the implementation we use in Insight3D. For hardware with programmable shaders, outlining can be done in a single pass.

One approach is to create extra triangles for the outline in the geometry shader. Instead of creating two triangles, six triangles are created: two for the interior and two on each side of the line for the outline (see Figure 7.12(a) and the implementation in OpenGlobe.Scene.Outlined PolylineGeometryShader). The downside to this approach is that it is subject to aliasing, unless the entire scene is rendered with antialiasing, as shown in Figure 7.12(c). The jagged edges are on both sides of the outline: when the outline meets the interior and when it meets the rest of the scene. We'll call these *interior* and *exterior aliasing*, respectively. Exterior aliasing is present in our line-rendering algorithm even when outlining is not used.

Aliasing can be avoided by using a 1D texture or a one-pixel high 2D texture, instead of creating extra triangles for the outline. The idea is to lay the texture across the line width-wise. A texture read in the fragment shader is then used to determine if the fragment belongs to the line's interior or outline. With linear texture filtering, the interior and outline colors are

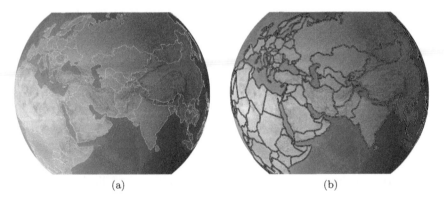

(a)   (b)

Figure 7.11. (a) No outline. (b) An outline makes polylines stand out.

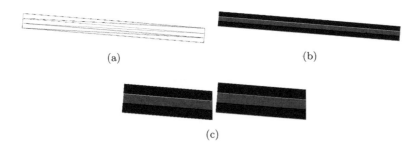

Figure 7.12. (a) Wireframe showing extra triangles for the outline using the geometry shader method. (b) With solid shading. (c) The left side shows aliasing using the geometry shader method; compare this to the right side, which uses the texture method.

smoothly blended without interior aliasing. Exterior aliasing is avoided by storing extra texels representing an alpha of zero on each end of the texture. With blending enabled, this provides exterior antialiasing.

The width of the texture is $interiorWidth + outlineWidth + outlineWidth + 2$. For example, a line with an interior width of three and an outline width of one would use a texture seven pixels wide (see Figure 7.13). The texture is created with two channels: the red channel, which is one for interior and zero for outline, and the green channel, which is zero on the left-most and right-most texels and one everywhere else. The texture only stores 0s and 1s, not actual line colors, so colors can change without rewriting the texture. Likewise, the same texture can be used for different line strips with different colors as long as they have the same interior and outline widths.

The geometry shader assigns texture coordinates when creating the triangle strip. Simply, points on the left side of the original line have a texture coordinate of zero and points on the right side have a texture

Figure 7.13. An example texture used for outlining. Here, the interior width is three pixels, and the outline width is one pixel. The red channel is one where the interior line is, and the green channel is one where the interior or outline is. Above, an example slice of a line width-wise is shown lined up with the texture.

```
flat in vec4 fsColor;
flat in vec4 fsOutlineColor;
in float fsTextureCoordinate;

out vec4 fragmentColor;

uniform sampler2D og_texture0;

void main()
{
  vec2 texel =
      texture(og_texture0, vec2(fsTextureCoordinate, 0.5)).rg;
  float interior = texel.r;
  float alpha = texel.g;

  vec4 color = mix(fsOutlineColor, fsColor, interior);
  fragmentColor = vec4(color.rgb, color.a * alpha);
}
```

Listing 7.6. Using a texture for outlining.

coordinate of one. In order to capture the left-most and right-most texels
used for exterior antialiasing without shrinking the line width, the triangle
strip is extruded by an extra half-pixel in each direction.

The fragment shader reads the texture using the interpolated texture
coordinate from the geometry shader. The red channel is then used to
blend between the interior and outline color for interior antialiasing. The
green channel is interpreted as an alpha value, which is multiplied by the
color's alpha for exterior antialiasing. The complete fragment shader is
shown in Listing 7.6.

The full implementation is in **OpenGlobe.Scene**.OutlinedPolyline
Texture. One downside to this approach is that overlapping lines can create artifacts since blending is used for exterior antialiasing. Of course,
this can be eliminated by rendering in back-to-front order at a potential
performance cost.

7.3.5 Sampling

Our polyline-rendering algorithm connects points using straight line segments in Cartesian space. Imagine what happens if two points are on the
ellipsoid but very far away; for example, what would a line segment between
Los Angles and New York look like? Of course, it would cut underneath the
ellipsoid. In fact, if averaged depth wasn't used, most of the line segment
would fail the depth test. The solution is to create a curve on the ellipsoid
that subsamples each line segment in a polyline using the algorithm presented in Section 2.4. Doing so makes the polyline better approximate the
curvature of the ellipsoid.

Some polylines will have a large number of points; for example, consider polylines bounding large countries such as Russia or China. After subsampling, these polylines will have even more points. Discrete LODs can be computed using an algorithm such as the popular Douglas-Peucker reduction [40, 73]. The LODs can then be rendered using traditional selection techniques (e.g., screen-space error).

LOD for individual polylines is not always useful. In virtual globes, polylines are typically rendered as a part of a layer that may only be visible for certain viewer heights or view distances. This can make LODs for individual polylines less important since a polyline's layer may be turned off before the polyline's lower LODs would even be rendered. Perhaps the best balance is a combination of layers and a handful of layer-aware discrete LODs.

> Run Chapter07VectorData to see polylines, polygons, and billboards rendered from shapefile data.

○○○○ **Try This**

7.4   Resources

To better understand vector data file formats, it is worth reading the *Esri Shapefile Technical Description* [49], OGC's KML 2.2 specification [131], and Google's "KML Reference" [64].

○○○○○ **8**

Polygons

Once we can render polylines, the next step is to render polygons on the globe or at a constant height above the globe. Examples of polygons in virtual globes include zip codes, states, countries, and glaciated areas, as shown in Figure 8.1. We define the interior of a polygon by a *line loop*, a polyline where the last point is assumed to connect to the first point.

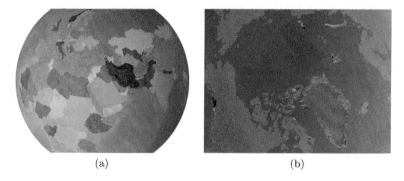

(a)   (b)

Figure 8.1. Randomly colored polygons for (a) countries and (b) glaciated areas.

8.1   Render to Texture

One approach to rendering polygons on the globe is to burn the polygons to a texture, we'll call this a *polygon map*, and render the globe with multi-texturing. The polygon map can be created using a rasterization technique such as flood fill or scan-line fill. In the simplest case, a single one-channel texture can be used, where a value of zero indicates the texel is outside of

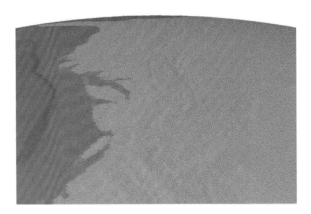

Figure 8.2. A polygon of North America rendered using a polygon map. The low resolution of the map leads to aliasing in the foreground.

a polygon and a value greater than one represents the interior of a polygon and its opacity. When the globe is rendered, the color of its base texture is blended with the polygon's color based on the polygon map. If multiple polygon colors are required, the polygon map can be expanded to four channels or multiple textures can be used. Multiple textures are also useful for supporting overlapping polygons.

A major benefit of this approach is that polygons automatically conform to the globe, even if the globe is rendered with terrain. Also, performance is independent of the number of polygons or number of points defining each polygon. Instead, performance is affected by the size and resolution of the polygon map—which leads to its weakness. Figure 8.2 shows a polygon of North America rendered with this approach. Since the resolution of the polygon map is not high enough, aliasing occurs in the foreground, leading to blocky artifacts. Using a higher-resolution map will, of course, fix the problem at the cost of using a significant amount of memory.

Ultimately, the polygon map can be treated like any other high-resolution texture and rendered with LOD techniques discussed in Part IV. This approach is used in Google Earth and appears to be used in ArcGIS Explorer, among others.

8.2   Tessellating Polygons

Instead of using a polygon map, we will detail an implementation based on tessellating a polygon into a triangle mesh, which is then rendered separately from the globe. This geometry-based approach does not suffer from

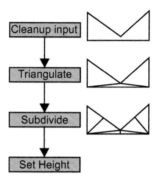

Figure 8.3. The pipeline used to convert a polygon's line loop into a triangle mesh that conforms to the ellipsoid surface of a globe.

large texture memory requirements or aliasing artifacts around its edges. It can also be implemented nearly independently of the globe rendering algorithm, leading to less coupling and clean engine design.

Figure 8.3 shows the pipeline to convert a line loop representing the boundary of a polygon on a globe to a triangle mesh that can be rendered with a simple shader. First, the line loop needs to be cleaned up so the rest of the stages can process it. Next, the polygon is divided into a set of triangles, a process called *triangulation*. In order to make the triangle mesh conform to the ellipsoid surface, the triangles are subdivided until a tolerance is satisfied. Finally, all points are raised to the ellipsoid's surface or a constant height.

8.2.1   Cleaning Up the Input

One thing that makes virtual globes so interesting is that a wide array of geospatial data is available for viewing. The flip side to this is that a plethora of file formats exist, each with their own conventions. For example, KML and shapefiles are two very popular sources of polygon data. Ignoring interior holes, in KML, the <Polygon> element's winding order is counterclockwise [64], whereas the winding order for polygons in shapefiles is clockwise [49]. To add to the fun, some files do not always follow the format's specifications.

To make our code as robust as possible, it is important to consider the following cleanup before moving on to the triangulation stage:

- *Eliminate duplicate points.* The algorithms later in our pipeline assume the polygon's boundary is a line loop; therefore, the last point should not be a duplicate of the first point. Consecutive duplicate

```
public static IList<T> Cleanup<T>(IEnumerable<T> positions)
{
  IList<T> positionsList =
      CollectionAlgorithms.EnumerableToList(positions);
  List<T> cleanedPositions =
      new List<T>(positionsList.Count);

  for (int i0 = positionsList.Count - 1, i1 = 0;
       i1 < positionsList.Count; i0 = i1++)
  {
    T v0 = positionsList[i0];
    T v1 = positionsList[i1];

    if (!v0.Equals(v1))
    {
       cleanedPositions.Add(v1);
    }
  }

  cleanedPositions.TrimExcess();
  return cleanedPositions;
}
```

Listing 8.1. Removing duplicate points from a list of points.

points anywhere along the boundary are allowed by some formats, such as shapefiles, but these should be removed as they can lead to degenerate triangles during triangulation. One method for removing duplicate points is to iterate through the points, copying a point to a new list only if it isn't equal to the previous point, as done in Listing 8.1.

This method illustrates a useful technique for iterating over polygon edges [13]. At each iteration, we want to know the current point, v1, and the previous point, v0. Since the last point implicitly connects to the first point, i0 is initialized to the index for the last point, and i1 is initialized to zero. After each iteration, i0 is set to i1 and i1 is incremented. This allows us to iterate over all the edges without treating the last point as a special case or using modulo arithmetic.

Try This ○○○○

Make Cleanup<T> more efficient: instead of creating a new list, modify the existing list and adjust its size. You will need to modify the method's signature.

- *Detect self-intersections.* Unless the triangulation step in the pipeline can handle self-intersections, it is important to verify that the polygon doesn't cross over itself, like the one in Figure 8.4(c) does.

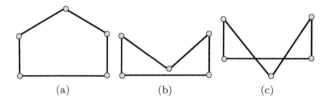

Figure 8.4. (a) Simple convex polygon. (b) Simple concave polygon. (c) Nonsimple polygon.

Even though some format specifications will "guarantee" no self-intersections, it can still be worth checking. One technique to check for self-intersections is to project the points on a plane tangent to the globe at the polygon's centroid. This allows us to treat the 3D polygon as a 2D polygon. Now it is a simple matter of verifying no edge intersects any other edge. The biggest pitfall with this technique is that large polygons that wrap around the globe may be distorted when projected onto the tangent plane. One approach to combating this is to only project the polygon if its bounding sphere's radius is below a threshold. See **OpenGlobe.Core**.EllipsoidTangentPlane for code that computes a tangent plane and projects points onto it.

- *Determine winding order.* Some formats store a polygon's points in clockwise order, others use counterclockwise order. Our algorithms can either handle both winding orders or only handle one and can reverse the order of the points if the winding orders do not match. Reversing can be efficiently implemented in $\mathcal{O}(n)$ by swapping the first and last points, then repeatedly swapping interior points until the middle of the list is reached. Most platforms provide a method for this, such as .NET's List<T>.Reverse.

  Algorithms later in our pipeline assume counterclockwise order, so if points are provided clockwise, they need to be reversed. One technique for determining winding order is to project the polygon onto a tangent plane as mentioned above, then compute the area of the 2D polygon. The sign of the area determines the winding order: nonnegative is counterclockwise and negative is clockwise (see Listing 8.2).

Eliminating duplicate points, checking for self-intersections, and handling winding order can chew up some CPU time. A naïve implementation that checks for self-intersection is $\mathcal{O}(n^2)$, and the other two steps are $\mathcal{O}(n)$. Although the vast majority of time will be spent in later stages of our pipeline, it would be nice to be able to skip these steps, especially considering that using a tangent plane can sometimes cause problems. However,

```
public static PolygonWindingOrder ComputeWindingOrder (
    IEnumerable<Vector2D> positions)
{
  return (ComputeArea(positions) >= 0.0)
           ? PolygonWindingOrder.Counterclockwise
           : PolygonWindingOrder.Clockwise;
}

public static double ComputeArea(IEnumerable<Vector2D> positions)
{
  IList<Vector2D> positionsList =
      CollectionAlgorithms.EnumerableToList(positions);

  double area = 0.0;
  for (int i0 = positionsList.Count - 1, i1 = 0;
       i1 < positionsList.Count; i0 = i1++)
  {
    Vector2D v0 = positionsList[i0];
    Vector2D v1 = positionsList[i1];

    area += (v0.X * v1.Y) - (v1.X * v0.Y);
  }
  return area * 0.5;
}
```

Listing 8.2. Using a polygon's area to compute its winding order.

we advise against skipping these cleaning steps unless you are using your own file format that only your application writes. This is one of the many challenges virtual globes face: handling a wide array of formats.

8.2.2 Triangulation

GPUs don't rasterize polygons—they rasterize triangles. Triangles make hardware algorithms simpler and faster because they are guaranteed to be convex and planar. Even if GPUs could directly rasterize polygons, the GPU is unaware of our need to render polygons on a globe. So once we've cleaned up a polygon, the next stage in the pipeline is triangulation: decomposing the polygon into a set of triangles without introducing new points. We will limit ourselves to *simple polygons*, which are polygons that do not contain consecutive duplicate points or self-intersections and each point shares exactly two edges.

A *convex polygon* is a polygon where each interior angle is less than 180°, as in Figure 8.4(a). Just because a polygon is simple does not mean it needs to be convex. Although some useful shapes like circles and rectangles are convex, most real-world geospatial polygons will have one or more interior angles greater than 180°, making them concave (see Figure 8.4(b)).

A nice property of simple polygons is that every simple polygon can be triangulated. A simple polygon with n points will always result in $n - 2$

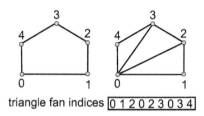

triangle fan indices $\boxed{0\ 1\ 2\ 0\ 2\ 3\ 0\ 3\ 4}$

Figure 8.5. Decomposing a convex polygon into a triangle fan.

triangles. For convex polygons, triangulation is easy and fast at $\mathcal{O}(n)$. As shown in Figure 8.5, a convex polygon can simply be decomposed into a triangle fan pivoting around one point on the boundary.

A fan can result in many long and thin triangles (e.g., consider a fan for a circle). These types of triangles have an adverse effect on rasterization because fragments along shared triangle edges are redundantly shaded. Persson suggests an alternative greedy triangulation algorithm based on creating triangles with the largest possible area [139]. If convex polygons are common in your application, consider using a separate code path for them because triangulating concave polygons is considerably more expensive.

There are several algorithms for triangulating concave polygons, with varying runtimes, from a naïve $\mathcal{O}(n^3)$ algorithm to a complex $\mathcal{O}(n)$ one [25]. We're going to focus on a simple and robust algorithm called *ear clipping* that we've found works well for a wide range of geospatial polygons: zip codes, states, country borders, etc. In order to understand the fundamentals, we're going to describe the $\mathcal{O}(n^3)$ implementation first, then briefly look at optimizations to make the algorithm very fast in practice.

Since the type of polygons we are interested in triangulating are defined on the face of an ellipsoid, they are rarely planar in three dimensions (i.e., they form a curved surface in the WGS84 coordinate system). Fortunately, ear clipping can be extended to handle this. We'll start by explaining ear clipping for planar 2D polygons, then we'll extend ear clipping to 3D polygons on the ellipsoid.

Ear clipping in two dimensions. An *ear* of a polygon is a triangle formed by three consecutive points, p_{i-1}, p_i, p_{i+1}, that does not contain any other points in the polygon, and the *tip* of the ear, p_i, is convex. A point is convex if the interior angle between the shared edges, $p_{i-1} - p_i$ and $p_{i+1} - p_i$, is not greater than $180°$.

The main operation in ear clipping is walking over a polygon's points and testing if three consecutive points form an ear. If so, the tip of the ear

(a)   (b)

Figure 8.6.   (a) A simple, planar polygon before ear clipping.   (b) Initial data structures before ear clipping. Only indices are shown in the `remaining Positions` list.

```
internal struct IndexedVector<T> : IEquatable<IndexedVector<T>>
{
  public IndexedVector(T vector, int index) { /* ... */ }
  public T Vector  { get; }
  public int Index { get; }

  // ...
}
```

Listing 8.3. A type used to hold a point and an index.

is removed and the ear's indices are recorded. Ear clipping then continues with the remaining polygon until a single triangle, forming the final ear, remains. As ears are removed and the polygon is reduced, new ears will appear until the algorithm finally converges. The union of the ears forms the triangulation of the polygon.

An implementation can simply take the polygon's boundary points as input and return a list of indices, where every three indices is a triangle in the polygon's triangulation. A full implementation is provided in **OpenGlobe.Core**.EarClipping. We will outline the major steps here by going through an example of triangulating the polygon in Figure 8.6(a).

Since we're not actually going to remove points from the input polygon, the first step is to copy the points to a new data structure. A doubly linked list is a good choice since it supports constant time removal. Given that we want to return a list of indices to the input positions, each node in the linked lists should contain the point and its index[1] using a type such as our IndexedVector<T> in Listing 8.3.

Using IndexedVector<T>, the linked list can be created with the following code:

---

[1]If the points are already stored in an indexable collection, just the index could be stored in each linked-list node. In our implementation, we follow the .NET convention of passing collections using IEnumerable, which does not support indexing.

```
public static class EarClipping
{
  public static IndicesUnsignedInt Triangulate(
     IEnumerable<Vector2D> positions)
  {
    LinkedList<IndexedVector<Vector2D>> remainingPositions =
        new LinkedList<IndexedVector<Vector2D>>();

    int index = 0;
    foreach (Vector2D position in positions)
    {
      remainingPositions.AddLast(
          new IndexedVector<Vector2D>(position, index++));
    }

    IndicesUnsignedInt indices = new IndicesUnsignedInt(
        3 * (remainingPositions.Count - 2));

    // ...
  }
}
```

The above code also allocates memory for the output indices using
the fact that triangulation of a simple polygon of n points creates $n - 2$
triangles. After this initialization, the contents of `remainingPoints` and
`indices` are shown in Figure 8.6(b). The rest of the algorithm operates on
these two data structures by removing points from `remainingPoints` and
adding triangles to `indices`.

First, the triangle formed by points (p_0, p_1, p_2), shown in Figure 8.7, is
tested to see if it is an ear. The first part of the test is to make sure the tip is
convex. Since we are assured the input points are ordered counterclockwise
due to the cleanup stage of our pipeline, we can test if a tip is convex by
checking if the sign of the cross product between two vectors along the
outside of the triangle is nonnegative, as shown in Listing 8.4.

If the tip is convex, the triangle containment test follows, which checks
if every point in the polygon, except the points forming the triangle, is
inside the triangle. If all points are outside of the triangle, the triangle is
an ear, and its indices are added to `indices` and its tip is removed from

```
private static bool IsTipConvex(Vector2D p0, Vector2D p1,
                                Vector2D p2)
{
  Vector2D u = p1 - p0;
  Vector2D v = p2 - p1;
  return ((u.X * v.Y) - (u.Y * v.X)) >= 0.0;
}
```

Listing 8.4. Testing if a potential ear's tip is convex.

Figure 8.7. Ear test for triangle (p_0, p_1, p_2).

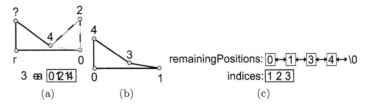

Figure 8.8. (a) Ear test for triangle (p_1, p_2, p_3). (b) Polygon after removing p_2. (c) Data structures after removing p_2.

`remainingPositions`. In our example, although the tip of triangle (p_0, p_1, p_2) is convex, p_3 is inside the triangle, making it fail the triangle containment test.

So ear clipping moves on to test triangle (p_1, p_2, p_3), shown in Figure 8.8(a). This triangle passes both the convex tip and triangle containment test, making it an ear that can be clipped, so p_2 is removed from `remainingPositions` and indices $(1, 2, 3)$ are added to `indices`. The updated polygon and data structures are shown in Figures 8.8(b) and 8.8(c).

The next triangle in question is triangle (p_1, p_3, p_4), shown in Figure 8.9, which fails the convex tip test. So ear clipping moves on to triangle (p_3, p_4, p_0), which is an ear, resulting in the changes shown in Figure 8.10.

After removing p_4, only one triangle remains, shown in Figure 8.10(b). The indices for this triangle are added to `indices` and ear clipping is complete. The final triangulation is shown in Figure 8.11.

Figure 8.9. Ear test for triangle (p_1, p_3, p_4).

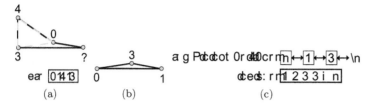

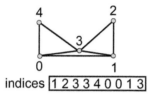

(a)   (b)   (c)

Figure 8.10. (a) Ear test for triangle (p_3, p_4, p_0). (b) Polygon after removing p_4. (c) Data structures after removing p_4.

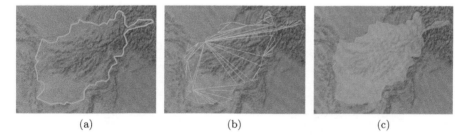

Figure 8.11. Final triangulation after ear clipping.

Figure 8.12 shows the triangulation of a more complex, real-world polygon. Creating a triangle fan for a convex polygon is basically ear clipping, except each triangle is guaranteed to be an ear, making the algorithm much faster since no ear tests are required.

(a)   (b)   (c)

Figure 8.12. (a) The line loop of a polygon for Afghanistan. (b) The wireframe triangulation of the polygon. (c) The triangulation rendered with translucency.

Ear clipping on the ellipsoid. Our approach to ear clipping thus far works well for planar 2D polygons, but it doesn't work for polygons defined on a globe by a set of geographic (longitude, latitude) pairs. We may be tempted to treat longitude as x and latitude as y, then simply use our 2D ear-clipping implementation. In addition to wraparound issues at the poles and the IDL, this approach has a more fundamental problem: our implementations for `IsTipConvex` and `PointInsideTriangle` are designed for Cartesian, not geographic, coordinates, which makes the primary operation in ear clipping, ear testing, incorrect. For example, consider three points at the same latitude with longitudes of 10°, 20°, and 30°. If the geodetic coordinates are treated as Cartesian coordinates, this is a straight line segment. In reality, in Cartesian WGS84 coordinates, it is two line segments with an angle where they meet. The result is the ear test can have false positives or negatives creating incorrect triangulations, like those shown in Figure 8.13(a).

There are two ways to achieve a correct triangulation, like that of Figure 8.13(b). One way is to project the boundary points onto a tangent plane, then use 2D ear clipping to triangulate the points on the plane. Imagine yourself at the center of the ellipsoid looking out towards the tangent plane. Consider two consecutive points on the polygon. Two rays cast from the center of the globe through each point lie in a plane whose intersection with the ellipsoid forms a curve (see Section 2.4) between the points. This creates the property that curves are projected to line segments on the plane, allowing our ear test to return correct results. Triangulating 3D polygons by projecting them onto a plane is a common approach and is the technique used in fast industrial-strength triangulation (FIST) [70].

An alternative approach is to modify ear clipping to handle points on an ellipsoid. The main observation is that the 2D point in the triangle test can be replaced with a 3D point in the infinite-pyramid test. An infinite

(a)   (b)

Figure 8.13. (a) Triangulating a polygon on the ellipsoid by naïvely treating longitude as x and latitude as y results in an incorrect triangulation. (b) A correct triangulation is achieved by projecting the polygon onto a tangent plane or replacing the 2D point in the triangle test with a 3D point in the infinite-pyramid test.

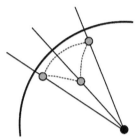

Figure 8.14. An infinite pyramid formed by the center of the ellipsoid and three points on its surface. When ear clipping polygons on an ellipsoid, the point-in-triangle test should be replaced with a point-in-infinite-pyramid test.

pyramid is formed with the ellipsoid center at its apex, extending through a triangular base formed by three points on the ellipsoid's surface, as shown in Figure 8.14. The inside of the pyramid is the space inside the intersection of three planes. Each plane contains the ellipsoid's center and two points in the triangle. The intersection of a plane and the ellipsoid forms a curve between the two points on the surface, shown as dashed lines in Figure 8.14.

The point-in-infinite-pyramid test is performed by first computing outward-facing normals for each plane. A point is in the pyramid if it is behind each plane. Since the pyramid's apex lies in all three planes, the sign of the dot product of the vector from the apex to the point in question with a plane's normal determines the side of the plane the point is on. If all three dot products are negative, the point is inside the pyramid (see Listing 8.5).

```
public static bool PointInsideThreeSidedInfinitePyramid(
    Vector3D point,
    Vector3D pyramidApex,
    Vector3D pyramidBase0,
    Vector3D pyramidBase1,
    Vector3D pyramidBase2)
{
  Vector3D v0 = pyramidBase0 - pyramidApex;
  Vector3D v1 = pyramidBase1 - pyramidApex;
  Vector3D v2 = pyramidBase2 - pyramidApex;
  Vector3D n0 = v1.Cross(v0);
  Vector3D n1 = v2.Cross(v1);
  Vector3D n2 = v0.Cross(v2);
  Vector3D planeToPoint = point - pyramidApex;

  return (planeToPoint.Dot(n0) < 0) &&
         (planeToPoint.Dot(n1) < 0) &&
         (planeToPoint.Dot(n2) < 0);
}
```

Listing 8.5. Testing if a point is inside an infinite pyramid.

```
private  static  bool  IsTipConvex(Vector3D  p0,  Vector3D  p1,
                                   Vector3D  p2)
{
    Vector3D  u = p1 - p0;
    Vector3D  v = p2 - p1;

    return u.Cross(v).Dot(p1) >= 0.0;
}
```

Listing 8.6. Testing if the tip of a potential ear on the ellipsoid is convex.

Replacing the point-in-triangle test with a point-in-infinite-pyramid test gets us halfway to ear clipping a polygon on an ellipsoid. The other half is determining if a potential ear's tip is convex. Given three points forming a triangle on the ellipsoid, (p_0, p_1, p_2), form two vectors along the outside of the triangle relative to the polygon: $\mathbf{u} = p_1 - p_0$ and $\mathbf{v} = p_2 - p_1$. If the tip, p_1, is convex, the cross product of $\mathbf{u}$ and $\mathbf{v}$ should yield an upward-facing vector relative to the surface. This can be verified by checking the sign of the dot product with $p_1 - 0$, as shown in Listing 8.6.

Ear clipping a polygon on an ellipsoid is now a simple matter of changing our 2D ear-clipping implementation to use Vector3D instead of Vector2D and using our new `PointInsideThreeSidedInfinitePyramid` and `IsTipConvex` methods. The input is a polygon's line loop in WGS84 coordinates. A complete implementation is provided in OpenGlobe. Core.EarClippingOnEllipsoid. Compare this to the 2D implementation in OpenGlobe.Core.EarClipping.

Besides creating a triangle mesh for rendering, triangulation is also useful for point in polygon testing. A point on an ellipsoid's surface is also in a polygon on the surface if it is in any of the infinite pyramids defined by each triangle in the polygon's triangulation. This can be made quite efficient by precomputing the normals for each pyramid's plane, reducing each point in pyramid test to three dot products. Geospatial analyses, such as point in polygon tests, are the bread and butter of GIS applications.

8.2.3   Ear-Clipping Optimizations

Given that virtual globe users commonly want to visualize a large number of geospatial polygons, and each polygon can have a large number of points, it is important to optimize ear clipping because a naïve implementation doesn't scale well at $\mathcal{O}(n^3)$. Thankfully, this is a worst-case runtime, and in practice, our naïve implementation will run much closer to $\mathcal{O}(n^2)$. But there is still plenty of room for optimization.

First, let's understand why the runtime is $\mathcal{O}(n^3)$; the triangle containment test is $\mathcal{O}(n)$, finding the next ear is $\mathcal{O}(n)$, and $\mathcal{O}(n)$ ears are needed. Although the triangle containment test is $\mathcal{O}(n)$, n decreases as ear tips are removed. In practice, the next ear is found much quicker than $\mathcal{O}(n)$, which is why the algorithm rarely hits its $\mathcal{O}(n^3)$ upper bound using real polygons.

At least $\mathcal{O}(n)$ ears are always required, but the other two steps can be improved. In fact, the $\mathcal{O}(n)$ search for the next ear can be eliminated, making the worst-case runtime $\mathcal{O}(n^2)$. The key observation is that removing an ear tip only affects the ear-ness of the triangle to its left and its right, so an $\mathcal{O}(n)$ search for the next ear is not required after ears in the initial polygon are found. This does require some extra bookkeeping.

Eberly describes an implementation that maintains four doubly linked lists simultaneously: lists for all polygon points, reflex points only, concave points only, and ear tips [44]. *Reflex points* are those points whose incident edges form an interior angle larger than 180°. The reason for keeping a list of these points is because only reflex points need to be considered during the triangle containment test, bringing the runtime down to $\mathcal{O}(nr)$, where r is the number of reflex points. For geospatial polygons including zip codes, states, and countries, Ohlarik found 40% to 50% of a polygon's initial points were reflex [127].

The final bottleneck to consider is the linear search, now $\mathcal{O}(r)$, used in the triangle containment test. This can be replaced by storing reflex points in leaf nodes of a spatial binary tree [127]. The linear search is then replaced with only tests for reflex points in the leaves whose AABB's overlap the triangle in question. Empirically, this optimization was found to scale very well on geospatial polygons, despite the need to create the tree.

Similarly, Held presents a detailed description and analysis of using geometric hashing, both bounding volume trees and regular grids, to optimize ear clipping [69]. Although geometric hashing does not improve the worst case runtime, in practice, it can obtain near $\mathcal{O}(n)$ runtimes.

Add any of the above optimizations to `OpenGlobe.Core.`EarClipp ing. Record timings before and after. Are the results always what you expect?

○○○○ Try This

8.2.4 Subdivision

If the globe were flat, a polygon would be ready for rendering after triangulation, but consider the wireframe rendering of the large triangle in Figure 8.15(a). From this view, the triangle appears to be on the ellipsoid's surface. By viewing the triangle edge, as in Figures 8.15(b) and 8.15(c), it becomes clear that only the triangle's endpoints are on the ellipsoid; the triangle itself is well under the surface. What we really want is the triangle projected onto the surface of the ellipsoid, as shown in Figure 8.15(d), or even at a constant height above the ellipsoid. In fact, we'd like to approximate the intersection of the ellipsoid and the infinite pyramid formed from the ellipsoid's center and the triangle (recall Figure 8.14). This will result in curves between triangle endpoints.

In order to approximate the triangle's projection, we use a subdivision stage immediately after triangulation (see Figure 8.16). This algorithm is very similar to the subdivision-surfaces algorithm discussed in Section 4.1.1, which subdivides a regular tetrahedron into an ellipsoid. Here, the idea is to subdivide each triangle into two new triangles until a stopping condition is satisfied.

We'll use *granularity* as our stopping condition. Consider two points, p and q, forming an edge of a triangle. The granularity is the angle between $\mathbf{p} - \mathbf{0}$ and $\mathbf{q} - \mathbf{0}$. For our stopping condition to be satisfied, the granularity of every edge in a triangle needs to be less than or equal to a given

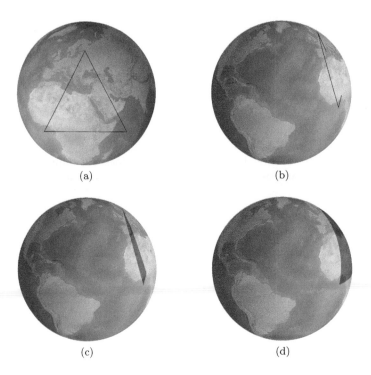

(a)

(b)

(c)

(d)

Figure 8.15. (a) From this view, the triangle appears to be on the globe. (b) In reality, only its endpoints are on the globe, as evident in this view. (c) When filled, the triangle is shown to be under the globe and partially clipped. (d) The desired result, which requires subdividing the triangle.

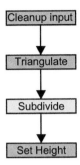

Figure 8.16. The subdivision stage immediately follows triangulation.

granularity (e.g., 1°). If the stopping condition is not satisfied, the triangle is divided into two by bisecting the edge with the largest granularity. This continues recursively until all triangles meet the stopping condition. At this point, the subdivided triangles are still in the same plane as the original triangles. The final stage of our pipeline will raise them to the ellipsoid's surface or above it.

In Section 4.1.2, we presented a recursive implementation to the tetrahedron subdivision algorithm. Here, we will present a queue-based subdivision implementation. The complete code is in **OpenGlobe.Core**.Tri angleMeshSubdivision. The method takes positions and indices computed from triangulation and a granularity and returns a new set of positions and indices that satisfy the stopping condition (see Listing 8.7).

```
public class TriangleMeshSubdivisionResult
{
    public ICollection<Vector3D> Positions  {get; }
    public IndicesUnsignedInt Indices { get; }

    // ...
}

public static class TriangleMeshSubdivision
{
    public static TriangleMeshSubdivisionResult Compute(
        IEnumerable<Vector3D> positions ,
        IndicesUnsignedInt indices , double granularity)
    {
        // ..
    }
}
```

Listing 8.7. Method signature for subdivision.

Figure 8.17. Subdividing one triangle into two by bisecting the edge with the largest granularity.

```
public struct Edge : IEquatable<Edge>
{
  public Edge(int index0, int index1) { /* ... */ }

  public int Index0 { get; }
  public int Index1 { get; }
  // ...
}
```

Listing 8.8. An edge of a triangle.

The algorithm works by managing two queues and a list of positions. The `triangle` queue contains triangles that may need to be subdivided, and the `done` queue contains triangles that have met the stopping condition. First, the indices for all input triangles are placed on the `triangles` queue, and the input positions are copied to a new positions list: `subdivided Positions`. While the `triangles` queue is not empty, a triangle is dequeued and the granularity for each edge is computed. If no granularity is larger than the input granularity, the triangle is enqueued on the `done` queue. Otherwise, the center of the edge with the largest granularity is computed and added to `subdividedPositions`. Then, two new subdivided triangles are enqueued onto `triangles`, and the algorithm continues. An example subdivision is shown in Figure 8.17.

This implementation can result in a lot of duplicate positions. For example, consider two adjacent triangles sharing an edge. Since each triangle is subdivided independently, each will subdivide the edge and add the same position to `subdividedPositions`. An easy way to avoid this duplication is to build a cache of subdivided edges and check the cache before subdividing an edge. To implement this, first create a type for a triangle edge. It simply needs to store two indices, as done in Listing 8.8.

The subdivision algorithm then needs a dictonary from an Edge to the index of the position that was created when the edge was subdivided, such as the following:

```
Dictionary<Edge, int> edges = new Dictionary<Edge, int>();
```

Consider a triangle that needs to be subdivided along an edge with indices `triangle.I0` and `triangle.I1`, which correspond to positions `p0` and `p1`. Without edge caching, this code would look as follows:

```
subdividedPositions.Add((p0 + p1) * 0.5);
int i = subdividedPositions.Count - 1;

triangles.Enqueue(
    new TriangleIndicesUnsignedInt(triangle.I0, i, triangle.I2));
triangles.Enqueue(
    new TriangleIndicesUnsignedInt(i, triangle.I1, triangle.I2));
```

With edge caching, first the cache is checked for a subdivided edge with the same indices. If such an edge exists, the index of the subdivided point is retrieved from the cache. Otherwise, the edge is subdivided, the new position is added to `subdividedPositions`, and an entry is made in the cache:

```
Edge edge = new Edge(Math.Min(triangle.I0, triangle.I1),
                     Math.Max(triangle.I0, triangle.I1));
int i;
if (!edges.TryGetValue(edge, out i))
{
    subdividedPositions.Add((p0 + p1) * 0.5);
    i = subdividedPositions.Count - 1;
    edges.Add(edge, i);
}
// ...
```

The edge is always looked up using the minimum index as the first argument and the maximum index as the second argument since edge (i_0, i_1) equals (i_1, i_0).

> Try implementing an adaptive subdivision algorithm on the GPU using the programmable tessellation stages introduced in OpenGL 4.

○○○○ **Try This**

8.2.5 Setting the Height

The final stage of our pipeline is also the simplest stage. Each position from the subdivision stage is scaled to the surface by calling Ellip soid.**ScaleToGeocentricSurface** (see Section 2.3.2). The wireframe result is shown in Figure 8.18.

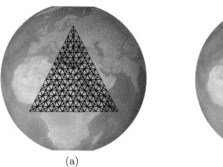

(a)   (b)

Figure 8.18. (a) The triangle from Figure 8.15(a) after subdivision with a 5°
granularity. (b) The same wireframe after scaling all the positions to the surface.

8.2.6 Rendering

Once a triangle mesh is produced from the polygon pipeline, rendering is
easy. Polygons are most commonly rendered with translucency, as shown in
figures throughout this chapter. This is achieved by using a blend equation
of add with a source factor of source$_\alpha$ and a destination factor of $1 -$
source$_\alpha$.

For lighting, normals for the polygon can be computed procedurally
in the fragment shader using the ellipsoid's geodetic surface normals (see
Section 2.2.2). This takes advantage of the fact that the polygon is on the
ellipsoid; we could not use the ellipsoid's normals for arbitrary polygons in
space.

Even though back-face culling is usually only used for closed opaque
objects, it can be used when rendering our translucent polygons if the
polygon is rendered directly on the ellipsoid, that is, with a height of zero.
This eliminates rasterizing fragments for triangles on the back face of the
globe.

A complete implementation of the polygon pipeline and rendering is in
OpenGlobe.Scene.Polygon.

8.2.7 Pipeline Modifications

The polygon pipeline is a subset of what we've used in STK and Insight3D
for years. There are additional stages and worthwhile modifications worth
considering. For starters, the triangulation stage can be modified to support polygons with holes. Typically, the convention is that the outer boundary has one winding order and boundaries for interior holes have the opposite winding order. The key to triangulating such polygons with ear

clipping is to first convert the polygon to a simple polygon by adding two
edges connecting mutually visible points in both directions for every hole.
Eberly describes such an algorithm [44].

There are several stages that can be added to improve rendering performance. Multiple LODs can be computed. One approach is to compute
discrete LODs after the height-setting stage. For a fast and easy-to-code
algorithm, see vertex clustering [106, 146]. For a good balance of speed
and visual fidelity, consider quadric error metrics [55]. LODs can also be
computed as part of the subdivision step by stopping subdivision early
to create lower LODs. The drawback is that every LOD will include all
triangles in the original triangulation.

Similar to polylines, polygons in virtual globes are typically rendered as
part of a layer that may only be visible for certain viewer heights or view
distances. This can make LODs for individual polygons less useful since a
polygon's layer may be turned off before the polygon's lower LODs would
even be rendered.

To improve rendering throughput, vertices and indices can be reordered
into a GPU vertex cache-coherent layout [50, 148]. This stage can occur
after subdivision or after LOD computation. If LODs are used, indices for
each LOD should be reordered independently.

If the polygon is going to be culled, a bounding-sphere or other bound-
ing-volume computation stage can be added to the pipeline. In many cases,
just the boundary points are necessary to determine the sphere. However,
one could conceive of polygons on really oblate ellipsoids where this is not
true, requiring a robust solution to consider all points after subdivision.

Finally, it is worth observing that performance improvements can be
made by combining pipeline stages. Fewer passes over the polygon result
in less memory access and fewer cache misses. In particular, it is easy to
combine the subdivision stage and the height-setting stage. The bounding-
volume stage can also be combined with other stages. Although combining
stages can result in a performance win, it comes with the maintenance cost
of less clear code.

8.3   Polygons on Terrain

The polygon pipeline is an excellent approach to rendering polygons on
an ellipsoidal globe. We've used it successfully in commercial products for
many years. What happens, though, if the globe is not a simple ellipsoid
but also includes terrain? The polygon's triangle mesh will render under
the terrain, or above it in the case of undersea terrain. This is because
the triangle mesh approximates the ellipsoid, not the terrain. Here, we
introduce modifications to the polygon pipeline and a new way to render

Figure 8.19. Polygon that conforms to terrain rendered using shadow volumes. (Image taken using STK.)

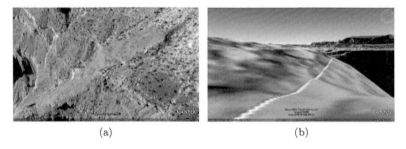

(a)   (b)

Figure 8.20. (a) A polygon conforming to terrain using render to texture. Image (C) 2010 DigitalGlobe. (b) Aliasing occurs along the boundary when zoomed in close. Image USDA Farm Service Agency, Image NMRGIS, Image (C) 2010 DigitalGlobe. (Figures taken using Google Earth.)

the triangle mesh based on shadow volumes that make a polygon conform to terrain, as shown in Figures 8.19 and 8.20(a).

8.3.1 Rendering Approaches

Before describing our approach in detail, let's consider a few alternative approaches:

- *Terrain-approximating triangle mesh.* Perhaps the most obvious way to make a triangle mesh conform to terrain is to modify the set-height stage of the pipeline (see Section 8.2.5). That is, instead of raising each position of the mesh to the ellipsoid surface, raise each position to the terrain's surface. Of course, this requires a function that can return the terrain's height given a geographic position.

Although this approach is simple, it does not produce satisfactory visual quality. Sometimes a polygon's triangles will be coplanar with terrain, and other times, they will be under the terrain creating obvious rendering artifacts. This cannot be resolved by a technique like averaged depth (see Section 7.2) because the triangle mesh needs to be depth tested against the true depth values of the terrain. Furthermore, most terrain is implemented using LOD; when the terrain LOD changes as the viewer zooms, different artifacts will appear.

- *Render to texture.* An alternative approach is to render a polygon to a texture, then render the texture on terrain using multitexturing. As mentioned in Section 8.1, a major drawback to this is that aliasing can occur if the texture is not of sufficient resolution, as shown in Figure 8.20(b). Also, creating the texture can be a slow process, making this approach less attractive for dynamic polygons.

A third approach is based on *shadow volumes* [36, 149]. A triangulated polygon is raised above terrain, duplicated, then lowered below terrain, forming the caps of a closed volume encompassing the terrain, as shown in Figure 8.23(b). Terrain intersecting the volume is shaded using shadow-volume rendering. This approach has several advantages:

- *Visual quality.* No aliasing occurs along the polygon boundary. The intersection of the shadow volume and terrain is accurate to the pixel. As the viewer zooms, the polygon's visual quality does not change.

- *Decoupled from terrain rendering.* Polygons are rendered separately and independently of terrain. It doesn't matter what terrain LOD algorithm is used, or if LOD is used at all. The only requirement is that the polygon's shadow volume needs to enclose the terrain and it needs to be rendered after terrain, when the terrain's depth buffer is available.

- *Low memory requirements.* For many polygons, this approach uses less memory than render to texture. Shadow-volume extrusion can be done in a geometry shader, further reducing the memory requirements.

- *Versatility.* This approach can be generalized to shade models or any geometry intersecting the shadow volume.

8.3.2 Shadow Volumes

Let's first look at shadow volumes in general before detailing how to render polygons with them. One approach to shadow rendering is to use shadow volumes and the stencil buffer to determine areas in shadow [68]. These areas are only shaded with ambient and emission components, while the rest of the scene is fully shaded, including diffuse and specular components.

Consider a point light source and a single triangle. Cast three rays, each from the point through a different vertex of the triangle. The truncated infinite pyramid formed on the side of the triangle opposite of the point light is a shadow volume, as shown in Figure 8.21(a).

Objects inside the shadow volume are in shadow. This is determined on a pixel-by-pixel basis. Imagine casting a ray from the eye through a pixel. A count is initialized to zero. When the ray intersects the front face of a shadow volume, the count is incremented. When the ray intersects the back face of a shadow volume, the count is decremented. If the count is nonzero when the ray finally intersects an object, the object is in shadow. Otherwise, the object is not in shadow (see Figure 8.21(b)). This works for nonconvex volumes and multiple overlapping volumes.

In practice, this is usually implemented with the stencil buffer. The algorithm for rendering the scene is as follows:

- Render the entire scene to the color and depth buffers with just ambient and emission components.

- Disable color buffer and depth buffer writes.

- Clear the stencil buffer and enable the stencil test.

- Render the front faces of all shadow volumes. Each object that casts a shadow will have a shadow volume. During this pass, increment the stencil value for fragments that pass the depth test.

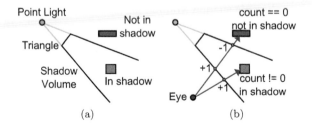

(a)   (b)

Figure 8.21. (a) A side view of a shadow volume formed from a point light and a triangle. Objects inside the volume are in shadow. (b) Conceptually, rays are cast from the eye through pixels to determine if an object is in shadow.

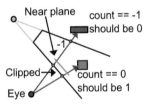

Figure 8.22. Z-pass shadow volumes lead to an incorrect count when shadow volumes are clipped by the near plane.

- Render shadow-volume back faces, decrementing the stencil value for fragments that pass the depth test.

- At this point, the stencil buffer is nonzero for pixels in shadow.

- Enable writing to the color buffer and set the depth test to less than or equal to.

- Render the entire scene again with additive diffuse and specular components and a stencil test that passes when the value is nonzero. Only fragments not in shadow will be shaded with diffuse and specular components.

Incrementing and decrementing the stencil value when rendering the shadow-volume front and back faces, respectively, implements casting a ray through the scene to determine areas in shadow. Although this is described using two passes, it can be implemented in a single pass using two-sided stencil, which allows different stencil operations for front- and back-facing triangles rendered in the same pass. This is a core feature since OpenGL 2.0.

The algorithm as described above is called z-pass because the stencil value is modified when a fragment passes the depth test. This algorithm fails when the viewer is inside the volume or the near plane intersects it. Since part of the volume is clipped by the near plane, the count becomes wrong, leading to an incorrect shadow test, as shown in Figure 8.22. The problem can be moved to the far plane, where it can be dealt with more easily by using z-fail shadow volumes [15, 89]. With z-fail, the front- and back-facing shadow-volume rendering passes are changed to:

- render shadow-volume back faces, incrementing the stencil value if the fragment fails a less-than depth test;

- render shadow-volume front faces, decrementing the stencil value if the fragment fails a less-than depth test.

This also requires that shadow volumes be closed at their far ends. An incorrect count can still be computed if the shadow volume is clipped by the far plane. This can be solved in hardware using depth clamping, a core feature since OpenGL 3.2. Depth clamping allows geometry clipped by the near or far plane to be rasterized and writes the depth value clamped to the far depth range.

8.3.3   Rendering Polygons Using Shadow Volumes

Since shadow volumes are used for rendering shadows, how do we use them to render polygons that conform to terrain? There are two key insights:

- Shadow volumes are used to determine the area in shadow. A shadow volume can be constructed however we choose; in fact, it could be a volume encompassing a polygonal region of terrain, like that shown in Figure 8.23(b).

- The area in shadow is only shaded with ambient and emission components, but we can actually shade areas in shadow however we please.

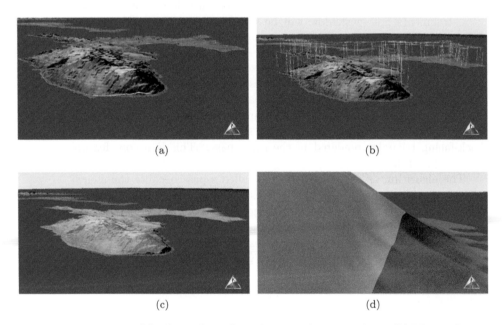

Figure 8.23. (a) The outline of a polygon is shown in white. (b) The outline is triangulated and extruded to form a shadow volume intersecting terrain. (c) The polygon is rendered filled on terrain using a shadow volume. (d) Zooming in close on the polygon does not create aliasing artifacts. (Images taken using STK.)

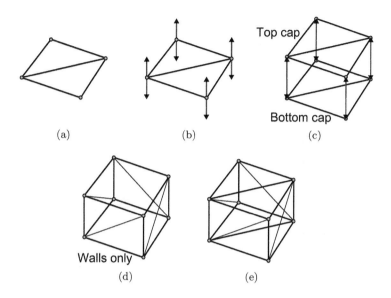

Figure 8.24. (a) A triangulated polygon before entering the shadow-volume creation stage. (b) Each vertex's normal and its opposite. (c) The top and bottom caps are formed by extruding each vertex. (d) A triangle strip forms a wall connecting the top and bottom cap. (e) The polygon's complete shadow volume with top and bottom caps and a wall.

We can render the globe and terrain as usual, build shadow volumes around polygons that encompass the terrain, then shade the area "in shadow," that is, the terrain inside the volume, by modulating the color buffer with that of the polygon's color. The result is shown in Figure 8.23(c).

The algorithm starts with a polygon's boundary, like that shown in Figure 8.23(a). It then goes through every stage of our polygon pipeline except the last: input cleanup, triangulation, and subdivision. The final stage, set height, is replaced with a stage that builds the shadow volume.

In this stage, every position is raised along the ellipsoid's geodetic surface normal, forming a mesh above the terrain. This is the top cap of the shadow volume. Likewise, every vertex is lowered along the direction opposite the geodetic surface normal, forming the bottom cap below terrain. Finally, a wall is created along the polygon's boundary that connects the bottom cap and top cap, forming a closed shadow volume. This process is shown in Figure 8.24.

Once a shadow volume is created for a polygon, rendering the polygon is similar to the shadow-volume rendering algorithm previously described. The polygon can be rendered on a globe with terrain or just an ellipsoidal globe. All that is required is that the depth buffer be written before the polygons are rendered. The steps are as follows:

- Render the globe and terrain to the color and depth buffers.

- Disable color buffer and depth buffer writes (same as Section 8.3.2).

- Clear the stencil buffer and enable the stencil test (same as Section 8.3.2).

- Render the front and back faces of each polygon's shadow volume to the stencil buffer using either the z-pass or z-fail algorithm described previously (same as Section 8.3.2).

- At this point, the stencil buffer is nonzero for globe/terrain pixels covered by a polygon.

- Enable writing to the color buffer and set the depth test to less than or equal to (same as Section 8.3.2).

- Render each shadow volume again with a stencil test that passes when the value is nonzero. The most common way to shade the fragments is to use blending to modulate the polygon's color with the globe/terrain color, based on the polygon's alpha value.

Unlike using shadow volumes for shadows, using shadow volumes for polygons does not require rendering the entire scene twice. Instead, the final pass renders just polygon shadow volumes, not the full scene. If all polygons are the same color, performance can be improved by replacing the second pass with a screen-aligned quad. Doing so reduces the geometry load and can reduce fill rate because the depth complexity of the shadow volumes can be high, especially for horizon views, whereas the depth complexity of the quad is always one.

8.3.4 Optimizations

Rendering polygons with shadow volumes uses a great deal of fill rate. In particular, there is a large amount of wasted rasterization for fragments that do not pass the stencil test on the final pass. A ray cast through these pixels goes through the shadow volume(s) without intersecting the terrain or globe. When looking straight down at the globe, there is no wasted rasterization overhead, but horizon views can have a large amount of overhead, as shown in Figure 8.25.

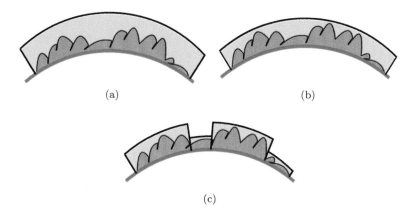

Figure 8.25. Shadow-volume rasterization overhead is shown in gray. (a) Using the globe's maximum terrain height for the shadow volume's top cap results in the most wasted rasterization. (b) Using the maximum terrain height of terrain encompassed by the polygon can decrease rasterization overhead. (c) Breaking the shadow volume into sections and using the maximum terrain height in each section can further reduce rasterization overhead.

One way to minimize the amount of rasterization overhead is to use tight-fitting shadow volumes. The simplest way to extrude the top and bottom caps is to move the top cap up along the maximum terrain height of the globe and to move the bottom cap down along the minimum terrain height. This can create a great deal of waste, creating tall, thin shadow volumes that yield very few shaded pixels for horizon views. A more sophisticated approach stores minimum and maximum terrain heights in a grid or quadtree covering the globe. The minimum and maximum heights for terrain approximately covered by a polygon is used to determine the height for the bottom and top caps. This can still lead to lots of wasted rasterization when terrain has high peaks and low valleys. To combat this, a shadow volume can be broken into sections, each with a different top cap and bottom cap height. Using multiple section trades increases vertex processing for decreased fill rate.

The amount of vertex memory required can be reduced by using a geometry shader. Instead of creating the shadow volume on the CPU, it can be created on the GPU using the original polygon triangulation. The geometry shader outputs the extruded top cap and bottom cap and the connecting wall. An additional benefit of this approach is that if terrain LOD requires the shadow volume to dynamically change its height, it can do so on the GPU.

Just like normal meshes, the shadow-volume mesh can be optimized for GPU vertex caches and have LOD applied. In addition, when the viewer is sufficiently far away, the shadow volume can be replaced with a simple mesh floating above terrain. At this view distance, the extra fill rate caused by shadow volumes usually isn't a problem, but the number of render passes is reduced by not using shadow volumes.

8.4   Resources

There are many discussions on triangulation on the web [44, 69–71, 127]. *Real-Time Rendering* contains an excellent survey of shadow volumes in general [3]. In recent years, using shadow volumes to render polylines and polygons on terrain has received attention in the research literature [36, 149].

9

Billboards

Screen-aligned *billboards* are textured quads rendered parallel to the viewport so that they are always facing the viewer. To support nonrectangular shapes, these quads are rendered with a texture containing an alpha channel, in combination with discard in the fragment shader (see Figure 9.1).[1]

Billboards used in this fashion are also called *sprites* and serve many important purposes in virtual globe rendering, the most prominent of which is rendering icons for places of interest: cities, landmarks, libraries, hospitals, etc.—pretty much anything interesting whose position can be defined or approximated by a single geographic position (see Figure 9.2(a)). The position doesn't always need to be static; icons may be used for the current position of an airplane in flight. Most virtual globes even allow users to place icons at positions of their choosing (e.g., Google Earth's placemarks and ArcGIS Explorer's point notes).

(a)   (b)

Figure 9.1. (a) Billboards are rendered using two screen-aligned triangles that form a quad. (b) A texture's alpha channel is used to discard transparent fragments.

---

[1]The alpha test could be used in place of discard, but it was deprecated in OpenGL 3.

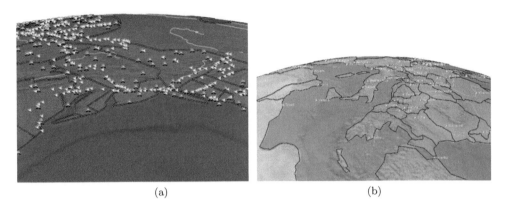

(a)   (b)

Figure 9.2. (a) Billboards used to render icons for airports and Amtrak stations in the United States. (b) Billboards used to render icons and labels for cities in Europe.

In addition to icons, billboards are also used to render text facing the viewer, as in Figure 9.2(b). In many cases, point features contain metadata, including a label that is displayed alongside its icon. Another type of text rendering seen in virtual globes is labels burned into imagery (e.g., the names of roads overlaid on satellite imagery). This type of text becomes hard for the user to read when the viewer is looking at an edge.

Billboards, in all their varieties, can serve many other uses: rendering vegetation, replacing 3D models with lightweight *imposters*, and rendering a wide range of image-based effects, including clouds, smoke, and fire. Our focus is on using billboards to render icons and text from GIS datasets.

9.1   Basic Rendering

Let's put aside textures for now and start our billboard-rendering algorithm by focusing on rendering screen-aligned quads of constant pixel size. The key observation is that since a billboard's position is defined in model (or world) coordinates and its size is defined in pixels, the size of a billboard in model space changes as the viewer zooms in and out. For example, a 32×32 pixel billboard for an airport is only a few meters large when zoomed in at street level. As the viewer zooms out, the size of the billboard in pixels does not change, but its size in meters increases. This means that when the viewer moves, our rendering algorithm needs to recompute the model-space size of every visible billboard in order to maintain its constant pixel size.

The basic approach is to transform the billboard's center point from model to window coordinates, then use simple 2D offsets, scaled by the bill-

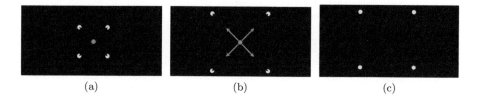

(a)   (b)   (c)

Figure 9.3. (a) Once transformed to window coordinates, a billboard's position is copied and moved towards each of the quad's corners. (b) The translation is scaled by one-half the texture's size, s, to compute the actual corner point. (c) The result is modeled with a two-triangle triangle strip.

board's texture size, to compute the corner points for a viewport-aligned quad. The quad, modeled with a two-triangle triangle strip, is then rendered using an orthographic projection (see Figure 9.3). Working in window coordinates makes it easy to align the billboard with the viewport and maintain its constant pixel size. The process is very similar to rendering lines of a constant pixel width discussed in Section 7.3.3. There are a variety of options for where exactly to do the transforms:

- *CPU.* Before programmable hardware, it was common to transform each billboard on the CPU, then render a quad either one at a time using OpenGL immediate mode or in an all-in-one call using client vertex arrays. On today's hardware this is horribly inefficient. In particular, it results in a lot of per-billboard CPU overhead, system bus traffic, and system-memory cache misses in the common case of rendering a large number of billboards. Furthermore, when the CPU has to touch each billboard and send it to the GPU, the CPU is not able to feed the GPU quickly enough to fully utilize it. The increasing gap between memory bandwidth and processing power will only make this approach worse over time. There must be a better way! In fact, since each billboard can be transformed independently of the others, a GPU implementation is ideal.

- *Point sprites.* In order to utilize the GPU for billboard rendering, both OpenGL and Direct3D exposed a fixed-function feature called *point sprites* [33]. Point sprites generalize point rendering to allow points to be rasterized as textured squares on the GPU, eliminating the need for per-billboard transforms on the CPU and significant system bus traffic, and only requiring one-fourth the amount of memory. Although point sprites can provide a significant performance improvement, they are less commonly used today in favor of the more flexible approach of using shaders. In fact, fixed-function point sprites were removed in Direct3D 10.

- *Vertex shader.* Billboards are straightforward to implement in a vertex shader. Each billboard is represented by two triangles forming a quad defined by four vertices. Each vertex stores the billboard's model position and an offset vector of either $(-1, -1)$, $(1, -1)$, $(1, 1)$, or $(-1, 1)$ (see Figure 9.4). The triangles are degenerate until the vertex shader transforms the billboard's model position into window coordinates and uses the offset vector and texture size to translate the vertex to its corner of the quad. Similar to point sprites, this moves all the transforms to the GPU, but it uses four times the number of vertices as point sprites and requires larger vertices due to the offset vector. Fortunately, the geometry shader can be used to keep the flexibility of a shader-based approach without the need to duplicate vertices.

- *Geometry shader.* Billboard rendering is the quintessential geometry-shader example, which is why it is mentioned in both the NVIDIA and ATI programming guides (under the name point sprites) [123, 135]. Its implementation is basically fixed-function point sprites meet vertex-shader billboards. A point is rendered for each billboard, and the geometry shader transforms each point into window coordinates and outputs a viewport-aligned quad. This has all of the flexibility of the vertex-shader version, with only one-fourth the number of vertices, and no need for offset vectors. In a minimal implementation, each billboard only needs its position stored in the vertex buffer. In addition, the geometry shader reduces vertex setup costs, eliminates duplicate model-to-window transforms, and does not require any index data, making it the most memory-efficient, flexible way to render billboards.

Patrick Says ○○○○

I've had the pleasure of implementing billboard rendering using CPU, vertex-shader, and geometry-shader implementations in Insight3D. The geometry shader was the easiest to implement (with the exception that debugging the CPU version was the easiest) and is also the most efficient on today's hardware. Interestingly enough, the vertex-shader version outperformed the geometry-shader version on hardware with first-generation geometry-shader support. I resisted the urge to remove the geometry-shader version in hopes of it becoming faster, and sure enough, it did!

Figure 9.4. The vertex buffer layout for a single billboard rendered with the vertex-shader method. The billboard's position, $p0$, is duplicated across each vertex for a billboard.

Given the strength of the geometry-shader version, let's take a closer look at its implementation. We'll build up to the shaders used in `Open Globe.Scene`.BillboardCollection. The vertex data are simply one vertex per billboard containing the billboard's model-space position. A vertex shader is used to transform the position to window coordinates (see Listing 9.1). This transform could also be done in the geometry shader, but for best performance, per-vertex operations should occur in the vertex shader so shared vertices are not redundantly processed in the geometry shader. When rendering point primitives, as we are here, this rarely matters, but we do it regardless for illustration purposes.

The geometry shader, shown in Listing 9.2, is where the bulk of the work happens. Its purpose is to expand the incoming point to a quad. This takes advantage of the fact that the input primitive type does not need to be the same as the output primitive type. In this case, a point is expanded

```
in vec4 position;

uniform mat4 og_modelViewPerspectiveMatrix;
uniform mat4 og_viewportTransformationMatrix;

vec4 ModelToWindowCoordinates(
    vec4 v,
    mat4 modelViewPerspectiveMatrix,
    mat4 viewportTransformationMatrix)
{
  v = modelViewPerspectiveMatrix * v; // clip coordinates
  v.xyz /= v.w; // normalized device coordinates
  v.xyz = (viewportTransformationMatrix *
          vec4(v.xyz, 1.0)).xyz; // window coordinates
  return v;
}

void main()
{
  gl_Position =
      ModelToWindowCoordinates(
          position,
          og_modelViewPerspectiveMatrix,
          og_viewportTransformationMatrix);
}
```

Listing 9.1. Vertex shader for billboard rendering.

```
layout(points) in;
layout(triangle_strip, max_vertices = 4) out;

out vec2 fsTextureCoordinates;

uniform mat4 og_viewportOrthographicMatrix;
uniform sampler2D og_texture0;

void main()
{
  vec2 halfSize = 0.5 * vec2(textureSize(og_texture0, 0));

  vec4 center = gl_in[0].gl_Position;

  vec4 v0 = vec4(center.xy - halfSize, -center.z, 1.0);
  vec4 v1 = vec4(center.xy + vec2(halfSize.x, -halfSize.y),
                 -center.z, 1.0);
  vec4 v2 = vec4(center.xy + vec2(-halfSize.x, halfSize.y),
                 -center.z, 1.0);
  vec4 v3 = vec4(center.xy + halfSize, -center.z, 1.0);

  gl_Position = og_viewportOrthographicMatrix * v0;
  fsTextureCoordinates = vec2(-1.0, -1.0);
  EmitVertex();

  gl_Position = og_viewportOrthographicMatrix * v1;
  fsTextureCoordinates = vec2(1.0, -1.0);
  EmitVertex();

  gl_Position = og_viewportOrthographicMatrix * v2;
  fsTextureCoordinates = vec2(-1.0, 1.0);
  EmitVertex();

  gl_Position = og_viewportOrthographicMatrix * v3;
  fsTextureCoordinates = vec2(1.0, 1.0);
  EmitVertex();
}
```

Listing 9.2. Geometry shader for billboard rendering.

into a triangle strip. Thankfully, all triangle strips will be composed of four vertices, one for each corner point, so it is easy to explicitly set the output's upper bound (`max_vertices = 4`). If the number of vertices is variable, it is important to keep the number as low as possible, for best performance on NVIDIA hardware [123].[2]

In order to expand the point into a triangle strip, first the geometry shader determines the size of the billboard's texture, in pixels, using the `textureSize` function. This value is divided by two and used to create four new vertices, one for each corner of the quad. These vertices are then output as a triangle strip with clockwise winding order. The winding order is not particularly important here, however, because the triangle strip is

_____

[2]ATI, on the other hand, states "the ATI Radeon HD 2000 series is largely insensitive to the declared maximum vertex count" (p. 9) and says it is OK to set this value to a safe upper bound when it is hard to determine at compile time [135].

always facing the viewer, so facet culling provides no benefit. Also, observe that the texture coordinates are easily generated on the fly when the vertex is output.

Since geometry shaders must preserve the order in which geometry was rendered, the GPU buffers geometry-shader output to allow several geometry-shader threads to run in parallel. If a geometry shader does a significant amount of amplification and, in turn, outputs a large number of vertices, the output can overflow from on-chip buffers to slow off-chip memory, which is why geometry-shader performance is commonly bound by the size of its output. Our geometry-shader implementation does not do too much amplification. Given it inputs one vertex and outputs four, its amplification rate is $1 : 4$, which just so happens to be a rate that ATI hardware includes special support for [135]. See their programming guide for additional restrictions.

The output size of a geometry shader is commonly measured in number of scalar components. In our geometry shader, each vertex output is composed of a four-scalar position and a two-scalar texture coordinate, for a total of six scalars per vertex. Since we are outputting four vertices, the total number of scalars output is 24. The lower this number, the faster the geometry shader will run. On an NVIDIA GeForce 8800 GTX, a shader runs at maximum performance when outputting 20 or less scalars. Performance then degrades in a nonsmooth fashion as the number of scalars increases. For example, a shader runs at 50% performance when outputting 27 to 40 scalars. Therefore, it is important to minimize the number of scalars output in the geometry shader even if it calls for doing additional computation in the fragment shader.

```
in vec2 fsTextureCoordinates;
out vec4 fragmentColor;

uniform sampler2D og_texture0;
uniform vec4 u_color;

void main()
{
  vec4 color = u_color * texture(og_texture0,
                                 fsTextureCoordinates);

  if (color.a == 0.0)
  {
      discard;
  }
  fragmentColor = color;
}
```

Listing 9.3. Fragment shader for billboard rendering.

Let us turn our attention to the fragment shader in Listing 9.3. In most virtual globes, lighting does not influence billboard shading, which makes fragment shading quite simple. In the simplest case, the texel from the billboard's texture is assigned as the output color. Since we want to support nonrectangular shapes, our shader also discards fragments based on the texture's alpha channel. In addition, the texel isn't used directly. Instead, it is modulated with a user-defined color, shown here as a uniform, but it could also be defined per vertex and passed through to the fragment shader. In many cases, this will be white to let the texel come through as is. This color is most often used when rendering billboards with white textures that should be shaded a different color. This eliminates the need for creating different color versions of the same texture.

These vertex, geometry, and fragment shaders are all that are needed for a basic billboard implementation. However, there are lots of things that will make our billboards more useful. Let's start with efficiently rendering a large number of billboards with different textures.

9.2   Minimizing Texture Switches

For best performance, we want to render as many billboards as possible using the fewest number of draw calls: ideally, only one. Batching positions for several billboards into a single vertex buffer usually isn't a problem, but it is only half the batching battle. How do you assign a potentially different texture to each billboard? The last thing you want is code like in Listing 9.4.

Here, the need to bind a new texture reduces the number of billboards that can be rendered with a single draw call. This problem extends well beyond billboard rendering; NVIDIA conducted an internal survey of four Direct3D 9 games and found `SetTexture` to be the most frequent render-state change or "batch breaker" [121]. The code above shows a worst case scenario; only a single billboard is rendered per draw call. The fact that the first texture is rebound for the third draw call hints toward one technique for improving batch size: sorting by texture. Let's consider this and other techniques:

- *Sorting by texture.* Sorting by state, whether it be shader, vertex array, texture, etc., is a common technique for minimizing driver CPU overhead and exploiting the parallelism of the GPU (see Section 3.3.6). In our code example, sorting by texture would allow us to combine the first and third draw calls. If the vertex buffer isn't also rearranged, a call like `glMultiDrawElements` can be used to render

```
context.TextureUnits[0].Texture2D = Texture0;
context.Draw(PrimitiveType.Points, 0, 1, _drawState,
    sceneState);
context.TextureUnits[0].Texture2D = Texture1;
context.Draw(PrimitiveType.Points, 1, 1, _drawState,
    sceneState);
context.TextureUnits[0].Texture2D = Texture0;
context.Draw(PrimitiveType.Points, 2, 1, _drawState,
    sceneState);
// ...
```

Listing 9.4. Texture binds inhibiting batching of billboards.

multiple sets of indices in a single call. If there are only a handful of unique textures and lots of billboards, sorting by texture can result in large batch sizes.

Be wary of sorting every frame, as the added CPU overhead can drown out the performance gains of batching. Ideally, sorting occurs once at initialization. However, this makes dynamically changing a billboard's texture more difficult. Another drawback to sorting is if a large number of unique textures is used, an equally large number of draw calls is required.

- *Texture atlases.* An alternative to sorting by texture is to use a single large texture, called a *texture atlas*, containing all the images for a given batch, as shown in Figure 9.5. To avoid texture-filtering artifacts, each image is separated by a small border. Of course, texture

Figure 9.5. A texture atlas composed of seven images.

coordinates need to be computed and stored for each billboard, as they no longer simply range from $(0,0)$ to $(1,1)$. Besides providing better texture memory management and coherence than sorting by texture, texture atlases allow for very large batches. The texture atlas is bound once, then a single draw call is made. A downside is texture coordinates can no longer be procedurally generated in the geometry, as done in Listing 9.2. Instead, texture coordinates bounding a subrectangle of the texture atlas are stored per billboard in the vertex buffer. Also, the texture atlas itself should be organized such that it doesn't contain a large amount of empty space. A downside to texture atlases is that texture addressing modes that use coordinates outside $[0,1]$ need to be emulated in a shader [121]. Fortunately, this is not a show-stopper for virtual globe billboards.

It is also important to keep mipmapping in mind when using a texture atlas. Generating a mipmap chain based on the entire atlas with a call like `glGenerateMipmap` will make images bleed together at higher levels of the mip chain. Instead, the mipmap chain should be generated individually for each image, then the texture atlas should be assembled. Other mipmapping considerations are covered in *Improve Batching Using Texture Atlases* [121].

- *Three-dimensional textures and texture arrays.* A third technique for minimizing texture switches is to use a 3D texture, where each 2D slice stores a unique image. Similar to texture atlases, this requires only a single texture bind and draw call. An advantage over texture atlases is that only a single texture-coordinate component needs to be stored per billboard, since this component can be used as the third texture coordinate to look up a 2D slice. For us, the most prominent downside to this technique is that all the 2D slices need to be the same dimensions.

Another downside is this technique is incompatible with mipmapping. The entire 3D volume is mipmapped, as opposed to each individual 2D slice. Because of this, a feature called *array textures* was added to OpenGL [22, 90]; a similar feature called texture arrays was added to Direct3D 10. This is basically a 3D texture where each 2D slice has its own mip chain, with no filtering between slices. So mipmapping works as expected for each individual slice. Array textures can also be 2D, storing a set of individually mipmapped 1D textures.

Since our billboards maintain a constant pixel size, mipmapping is not needed.[3] Therefore, the advantages of array textures over 3D textures are

---

[3]If billboards are sized dynamically, perhaps based on their distance to the viewer,

not important to us. Sorting by texture can lead to lots of small batches when many textures are used, and it is hard to come up with an efficient and flexible implementation. For us, the contest is between texture atlases and 3D textures. Since each 2D slice in a 3D texture needs to be the same size, either each billboard texture needs to be the same size or additional per-billboard texture coordinates need to be stored to access the subrectangle of a slice. If each slice is not packed with multiple images, using 3D textures can waste a lot of memory if the images vary significantly in size. For our needs, the most flexible approach is to use texture atlases.

9.2.1 Texture-Atlas Packing

Before changing our shaders to support texture atlases, we need to implement code to create a texture atlas. Texture-atlas creation can be done during initialization, at runtime in a separate resource-preparation thread, or as part of an offline tool-chain process, common in games. NVIDIA provides a tool for this purpose.[4] In virtual globes, it is not always known in advance what images will be used to create a texture atlas, so we'll focus on designing a class that can be used at runtime. Specifically, given a list of input images, we want to create a single atlas containing all the images and texture coordinates for each of the input images (see Listing 9.5).

Example input images and the output atlas and texture coordinates are shown in Figure 9.6.

A primary concern in texture-atlas creation is minimizing empty space in the new texture atlas. Naïve algorithms can result in a significant amount of wasted space. For example, consider the simple approach of making the atlas as high as the highest input image and as wide as the sum of each image's width. If one image is tall and narrow and every other image is short and wide, the majority of the atlas will be wasted space! Also, this approach will quickly overrun the OpenGL maximum texture-width limitation,[5] requiring more texture atlases be created.

Minimizing the amount of wasted space in a texture atlas is an application of the *2D rectangle bin-packing problem*, which turns out to be NP-hard [54]. Here, we will outline a simple heuristic-based greedy algorithm that is easy to implement; runs quickly; doesn't have input image size restrictions, such as requiring power of two dimensions or squares; and doesn't require a shader to later rotate texture coordinates. The resulting atlas's wasted space is far from minimal, but this algorithm will perform

---

mipmapping is useful. For example, as the viewer zooms out, the billboard may shrink, making mipmapping useful.

[4]http://developer.nvidia.com/legacy-texture-tools

[5]The OpenGL maximum texture width and height are usually at least 8,192 texels on today's hardware.

Figure 9.6. Four images (left) used to create a texture atlas (center) and texture coordinates for the original images (right). In the coordinates for the .NET Bitmap, y increases from the top, as opposed to OpenGL, where y increases from the bottom.

well on common virtual globe inputs, especially in the easy cases, for example, when all input images are 32×32. Full source code is in **OpenGlobe**. **Renderer**.TextureAtlas.

The first step of our packing algorithm is to determine the width for the texture atlas. We'd like the atlas to be reasonably square so we don't run into an OpenGL width or height limitation when the other dimension still has plenty of room. Our approach doesn't guarantee the atlas is square but provides sensible results for most realistic inputs. We simply take the square root of the sum of the areas of each input image, including a border (see Listing 9.6).

This function also tracks the maximum input image's width and is sure to create an atlas with at least this width. Next, we sort the input images in descending order by height. The result of sorting an example set of input

```
IList<Bitmap> bitmaps = new List<Bitmap>();
bitmaps.Add(new Bitmap("image0.png"));
bitmaps.Add(new Bitmap("image1.png"));
// ...

TextureAtlas atlas = new TextureAtlas(bitmaps);
Texture2D texture = Device.CreateTexture2D(atlas.Bitmap,
    TextureFormat.RedGreenBlueAlpha8, false);

// image0 has texture coordinate atlas.TextureCoordinates[0]
// image1 has texture coordinate atlas.TextureCoordinates[1]
// ...
atlas.Dispose();
```

Listing 9.5. Using the TextureAtlas class.

```
private static int ComputeAtlasWidth(IEnumerable<Bitmap> bitmaps,
                                     int borderWidthInPixels)
{
  int maxWidth = 0;
  int area = 0;
  foreach (Bitmap b in bitmaps)
  {
    area += (b.Width + borderWidthInPixels) *
            (b.Height + borderWidthInPixels);
    maxWidth = Math.Max(maxWidth, b.Width);
  }

  return Math.Max((int)Math.Sqrt((double)area),
                  maxWidth + borderWidthInPixels);
}
```

Listing 9.6. Computing the width for a texture atlas.

images is shown in Figure 9.7(b). At expected $\mathcal{O}(n \log n)$ complexity,[6] this is theoretically the most expensive part of our algorithm, given that every other pass is $\mathcal{O}(n)$ with respect to the number of input images. In practice, the bottleneck can be memory copies from the input images to the atlas, especially if the number of images is small and each image is large.

Now that the images are sorted by height, we make a pass over the images to determine the atlas's height and subrectangle bounds for each image. Starting at the top of the atlas and proceeding left to right, we iterate over the images in sorted order and temporarily save subrectangles bounds for each as if we're copying the image to the atlas without overlapping previous images (see Figures 9.7(c)–9.7(f)). When a row of images exceeds the atlas's width, a new row is created, and the process continues until all images have been accounted for. For a given row, the height is determined by the height of the first image since that will be the tallest image, given the images were sorted descending by height. The height of the atlas is the sum of the height of the rows.

Now that we know the width and height of the atlas, the atlas can be allocated. In a final pass, the images are copied to the atlas using the subrectangle bounds computed in the previous pass, and texture coordinates are computed using a simple linear mapping. The final atlas for the images in Figure 9.7(a) is shown in Figure 9.7(g). This implementation stores the atlas in a Bitmap so it is suitable for offline use (e.g., writing to disk) or runtime use (e.g., as the source for a texture). If atlas creation will be done exclusively at runtime, a Texture2D can be allocated instead of a Bitmap, and each subrectangle can be uploaded individually.

---

[6]In .NET, List<T>.Sort is implemented using Quicksort. For many inputs, it may be worth coding a radix sort, which runs in $\mathcal{O}(nk)$; in our case, n is the number of images and k is the number of digits required for their height. In most cases, k will only be two or three.

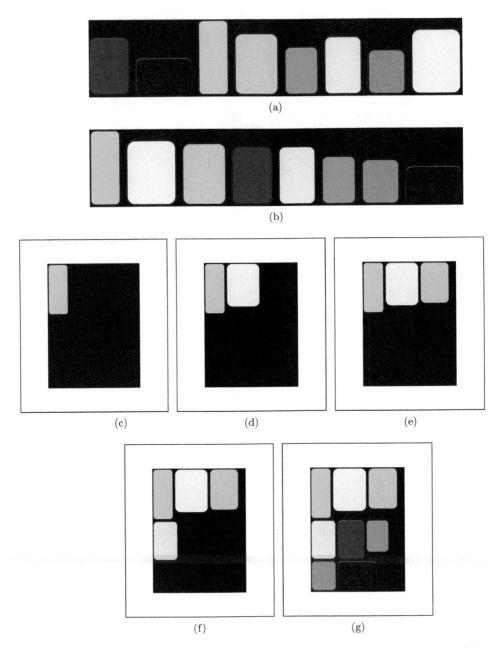

Figure 9.7. (a) A randomly ordered set of images. (b) The images sorted descending by height. (c)–(f) The first four steps of adding images to the atlas. (g) The final atlas.

Modify **OpenGlobe.Renderer**.TextureAtlas to have the option to write the atlas to a Bitmap or Texture2D object.

○○○○ **Try This**

An atlas with less wasted space can be created in a few ways, the simplest of which is to trim the width of the atlas so it is flush against the right most subrectangle. Our algorithm can also be extended to the approach presented by Igarashi and Cosgrove [75]. Before sorting the input images, images that are wider than they are tall are rotated 90° so all images are "stood up." The images are then sorted by height and placed in the atlas starting at the top row. The first row is packed left to right, the second row is packed right to left, and so on. As images are added, they are "pushed upward" in the atlas to fill any empty space between rows, like the empty space between the first two rows in Figure 9.7(g). When rendering with a texture atlas created in this fashion, the texture coordinates for images that were rotated also need to be rotated.

There are other alternatives for creating texture atlases. In some cases, texture atlases are created by hand, but that is rarely an option for virtual globes given their dynamic nature. Scott uses recursive spatial subdivision to pack light maps, which is essentially the same as creating texture atlases [151]. John Ratcliff provides C++ code for an approach that selects images with the largest area and largest edge first, filling the atlas starting at the bottom left using a free list and rotating images 90° if needed [143]. For comprehensive coverage of 2D rectangle bin-packing algorithms, see Jylänki's survey [78].

Modify **OpenGlobe.Renderer**.TextureAtlas to implement the packing algorithm described by Igarashi and Cosgrove [75] or another approach described above. Compare the amount of wasted space between your implementation and the original implementation for various inputs.

○○○○ **Try This**

9.2.2 Rendering with a Texture Atlas

Rendering billboards using a texture atlas is only slightly more involved than rendering with individual textures. The good news is we can just bind the texture atlas once, then issue a draw call for all our billboards.

After all, this was the ultimate goal—rendering large batches. In order
for each billboard to know what subrectangle in the atlas is its texture,
texture coordinates need to be stored per billboard. In our geometry-
shader implementation, each vertex needs two 2D texture coordinates, one
for the lower-left corner and one for the upper right of its subrectangle.
These two texture coordinates can be stored in a single vec4. Since we
are not using mirror or repeat texture addressing, the texture coordinates
will always be in the range $[0, 1]$ so the VertexAttributeComponentType
.`HalfFloat` vertex data type provides enough precision and only consumes
2 bytes per component.

To support texture atlases, the only addition needed to the billboard
vertex shader in Listing 9.1 is to pass through the texture coordinates:

```
in vec4 position;
in vec4 textureCoordinates;

out vec4 gsTextureCoordinates;

uniform mat4 og_modelViewPerspectiveMatrix;
uniform mat4 og_viewportTransformationMatrix;

vec4 ModelToWindowCoordinates(/* ... */) { /* ... */ }

void main()
{
  gl_Position =
      ModelToWindowCoordinates(position,
                               og_modelViewPerspectiveMatrix,
                               og_viewportTransformationMatrix);
  gsTextureCoordinates = textureCoordinates;
}
```

Changes to the original geometry shader in Listing 9.2 are only slightly
more involved. The pixel size of the billboard is no longer the size of the
entire texture, it is the size of the texture divided by the size of the billboard's subrectangle, or, equivalently, scaled by the subrectangle's span in
texture-coordinate space. Also, texture coordinates are no longer procedurally generated; instead, they are taken from the vertex input:

```
layout(points) in;
layout(triangle_strip, max_vertices = 4) out;

in vec4 gsTextureCoordinates[];
out vec2 fsTextureCoordinates;

uniform mat4 og_viewportOrthographicMatrix;
uniform sampler2D og_texture0;

void main()
```

```
{
  vec4 textureCoordinate = gsTextureCoordinates[0];
  vec2 atlasSize = vec2(textureSize(og_texture0, 0));
  vec2 subRectangleSize = vec2(
      atlasSize.x * (textureCoordinate.p - textureCoordinate.s),
      atlasSize.y * (textureCoordinate.q - textureCoordinate.t));
  vec2 halfSize = subRectangleSize * 0.5;

  vec4 center = gl_in[0].gl_Position;

  vec4 v0 = vec4(center.xy - halfSize, -center.z, 1.0);
  vec4 v1 = vec4(center.xy + vec2(halfSize.x, -halfSize.y),
                 -center.z, 1.0);
  vec4 v2 = vec4(center.xy + vec2(-halfSize.x, halfSize.y),
                 -center.z, 1.0);
  vec4 v3 = vec4(center.xy + halfSize, -center.z, 1.0);

  gl_Position = og_viewportOrthographicMatrix * v0;
  fsTextureCoordinates = textureCoordinate.st;
  EmitVertex();

  gl_Position = og_viewportOrthographicMatrix * v1;
  fsTextureCoordinates = textureCoordinate.pt;
  EmitVertex();

  gl_Position = og_viewportOrthographicMatrix * v2;
  fsTextureCoordinates = textureCoordinate.sq;
  EmitVertex();

  gl_Position = og_viewportOrthographicMatrix * v3;
  fsTextureCoordinates = textureCoordinate.pq;
  EmitVertex();
}
```

9.3 Origins and Offsets

At this point, our billboard implementation is pretty useful: it can efficiently render a large number of textured billboards. There are still several small features that can be implemented in the geometry shader to make it more useful, especially when billboards are lined up next to each other. For example, how would you render a label billboard to the right of an icon billboard, as done in Figure 9.8?

One solution is to treat the icon and label as one image. This can lead to a lot of duplication because unique images need to be made for each icon/label pair. Our geometry shader treats a billboard's 3D position as the center of the billboard. If an icon and label are treated as one billboard, its center point will be towards the middle of the label, as opposed to the more common center of the icon.

Our solution is to add per-billboard horizontal and vertical origins and a pixel offset. These are easily implemented in the geometry shader by translating the billboard's position in window coordinates.

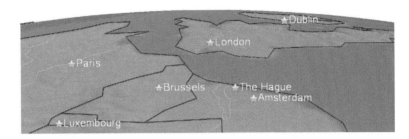

Figure 9.8. Placing an icon and label billboard next to each other as done here requires left aligning and applying a pixel offset to the label billboard.

the origins determine how the billboard is positioned relative to the transformed position (see Figures 9.9 and 9.10). Our implementation thus far implicitly uses a center horizontal and vertical origin. In order to obtain a different origin, a vec2 is stored per vertex. Each component is either -1, 0, or 1. For the x-component, these correspond to left, center, and right horizontal origins, respectively. For the y-component, these correspond to bottom, center, and top vertical origins. The geometry shader simply translates the billboard's position in window coordinates by the origins multiplied by half the billboard's width and height (see Listing 9.7).

```
// ...
in vec2 gsOrigin [];

void main ()
{
  // ...
  vec2 halfSize = /* ... */;

  vec4 center = gl_in [0]. gl_Position;
  center.xy += (halfSize * gsOrigin [0]);
  // ...
}
```

Listing 9.7. Geometry-shader snippet for horizontal and vertical billboard origins.

(a)   (b)   (c)

Figure 9.9. Billboard horizontal origins. (a) Left. (b) Center. (c) Right.

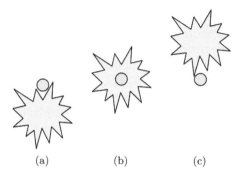

(a)   (b)   (c)

Figure 9.10. Billboard vertical origins. (a) Bottom. (b) Center. (c) Top.

In order the get the effect of Figure 9.8, the label billboard has the same position as the icon billboard and uses a left horizontal origin. The only additional step is to translate the label to the right in window coordinates so it is not on top of the icon. This can easily be achieved by storing a vec2 per vertex that is a pixel offset so the label can be offset to the right by half the width of the icon. The geometry-shader changes, shown in Listing 9.8, are trivial.

There are many other simple but useful operations: a rotation in window coordinates can be used, for example, to orientate a plane icon along the direction the plane is flying; scaling can be applied based on the distance to the billboard; or the billboard can be translated in eye coordinates, for example, to always keep it "on top" of a 3D model. There are a few things to keep in mind when adding features:

- Although these features generally do not increase the size of vertices output from the geometry shader, they tend to increase the

```
// ...
in vec2 gsPixelOffset[];

void main()
{
    // ...
    vec4 center = gl_in[0].gl_Position;
    center.xy += (halfSize * gsOrigin[0]);
    center.xy += gsPixelOffset;
    // ...
}
```

Listing 9.8. Geometry-shader snippet for pixel offsets.

size of vertices entering the vertex and geometry shader. Try to use the least amount of memory possible per vertex. For example, instead of storing horizontal and vertical components in a vec 2, they can be stored in a single component and extracted in the geometry shader using bit manipulation. See the geometry shader for **OpenGlobe.Scene**.BillboardCollection. Although it is less flexible, you can also combine things like the resulting pixel offset from the origins and the user-defined pixel offset into one pixel offset, which results in less per-vertex data and fewer instructions in the geometry shader.

- Careful management of vertex buffers can reduce the amount of memory used and improve performance. In particular, if an attribute has the same value across all billboards (e.g., the pixel offset is $(0,0)$), do not store it in the vertex buffer. Likewise, choose carefully between static, dynamic, and stream vertex buffers. For example, if billboards are used to show the positions for all the current flights in the United States, a static vertex buffer may be used for the texture coordinates since a billboard's texture is unlikely to change, and a stream vertex buffer may be used for the positions, which change frequently.

- If you cull billboards in the geometry shader, be careful to cull after all the offsets, rotations, etc., have been applied. Otherwise, a billboard may be culled when it should be visible. One reason OpenGL point sprites are less commonly used is because the entire sprite is culled based on its center point, which makes large sprites pop in and out along the edges of the viewport.

With respect to performance, it is also important to consider the fill rate. Depending on the texture, a large number of fragments will be dis-

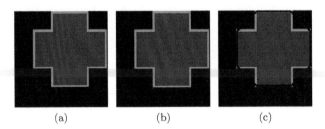

(a)   (b)   (c)

Figure 9.11. (a) A quad only requires four vertices but has a large amount of empty space that will result in discarded fragments. (b) Tightly fitting a bounding rectangle reduces the amount of empty space without increasing the number of vertices. (c) Using the convex hull decreases the amount of empty space even more but increases the vertex count to eight for this image.

carded, wasting rasterization resources. When billboards overlap, which is common in particle systems, overdraw makes the problem even worse. Persson presents a method and a tool for reducing the fill rate by using more vertices per billboard to define a tightly fit convex hull around an image's nonempty space [138] (see Figure 9.11). Of course, a balance needs to be struck between increasing vertex processing and decreasing fill rate.

9.4   Rendering Text

Given our billboard implementation, including the texture atlas and pixel-offset support, it is an easy task to render text for labels. In fact, no new rendering code needs to be written. Text can be implemented as a thin layer on top of billboards.

There are two common methods for rendering text. In both cases, we need a way to write a string to a Bitmap using a particular font. Open-Globe contains a helper method, Device.CreateBitmapFromText, shown in Listing 9.9.

One method for rendering text is to store each word, phrase, or sentence in a texture atlas and render each using a billboard. For example, creating the texture atlas may look like Listing 9.10.

The advantages to this method are that it is easy to implement and uses very little vertex data. The downside is the texture atlas can have a lot of duplication. In the example, ", CA" is duplicated twice. If we consider a finer granularity, "o" is duplicated four times. One technique to minimize duplication is to cache and reuse words or phrases. In the above example, if we chose to cache words, "CA" would only be stored once in the texture atlas but six billboards would be required (assuming commas

```
public static Bitmap CreateBitmapFromText(string text, Font font)
{
    SizeF size = /* ... */

    Bitmap bitmap = new Bitmap(
        (int)Math.Ceiling(size.Width),
        (int)Math.Ceiling(size.Height),
        ImagingPixelFormat.Format32bppArgb);
    Graphics graphics = Graphics.FromImage(bitmap);
    Brush brush = new SolidBrush(Color.White);
    graphics.DrawString(text, font, brush, new PointF());

    return bitmap;
}
```

Listing 9.9. Writing a string to a Bitmap in OpenGlobe.

```
Font font = new Font("Arial", 24);
IList<Bitmap> bitmaps = new List<Bitmap>();
bitmaps.Add(Device.CreateBitmapFromText("Folsom, CA", font));
bitmaps.Add(Device.CreateBitmapFromText("Endicott, NY", font));
bitmaps.Add(Device.CreateBitmapFromText("San Jose, CA", font));

TextureAtlas atlas = new TextureAtlas(bitmaps);
Texture2D texture = Device.CreateTexture2D(atlas.Bitmap,
    TextureFormat.RedGreenBlueAlpha8, false);
```

Listing 9.10. Creating a texture atlas for text rendering.

and spaces are considered part of their preceding word). For larger input sets, caching can be even more effective.

If we take caching to the extreme, we arrive at the other text-rendering method: storing individual characters in the texture atlas (see Figure 9.12) and rendering each character with a single billboard. The amount of texture memory used is small, but the amount of vertex data can become large. For a given string, each billboard stores the same 3D position but a different pixel offset so the correct position for each character is reconstructed in window coordinates in the geometry shader. In addition to storing the texture coordinates for each character with the texture atlas, the pixel offset (i.e., the width and height) of each character also needs to be stored. A pitfall to watch for when implementing this method is computing the texture atlas for an entire character set before knowing what characters are actually used. For ANSI character sets, the atlas will not be very large,

Figure 9.12. A texture atlas storing individual characters for text rendering. The order of the characters was determined by the atlas-packing algorithm, not their position in the alphabet.

depending on the font used, but some Unicode character sets contain lots of characters, most of which will go unused! Regardless of if we are storing individual characters or entire strings, we should be mindful of the size of the created texture. In some cases, multiple textures will be required.

Storing individual characters increases the amount of work on the vertex and geometry shaders but decreases the amount of fill rate. Fewer fragments are shaded since differences among character heights, spacing, and multiline text are taken into account. For example, in the string "San Jose," the geometry around the lowercase characters does not have to be as tall as the geometry around the uppercase ones, and no fragments are shaded for the space between "San" and "Jose" since it can be accounted for with pixel offsets. Texture-caching opportunities are also better when storing individual characters since characters are frequently repeated. Storing individual characters also provides the flexibility to transform and animate individual characters.

So, which method is better? Storing entire strings, individual characters, or a hybrid caching approach? Storing entire strings can be simpler to implement but wastes texture memory. A caching scheme can be efficient for some inputs but not for others; for example, imagine a simulation with objects named "satellite0," "satellite1," etc.—unless words are broken apart, caching will perform poorly. In Insight3D, we store individual characters in textures, as do most other text-rendering implementations we are aware of.

9.5 Resources

Considering that our implementations for polylines and billboards use geometry shaders, it is important to understand geometry-shader performance characteristics. A great place to start is the NVIDIA [123] and ATI [135] programming guides. Our polyline and billboard implementations require a solid understanding of transformations, which are nicely covered in "The Red Book" [155].

NVIDIA's white paper, *Improve Batching Using Texture Atlases*, is an excellent source of information [121]. In a *Gamasutra* article, Ivanov explains how they integrated texture atlases into a game's art pipeline and renderer [76]. For creating texture atlases, Jylänki provides a recent survey on 2D rectangle bin packing [78]. Jylänki's website includes example C++ code [79].

Matthias Wloka's slides go into great detail on using batching to reduce CPU overhead [183].

10

Exploiting Parallelism in Resource Preparation

The rendering of virtual globes requires constant reading from disk and network connections to stream in new terrain, imagery, and vector data. Before these new data are rendered, they usually go through CPU-intensive processing and are made into a renderer resource like a vertex buffer or texture. We call these tasks as a whole resource preparation. Doing resource preparation in the rendering thread dramatically reduces performance and responsiveness because it frequently blocks the thread on I/O or keeps it busy number crunching instead of efficiently feeding draw commands to the GPU.

This chapter explores ways to exploit the parallelism of modern CPUs via threading to offload resource preparation to other threads. We briefly review the parallelism of modern CPUs and GPUs, then consider various engine architectures for multithreaded resource preparation and look at the sometimes intimidating world of multithreading with OpenGL. Along the way, we learn how to use a message queue to communicate between threads.

If you are building a high-performance graphics engine, virtual globe or not, this material is for you!

10.1 Parallelism Everywhere

Modern CPUs and GPUs exploit a great deal of parallelism when executing our seemingly sequential code.

10.1.1   CPU Parallelism

Consider code written in a high-level programming language for a CPU. This code is compiled into lower-level instructions that the CPU executes. To boost performance, CPUs exploit instruction-level parallelism (ILP) to execute more than one instruction at the same time. Perhaps the most well-known form of ILP is pipelining, which allows new instructions to start executing before previous instructions finish. Suppose each instruction is executed in five stages: fetch, decode, execute, memory, and write back. Without pipelining, each instruction requires five cycles to execute. During those cycles, most of the hardware resources are not utilized. In an ideal pipelining scenario, a new instruction enters the fetch stage every clock cycle so multiple instructions are in flight at the same time. This is the same idea as having one load of laundry in the washer while having another load in the dryer. If we have several loads of laundry (instructions), we don't want to leave the washer idle until the dryer is finished. If there are no stalls in the pipeline, meaning no instructions are paused due to dependencies on previous instructions, an instruction can be completed every cycle, making the speedup directly proportional to the depth of the pipeline—five in our example.

Pipelining isn't the only form of ILP used to speed up our sequential code. Superscalar CPUs can execute more than one instruction at the same time. Combined with out-of-order execution, multiple instructions from the same instruction stream are executed in parallel using a scheduling algorithm that handles instruction dependencies and assigns stages of instructions to duplicated hardware units. Most CPUs are not more than four-way superscalar, meaning they can execute up to four instructions from the same instruction stream in parallel. Given branch-prediction accuracy and all the dependencies in real-world code, making processors any wider isn't currently effective [72].

In addition to ILP, many CPUs provide vector instructions to exploit data-level parallelism (DLP). These include SIMD operations that are widely used in vector math for computer graphics. For example, using a single instruction, each multiply in a dot product can be executed in parallel. It is called SIMD because there is a single instruction, multiply in our example, and multiple data since vectors, not scalars, are operated on. Superscalar is interinstruction parallelism, while vector instructions are intrainstruction parallelism.

What does all this computer architecture have to do with virtual globes? On the surface, it explains how CPUs use creative techniques to execute our single-threaded code efficiently. In addition, we can borrow ideas from a CPU's parallelism to help design and exploit parallelism in engine code. Before we do so, let's look at the parallelism we get for free with GPUs.

10.1.2   GPU Parallelism

The incredible triangle throughput and fill rates of today's GPUs are thanks to the amount of parallelism that can be exploited during rasterization and thanks to Moore's law providing more and more transistors to realize this parallelism. There is so much parallelism in rasterization that it is considered embarrassingly parallel. First, consider vertex and fragment shaders. Since transforming and shading a vertex does not depend on another vertex, a shader can execute on different vertices in parallel. The same is also true for fragments.

Similar to how CPUs use SIMD to exploit DLP, GPUs use single instruction, multiple threads (SIMT) to execute the same instruction on different data. Each vertex or fragment is shaded using a lightweight thread. When the same shader is used for several vertices or fragments, each instruction is executed at the same time (up to the number of lanes) on a different vertex or fragment using SIMT. For best performance, SIMT requires reasonable code-path coherency between threads so threads running on a single hardware unit follow the same code path. Divergent branches, for example, an if ... else statement where every other fragment takes a different part of the branch, can adversely affect performance.

A triangle does not need to go through all the stages of the pipeline and be written to the framebuffer before the next triangle enters the pipeline. Just like a pipelined CPU has multiple instructions in flight at the same time, multiple triangles are in flight on a GPU.

GPUs are designed to exploit the embarrassingly parallel nature of graphics. Today's GPUs have hundreds of cores that can host thousands of threads. When we issue `context.Draw`, we are invoking a massively parallel computer for rendering! Not only is the GPU itself parallel, but the CPU and GPU can work together in parallel. For example, while the GPU is executing a draw command, the CPU may be processing geometry that will be drawn in the future.

10.1.3   Parallelism via Multithreading

Given all the parallelism under the hood of a CPU and GPU, why do we need to write multithreaded code? After all, we are already writing massively parallel code by writing shaders—and this parallelism comes for free without the error-prone development and painful debugging stereotypically associated with multithreading. To add to the argument, we frequently hear claims that applications written using OpenGL cannot benefit from multithreading. So why embrace multithreading? Performance.

Given the trend of increasing CPU cores, software developers can no longer write sequential code and expect it to run faster every time a new

CPU is released. Although the increase in transistor density predicted by Moore's law has continued to hold true, this does not imply single-instruction stream performance continues to climb at the same pace. Due to a limited amount of ILP in real-world code and physical issues including heat, power, and leakage, these extra transistors are largely used to improve performance for multiple instruction streams.[1] To create multiple instruction streams, we need multiple threads. As Herb Sutter put it: "the free lunch is over" [161].

CPUs that exploit multiple instruction streams are said to exploit thread-level parallelism (TLP). CPUs can do this in many ways. Multithreaded CPUs support efficient switching between multiple instruction streams on a single CPU. This is particularly useful when a thread misses the L2 cache; instead of stalling the thread for a large number of cycles, a new thread is quickly swapped in. Although multithreaded CPUs do not improve the latency of any single instruction stream, they can increase the throughput of multiple instruction streams by tolerating cache misses.

Hyper-threaded CPUs turn TLP into ILP by executing instructions from different instruction streams at the same time, similar to how a superscalar processor executes instructions from the same instruction stream at the same time. The difference is the amount of parallelism in a single instruction stream is limited compared to the amount of parallelism in multiple instruction streams.

To utilize large transistor budgets, CPUs are exploiting multiple instruction streams by going multicore; loosely speaking, putting multiple CPUs on the same chip. This allows multiple instruction streams to truly execute at the same time, similar to a multiprocessor system with multiple CPUs. Considering that today's desktops can have six cores and even laptops can have four, we would be missing out on an awful lot of transistors if we did not write code to utilize multiple cores.

10.2   Task-Level Parallelism in Virtual Globes

If we write multithreaded code that runs on a multithreaded, hyper-threaded, multicore CPU, we can expect lots of fancy things to happen: threads execute at the same time on different cores, instructions from different threads execute at the same time on the same core, and threads are quickly switched on a core in the face of an L2 cache miss.

But how do we take advantage of this? Isn't most of our performance-critical code sequential: **bind vertex array**, **bind textures**, **set render**

---

[1] The most prominent exception is that larger caches can improve single instruction stream performance.

`state, issue draw call`, repeat? There is only one GPU, on most systems anyway. Well, yes, we spend a great deal of time issuing draw calls and multithreading has been shown to improve performance here [6, 80, 133], but multithreading can also be used to improve the responsiveness of our application. Considering that virtual globes are constantly loading and processing new data, multithreading is critical for smooth frame rates. Specifically, the following tasks are great candidates to move off the main rendering thread and onto one or more worker threads:

- *Reading from secondary storage.* Reading from disk or a network connection is very slow and can stall the calling thread.[2] Doing so in the rendering thread makes the application appear to freeze. By moving secondary storage access to a worker thread, the rendering thread can continue issuing draw calls while the worker thread is blocked on I/O. Doing so utilizes a multithreaded processor's ability to keep its core busy with unblocked threads.

- *CPU-intensive processing.* Virtual globes are not short on CPU-intensive processing, including triangulation (see Section 8.2.2), texture-atlas packing (see Section 9.2.1), LOD creation, computing vertex cache-coherent layouts, image decompression and even recompression, generating mipmaps, and other processing mainly used to prepare vertex buffers and textures. By moving these tasks to a worker thread, we are utilizing a hyper-threaded and multicore processor's ability to execute different threads at the same time.

- *Creating renderer resources.* Creating resources such as vertex buffers, textures, shaders, and state blocks can be moved to a separate thread to reduce the associated CPU overhead. Of course, data for vertex buffers and textures still need to travel over the same system bus, an operation that is generally considered asynchronous, even when executed from the main rendering thread. By moving these operations into a worker thread, we can simplify our application's architecture and avoid holding onto temporary copies of data prepared by the worker thread in system memory because renderer-resource creation isn't restricted to the rendering thread.

Delegating tasks like these to worker threads is an example of TLP, which is the focus of this chapter. Using this type of parallelism, tasks are the units of work and are executed concurrently on different CPU cores.

The goal is to reduce the amount of work done by the rendering thread so it can continue to issue draw commands. If the rendering thread spends

---

[2]With the exception of asynchronous I/O, which returns immediately and later issues an event when the I/O is complete.

too much time waiting on I/O or on CPU-intensive processing, it can starve the GPU. A primary goal of any 3D engine is to allow the CPU and GPU to work in parallel, fully utilizing both.

Resource preparation, mentioned in the introduction, is the pipeline of these three tasks: I/O, processing, and renderer-resource creation. Resource preparation is an excellent candidate for multithreading.

10.3   Architectures for Multithreading

In order to explore software architectures that move resource preparation to worker threads, let's build on Chapter07VectorData. At startup, the application loads several shapefiles on the main thread with code similar to Listing 10.1.

This involves going to disk, using CPU-intensive computations like triangulation, and creating renderer resources—all excellent candidates for multithreading! Chapter07VectorData takes a noticeable amount of time to start. By moving these tasks to a worker thread or threads, the application's responsiveness improves: the application starts quickly, the user can interact with the scene and spin the globe while shapefiles are loading, and a shapefile appears as soon as it is ready. In short, the main thread can focus on rendering while worker threads prepare shapefiles.

The usefulness of this goes far beyond application startup. The same responsiveness can be gained after the user selects files in a file open dialog. Better yet, shapefiles may be automatically loaded based on view parameters such as viewer altitude. If we combine automatic loading with a replacement policy to remove shapefiles, we are awfully close to an out-of-core rendering engine. This idea doesn't just apply to shapefiles, it is true for terrain and imagery as well.

```
_countries = new ShapefileRenderer(
    "110m_admin_0_countries.shp", /* ... */);
_states = new ShapefileRenderer(
    "110m_admin_1_states_provinces_lines_shp.shp", /* ... */);
_rivers = new ShapefileRenderer(
    "50m-rivers-lake-centerlines.shp", /* ... */);
_populatedPlaces = new ShapefileRenderer(
    "110m_populated_places_simple.shp", /* ... */);
_airports = new ShapefileRenderer("airprtx020.shp", /* ... */);
_amtrakStations = new ShapefileRenderer(
    "amtrakx020.shp", /* ... */);
```

Listing 10.1. Shapefile loading in the main thread in Chapter07VectorData.

10.3.1   Message Queues

When moving to multithreading, we need to consider how threads communicate with each other. The rendering thread needs to tell a worker thread to prepare a shapefile, and a worker thread needs to tell the rendering thread that a shapefile is ready for rendering. Depending on the division of labor, worker threads may even need to communicate with each other. In all cases, the communication is a request to do work or a notification that work is complete. These messages also contain data (e.g., the file name of the shapefile or the actual object for rendering).

An effective way for threads to communicate is via shared message queues. A message queue allows one thread to post messages (e.g., "load this shapefile" or "I finished loading a shapefile, here is the object for rendering") and another thread to process messages. Message queues do not require client code to lock; ownership of a message determines which thread can access it.

The code for our message queue is in **OpenGlobe.Core**.MessageQueue. Appendix A details its implementation. Here, we consider just its most important public members:

```
public class MessageQueueEventArgs : EventArgs
{
  public MessageQueueEventArgs(object message);
  public object Message { get; }
}

public class MessageQueue : IDisposable
{
  public event EventHandler<MessageQueueEventArgs>
    MessageReceived;
  public void StartInAnotherThread();
  public void Post(object message);
  public void TerminateAndWait();
  // ...
}
```

Post is used to add new messages to the queue. Messages are of type object,[3] so they can be simple intrinsic types, structs, or classes. Post is asynchronous; it adds a message to the queue, then returns immediately without waiting for it to be processed. For example, the rendering thread may post shapefile file names, then immediately continue rendering. Since Post locks under the hood, multiple threads can post to the same queue.

MessageReceived is the event that gets called to process a message in the queue. The message is received as MessageQueueEventArgs.Message. This

---

[3]If you were to implement this type of message queue in C or C++, you might use void * in place of object.

is where the work we want to move off the rendering thread is performed. The method `StartInAnotherThread` creates a new worker thread to process messages posted to the queue. This method only needs to be called once. Messages in the queue are automatically removed in the order they were added (i.e., first in, first out (FIFO)), and `MessageReceived` is invoked on the worker thread to process each. Finally, `TerminateAndWait` blocks the calling thread until the queue processes all of its messages. Our message queue contains other useful public members that we'll introduce as needed.

The following console application uses a message queue and a worker thread to square numbers:

```
class Program
{
    static void Main(string[] args)
    {
        MessageQueue queue = new MessageQueue();
        queue.StartInAnotherThread();

        queue.MessageReceived += SquareNumber;

        queue.Post(2.0);
        queue.Post(3.0);
        queue.Post(4.0);

        queue.TerminateAndWait();
    }

    private static void SquareNumber(object sender,
                                     MessageQueueEventArgs e)
    {
        double value = (double)e.Message;
        Console.WriteLine(value * value);
    }
}
```

It outputs 4.0, 9.0, and 16.0. First, it creates a message queue and immediately starts it in a worker thread by calling `StartInAnotherThread`. The thread will wait until it has messages to process. Each message posted to the queue is processed by the `SquareNumber` function in the worker thread. The main thread then posts three doubles to the queue. Since the main thread has no more work to do, it calls `TerminateAndWait` to wait for each double on the queue to be processed. This is a bit of a toy example because it doesn't have much, if any, parallelism between the main thread and the worker thread because the main thread doesn't really have anything else to do. In a 3D engine, the main thread continues rendering while worker threads prepare resources.

Given our understanding of message queues, let's explore architectures for moving shapefile preparation off the main thread.

10.3.2 Coarse-Grained Threads

Chapter07VectorData uses only a single thread. It first prepares shapefiles, then enters the draw loop. To make this application multithreaded, the simplest design is to add one worker thread responsible for preparing shapefiles, which allows the main thread to enter the draw loop immediately, instead of waiting for all shapefiles to be ready. The main thread spends its time issuing draw calls while the worker thread prepares shapefiles.

We call this a coarse-grained design because the worker thread is not divided into fine-grained tasks; it does everything that is required to prepare shapefiles for rendering: loading, CPU-intensive processing, and renderer-resource creation. Two message queues are used to communicate between the rendering and worker threads: one for requesting the worker thread to prepare a shapefile and another for notifying the rendering thread that a shapefile is ready (see Figure 10.1).

Chapter10Multithreading extends Chapter07VectorData to use coarse-grained threading with a single worker thread. It introduces a small class called ShapefileRequest to represent the "load shapefile" message that the rendering thread posts on the request queue for consumption by the worker thread. It contains the shapefile's file name and visual appearance, as seen in Listing 10.2.

```
internal class ShapefileRequest
{
    public ShapefileRequest(string filename,
                            ShapefileAppearance appearance);

    public string Filename { get; }
    public ShapefileAppearance Appearance { get; }
}
```

Listing 10.2. Shapefile request public interface.

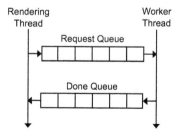

Figure 10.1. Coarse-grained threading with a single worker thread. The rendering thread and worker thread communicate using message queues.

The rendering thread has references to both the request and done queues. It addition, it keeps a list of shapefiles for rendering:

```
private readonly MessageQueue _requestQueue = new MessageQueue();
private readonly MessageQueue _doneQueue = new MessageQueue();
private readonly IList<IRenderable> _shapefiles =
    new List<IRenderable>();
```

As shapefiles come off the done queue, _doneQueue, they will be placed onto the shapefile list, _shapefiles, for rendering. When the application starts, instead of preparing the shapefiles immediately, as shown earlier in Listing 10.1, shapefile requests are added to the request queue, _requestedFiles, to be processed in the worker thread:

```
_requestQueue.Post(new ShapefileRequest(
    "110m_admin_0_countries.shp", /* ... */));
_requestQueue.Post(new ShapefileRequest(
    "110m_admin_1_states_provinces_lines_shp.shp", /* ... */));
_requestQueue.Post(new ShapefileRequest(
    "airprtx020.shp", /* ... */));
_requestQueue.Post(new ShapefileRequest(
    "amtrakx020.shp", /* ... */));
_requestQueue.Post(new ShapefileRequest(
    "110m_populated_places_simple.shp", /* ... */));
_requestQueue.StartInAnotherThread();
```

Compared to Listing 10.1 used in Chapter07VectorData, this code executes very quickly—it just adds messages to a queue. Shapefile preparation is moved to the worker thread using a small class:

```
internal class ShapefileWorker
{
  public ShapefileWorker(MessageQueue doneQueue)
  {
    _doneQueue = doneQueue;
  }

  public void Process(object sender, MessageQueueEventArgs e)
  {
    ShapefileRequest request = (ShapefileRequest)e.Message;
    _doneQueue.Post(
        new ShapefileRenderer(request.Filename, /* ... */));
  }

  private readonly MessageQueue _doneQueue;
}
```

The type ShapefileWorker represents the worker thread. It is constructed by the rendering thread, which passes it to the done queue:

```
_requestQueue.MessageReceived +=
    new ShapefileWorker(_doneQueue).Process;
```

Its trivial constructor executes in the rendering thread, but Process is invoked for each shapefile in the message queue on the worker thread. The Process call simply uses the message to construct the appropriate shapefile object for rendering. Once constructed, the object is placed on the done queue so it can be picked up by the rendering thread. The rendering thread uses a method called ProcessNewShapefile to respond to messages posted on the done queue:

```
_doneQueue.MessageReceived += ProcessNewShapefile;
```

It simply adds the shapefile to the list of shapefiles for rendering:

```
public void ProcessNewShapefile(object sender,
                                MessageQueueEventArgs e)
{
   _shapefiles.Add((IRenderable)e.Message);
}
```

The rendering thread explicitly processes the done queue once per frame using the ProcessQueue message-queue method, not yet introduced. This method empties the queue and processes each message synchronously in the calling thread. In our case, this is a very quick operation since the messages are simply added to the shapefiles list. The entire render-scene function is as follows:

```
private void OnRenderFrame()
{
   _doneQueue.ProcessQueue();

   Context context = _window.Context;
   context.Clear(_clearState);
   _globe.Render(context, _sceneState);

   foreach (IRenderable shapefile in _shapefiles)
   {
      shapefile.Render(context, _sceneState);
   }
}
```

An alternative to using the done queue is to allow the worker thread to add a shapefile directly to the shapefile list. This is less desirable because it decreases the parallelism between threads. The shapefile list would need to be protected by a lock. The rendering thread would need to hold this lock while it is iterated over the list for rendering. During this time, which is likely to be the majority of the rendering thread's time, the worker thread will not be able to add a shapefile to it. Furthermore, providing the worker thread access to the shapefile list increases the coupling between threads.

The code in Chapter10Multithreading is slightly more involved than the code presented in this section. Since our renderer is implemented with OpenGL, additional considerations (coming up in Section 10.4) need to be taken into account when renderer resources are created on the worker thread.

Try This ○○○○

Change Chapter10Multithreading to load shapefiles from a file open dialog instead of at startup.

Try This ○○○○

After adding a file open dialog to Chapter10Multithreading, add the ability to cancel loading a shapefile. How does this affect synchronization?

Try This ○○○○

Change the synchronization granularity in Chapter10Multithreading to be more frequent than per shapefile. Have the rendering thread incrementally render a shapefile; for example, start once 25% of it is prepared, then update again at 50%, and so on. This will improve the responsiveness, especially for large shapefiles, since features start appearing on the globe sooner. How does this affect the overall latency of preparing a shapefile? How much complexity does this add to the design?

Multiple threads. Coarse-grained threading with a single worker thread does an excellent job of keeping resource preparation off the rendering thread. But does it fully utilize a multicore CPU? How responsive is it? Consider a dual-core CPU, with one core running the rendering thread and another running the worker thread. How often is the worker thread blocked on I/O?

How often is the worker thread doing CPU-intensive work? It depends, of course; the bottleneck for large polyline shapefiles preparation may be I/O, while the bottleneck for preparing polygon shapefiles may be CPU-intensive triangulation, LOD creation, and computing vertex cache-coherent layouts.

Using a single worker thread does not fully utilize the second core nor does it fully utilize the I/O bandwidth. A single worker also does not scale to more cores. Since our message queue processes messages in the order they were received, a single worker does not provide ideal responsiveness. If a large shapefile is placed on the queue, no other shapefiles will be rendered until the large one is prepared. This increases the latency for rendering other shapefiles that can be prepared quickly. The problem becomes more prominent if instead of going to disk, the I/O is going to an unreliable network. In this case, the worker thread will be blocked until the connection times out.

A simple solution to these scalability and responsiveness problems is to use multiple coarse-grained worker threads. Figure 10.2 shows an example with three workers. Compare this to Figure 10.1 with only a single worker. When using our message queue without modifications, each worker thread can post to the same done queue, but a different request queue is required per worker. The simplest request-queue-scheduling algorithm is round-robin: just rotate through the request queues.

The ideal number of workers depends on many factors, including the number of CPU cores, amount of I/O, and type of I/O. For example, if the I/O is always over a network, more workers will be useful. This is similar to how a web browser may download each image of a webpage in

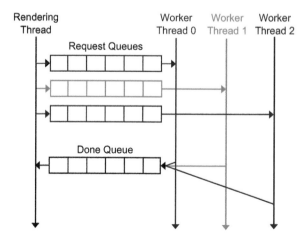

Figure 10.2. Coarse-grained threading with multiple worker threads.

a different thread. On the other hand, if the I/O is from a single optical disk, like a DVD, multiple threads can actually degrade performance due to excessive seeking. Likewise, multiple threads reading from a hard disk may hurt performance by polluting the operating system's file cache. More worker threads also means more memory, task switching, and synchronization overhead.

When a single worker thread is used, shapefiles are prepared one at a time in a FIFO order. With multiple workers, shapefiles are prepared concurrently, making the order that shapefiles are ready nondeterministic. Different shapefiles may complete in a different order from run to run.

Using multiple coarse-grained workers requires some attention to detail to avoid race conditions. In particular, it should be safe to create multiple independent instances of our shapefile types in separate threads. Fortunately, this is pretty straightforward since tasks like triangulation rarely have any threading problems; triangulating different polygons in different threads is completely independent. Shared data structures, such as a shader program cache, should be protected by a lock that is held for the least amount of time necessary.

Try This ○○○○

> Modify Chapter10Multithreading to use multiple coarse-grained workers using round-robin scheduling. Each worker will need its own GraphicsWindow (see Section 10.4). What are the downsides to round robin? What are the alternatives?

10.3.3   Fine-Grained Threads

An alternative to multiple coarse-grained threads is a pipeline of more finely grained threads connected via message queues, like that shown in Figure 10.3. Here, two worker threads are used. The rendering thread sends requests to the load thread via the load queue. This is the same request used with coarse-grained threads. The load thread is just responsible for reading from disk or the network. The load thread then sends the data to the processing thread for CPU-intensive processing algorithms and renderer-resource creation. Finally, the processing thread posts ready-to-render objects to the done queue.

In essence, we've broken our coarse-grained thread into two more finely grained threads: one I/O bound and another CPU bound. The rationale is that breaking up the tasks in this manner leads to better hardware utilization. With multiple coarse-grained threads, every thread could be

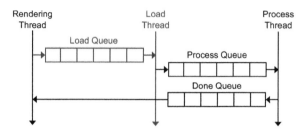

Figure 10.3. Fine-grained threading with separate workers for loading and processing (CPU intensive algorithms and renderer-resource creation).

waiting on I/O while CPU cores are idle, or vice versa. Similar to CPU pipelining, a pipeline of fine-grained threads results in longer latencies while trying to achieve better throughput. The latency is introduced by more message queues and the potential need to make intermediate copies. For example, in our pipeline, data from disk cannot be read directly into a mapped renderer buffer because the load thread and processing thread are independent.

It is generally worthwhile to use more than two worker threads. The number of processing threads can be equal to the number of available cores. As mentioned earlier, the number of load threads should be based on the type of I/O. If all I/O is from a local hard drive, too many load threads will contend for I/O. If the I/O is from different network servers, many load threads will help improve throughput.

> Modify Chapter10Multithreading to use fine-grained threads. How many load and processing threads are ideal in your case?

○○○○ Try This

Texture-streaming pipeline. Van Waveren presents a pipeline for streaming massive texture databases using fine-grained threads [172,173]. It is similar to our fine-grained pipeline for shapefiles in that it has dedicated threads for I/O and processing, although only the rendering thread talks to the graphics driver.

A texture-streaming thread reads lossy compressed textures in a format similar to JPEG. The compressed texture is then sent to the transcoding thread for de-recompression, while the texture-streaming thread continues reading other data. The transcoding thread decompresses the texture using an SIMD-optimized decompresser, then recompresses the texture to DXT.

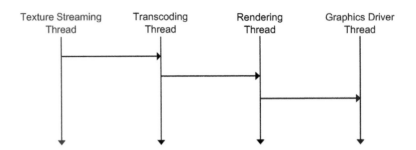

Figure 10.4. A pipeline for texture streaming that separates I/O and CPU processing, and assumes a threaded graphics driver (see Section 10.4.4).

DXT-compressed textures can be efficiently decoded in hardware so they result in using less graphics memory, generally faster rendering, and lower bandwidth. DXT is lossy so some quality may be lost, but when the same amount of memory is used for a DXT-compressed texture compared to a noncompressed texture, the DXT-compressed texture has better quality. Textures are not stored DXT compressed on disk because higher compression ratios can be achieved with other algorithms, reducing the amount of I/O. After DXT compression, the texture is passed to the rendering thread, which interacts with the graphics driver (see Figure 10.4).

Van Waveren observed that transcoding work was typically completely hidden by the time the streaming texture thread had read the data. Since the streaming thread spends most of its time waiting on I/O, a lot of CPU time is available for the transcoding thread.

10.3.4   Asynchronous I/O

We've presented fine-grained threads as a way to better utilize hardware by separating I/O- and CPU-bound tasks into separate threads. With fine-grained threads, I/O threads use blocking I/O system calls. Since these threads spend most of their time waiting on I/O, the CPU is free to execute the processing and rendering threads.

There is another way to achieve this hardware utilization without fine-grained threads: coarse-grained threads with asynchronous I/O. This is done by replacing the blocking I/O system calls with nonblocking, asynchronous I/O system calls that return immediately and trigger a callback when an I/O operation is complete. This lets the operating system manage pending I/O requests, instead of the application managing fine-grained I/O threads, resulting in less memory and CPU usage due to fewer threads and fewer potential data copies.

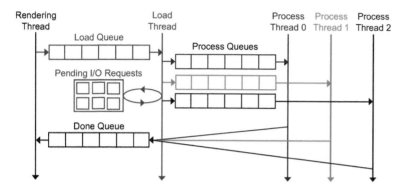

Figure 10.5. Fine-grained threads with a single load thread using asynchronous I/O, allowing the operating system to handle multiple pending I/O requests.

Asynchronous I/O can also be used with fine-grained threads. In this architecture, a single load thread uses asynchronous I/O, which then feeds multiple processing threads, as shown in Figure 10.5. Van Waveren suggests this method over coarse-grained threads with asynchronous I/O because a separate I/O thread allows sorting of multiple data requests to minimize seek times [173]. This is important for hard disks and DVDs but not for network I/O.

One drawback to threading architectures relying on asynchronous I/O is that they are not compatible with some third-party libraries.

10.3.5   Single-Threaded Test/Debug Mode

Regardless of the multithreading architecture used, it is worth including an option to run in a single thread. This is useful for debugging and performance comparisons. In fact, it is sometimes easier to implement a multithreaded architecture on a single thread and then, after verifying it works, add multithreading.

Chapter10Multithreading has a preprocessor variable, SINGLE_THREADED, that enables the application to run in a single thread. The major difference when running in a single thread is that the worker "thread" does not need a separate GraphicsWindow (see Section 10.4), and instead of creating a thread to process the request queue, it is processed immediately at startup:

```
#if SINGLE_THREADED
  _requestQueue.ProcessQueue();
#else
  _requestQueue.StartInAnotherThread();
#endif
```

Run Chapter10Multithreading both with and without `SINGLE_`
`THREADED` defined. Which one starts faster? Which one loads all the
shapefiles faster? Why?

Try This ○○○○

10.4   Multithreading with OpenGL

We've detailed using worker threads to offload I/O and CPU-intensive
computations from the rendering thread, but up until this point, we've
ignored how exactly the renderer resources are finally created. This section explains how to create renderer resources on a worker thread by taking a closer look at multithreading with OpenGL and its abstraction in
`OpenGlobe.Renderer`. Similar multithreading techniques can also be implemented with Direct3D [114].

Let's consider three threading architectures for creating GL resources
such as vertex buffers and textures.

10.4.1   One GL Thread, Multiple Worker Threads

In this architecture, worker threads do not make any GL calls. Instead,
they prepare data that are used to create GL resources in the rendering
thread, as shown in Figure 10.6. This is the simplest approach because
all GL calls are made from the same thread. It does not require multiple
contexts, locks around GL calls, or GL synchronization. On the downside,

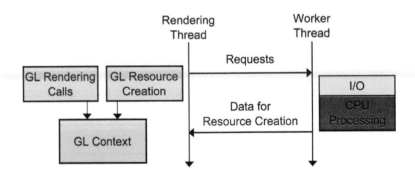

Figure 10.6. When only the rendering thread issues GL calls, worker threads
prepare data in system memory or driver-controlled memory using buffer objects.

it may require some engine restructuring since GL resources cannot be created until the data are in the rendering thread.

A variation of this approach is to create vertex buffer objects (VBOs) or pixel buffer objects (PBOs) in the rendering thread, then hand off memory-mapped pointers to them to a worker thread. The worker thread can then write into driver-controlled memory using the mapped pointer, without issuing any GL calls. This variation also has the pitfall of requiring some restructuring because buffer objects need to be created in advance by the rendering thread.

10.4.2 Multiple Threads, One Context

Another approach to GL threading is to access a single context using multiple threads. This architecture allows worker threads to create GL resources, as shown in Figure 10.7. A coarse-grained lock around every GL call is required to ensure multiple threads don't call into GL at the same time. Furthermore, a context can only be current in one thread at a time. Each GL call needs to be wrapped, similar to Listing 10.3. It is possible to reduce the number of calls to `MakeCurrent` by using more coarse-grained synchronization.

This approach is not recommended because the significant amount of locking eliminates the potential parallelism gained by calling GL from multiple threads.

```
public class ThreadSafeContext
{
  public void GLCall()
  {
    lock (_context)
    {
      _context.MakeCurrent(_window.WindowInfo);
      // GL call
      _context.MakeCurrent(null);
    }
  }

  private NativeWindow _window;
  private GraphicsContext _context;
}
```

Listing 10.3. Accessing a context from different threads requires locking so only one thread has a current context.

10.4.3 Multiple Threads, Multiple Contexts

Our preferred architecture is to use multiple threads, each with its own context that is shared with the contexts of other threads, as shown in

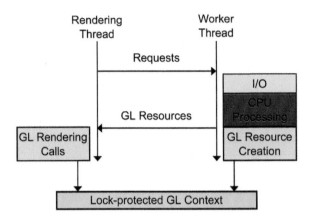

Figure 10.7. A single context accessed from multiple threads allows GL resource creation on a worker thread but requires a coarse-grained lock around GL calls and careful tracking of each thread's current context, as done in Listing 10.3.

Figure 10.8. This allows creating a GL resource in a worker thread, then using it in the rendering thread, without protecting each GL call with a lock. Each thread can make GL calls to its own context at the same time. GL synchronization is required so that the rendering thread's context knows when a GL resource created in a worker thread's context is completed. Since only a single context was used in the previous two approaches, GL synchronization was not required, making them easier to implement in this regard.

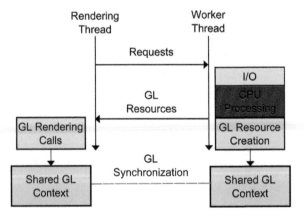

Figure 10.8. Multiple threads, each with a shared context. GL synchronization is required so the rendering thread knows when a GL resource created on a worker thread completed creation.

Once context sharing and synchronization are designed, this approach
leads to a clean architecture because most GL calls can be done at a natural
time. This can also be more efficient than making all GL calls in a single
thread because driver CPU overhead can be moved off the rendering thread
to a worker thread. Although CPU overhead is unlikely to be the bottleneck for generally bandwidth-limited tasks such as vertex buffer uploads,
tasks such as compiling and linking shaders can have a significant amount
of CPU overhead. Even texture uploads can have CPU overhead if the
data need to be compressed or converted to an internal format. Be aware
that the performance gain is potentially limited by the driver, which may
hold coarse-grained locks. The main motivation for this multithreading
architecture is that it lends itself to a natural design.

Sharing per-thread contexts. Our one-context-per-thread approach requires
that each thread contain an OpenGL context that shares its resources with
other OpenGL contexts. In Chapter10Multithreading, the context for the
rendering thread and worker thread are created at startup by creating a
window for each:

```
_workerWindow = Device.CreateWindow(1, 1);
_window = Device.CreateWindow(800, 600, "Multithreading");
```

Under the hood, Device.CreateWindow calls wglCreateContextAttribs and
wglMakeCurrent on Windows. This creates a context shared with existing
contexts, and makes it current in the calling thread.

The window for the worker thread is never shown. The worker-thread
context is created first because creating a context makes the context current
in the calling thread, and both contexts are created in the rendering thread.
Alternatively, this could be coded as follows:

```
_window = Device.CreateWindow(800, 600, "Multithreading");
_workerWindow = Device.CreateWindow(1, 1);
_window.Context.MakeCurrent();
```

The worker-thread context can be passed to a worker thread like the
one in Listing 10.4. The worker thread must then set the context current
at least once before any GL calls are made.

Shared contexts share some but not all resources. Care needs to be
taken so that a nonshared resource created in one context is not used in another context. The OpenGL 3.3 specification defines the following objects

```
internal class Worker
{
  public Worker(GraphicsWindow window)
  {
    _window = window; // Executed in rendering thread
  }

  public void Process(object sender, MessageQueueEventArgs e)
  {
    _window.Context.MakeCurrent(); // Executed in worker thread
    // GL calls ...
  }

  private readonly GraphicsWindow _window;
}
```

Listing 10.4. Setting a context current in a worker thread.

as not shared across contexts: framebuffer objects (FBOs), query objects, and vertex-array objects (VAOs) [152]. It makes sense not to share these objects because they are all lightweight; the overhead to make them shared may be significant relative to the object itself. FBOs and VAOs may sound heavyweight, but they are just containers for textures/render buffers and vertex/index buffers, respectively.

Shared objects in the OpenGL specification include PBOs, VBOs, shader objects, program objects, render buffer objects, sync objects, and texture objects.[4] Relative to nonshared objects, shared objects are usually much larger. For example, a VBO may contain a significant number of vertices, whereas a VAO is just a container whose size is independent of the size of the VBOs it contains.

The code in `OpenGlobe.Renderer` provides a natural separation between shared and nonshared objects. Nonshared objects, which can only be used in the context in which they are created, are created using factory methods on `OpenGlobe.Renderer`.Context:

```
public abstract class Context
{
  public abstract VertexArray CreateVertexArray();
  public abstract Framebuffer CreateFramebuffer();
  // ...
}
```

Shared objects are created using factory methods on `OpenGlobe.Render`er.Device:

---

[4]The texture object named zero is not shared.

```
public static class Device
{
  public static ShaderProgram CreateShaderProgram();
  public static VertexBuffer CreateVertexBuffer(/* ... */);
  public static IndexBuffer CreateIndexBuffer(/* ... */);
  public static MeshBuffers CreateMeshBuffers(/* ... */);
  public static UniformBuffer CreateUniformBuffer(/* ... */);
  public static WritePixelBuffer CreateWritePixelBuffer(
      /* ... */);
  public static Texture2D CreateTexture2D(/* ... */);
  public static Fence CreateFence();
  // ...
}
```

This is a natural separation; anyone can access the static Device to create a shared object, but only those with access to a Context can create a nonshared object. The trouble point is when a worker thread creates a nonshared object using its context, then passes it back to the rendering thread, which cannot use the object. In our experience, we are rarely tempted to create FBOs on worker threads, but VAOs are more seductive.

> Add a debugging aid that catches when a nonshared object is used in a context other than the one it was created in.

○○○○ Try This

Of course, the solution is to delay creating the VAO until the rendering thread can do it. This is similar to the first OpenGL multithreading architecture presented, in which all GL calls were done from the rendering thread, but now, only required GL calls are done on the rendering thread.

Using our abstractions in **OpenGlobe.Renderer** with only a single thread and single context, a class that renders a triangle mesh may use the code in Listing 10.5. The constructor creates a shader program and vertex array for a mesh. The call to **CreateVertexArray** creates vertex and index buffers for the mesh and attaches them to a newly created vertex array in one convenient call. The class's **Render** method is a simple one-liner that draws using the objects created in the constructor.

This class cannot be created in the same way on a worker thread. What context would be passed to the constructor? The worker thread cannot use the rendering thread's context without locking. It also cannot use its own context to create the vertex array. The class in Listing 10.6 shows the solution. The worker's context is used to create the shader program and vertex and index buffers, which are stored in an intermediate

```
public class RenderableTriangleMesh
{
  public RenderableTriangleMesh(Context context)
  {
    Mesh mesh = /* ... */;

    _drawState = new DrawState();
    _drawState.ShaderProgram =
        Device.CreateShaderProgram(/* ... */);
    _drawState.VertexArray = context.CreateVertexArray(
        mesh, _drawState.ShaderProgram.VertexAttributes,
        BufferHint.StaticDraw);
  }

  public void Render(Context context, SceneState sceneState)
  {
    context.Draw(PrimitiveType.Triangles,
                _drawState, sceneState);
  }

  private readonly DrawState _drawState;
}
```

Listing 10.5. If an object is used from only a single context, all renderer resources can be created at the same time.

MeshBuffers object, which is a simple container—like an engine-level vertex array instead of a GL vertex array:

```
public class MeshBuffers : Disposable
{
  public VertexBufferAttributes Attributes { get; }
  public IndexBuffer IndexBuffer { get; set; }

  // ...
}
```

In the rendering thread, before the object is rendered, MeshBuffers is used to create the actual vertex array using an overload to **CreateVertex Array**. This is expected to be a quick operation since the vertex/index buffers were already created in the worker thread.

Question ○○○○

Is it a good idea to create a vertex array immediately before rendering with it? Why or why not? What could you change to allow some time to pass between creating the vertex array and rendering?

```
public class RenderableTriangleMesh
{
  public RenderableTriangleMesh(Context context)
  {
    // Executes in worker thread

    Mesh mesh = /* ... */;

    _drawState = new DrawState();
    _drawState.ShaderProgram =
        Device.CreateShaderProgram(/* ... */);
    _meshBuffers = Device.CreateMeshBuffers(
        mesh, _drawState.ShaderProgram.VertexAttributes,
        BufferHint.StaticDraw);
  }

  public void Render(Context context, SceneState sceneState)
  {
    // Executes in rendering thread
    if (_meshBuffers != null)
    {
      _drawState.VertexArray =
          context.CreateVertexArray(_meshBuffers);
      _meshBuffers.Dispose();
      _meshBuffers = null;
    }

    context.Draw(PrimitiveType.Triangles,
                 _drawState, sceneState);
  }

  private readonly DrawState _drawState;
  private readonly MeshBuffers _meshBuffers;
}
```

Listing 10.6. If an object is created in one context and rendered in another, nonshared renderer resources need to be created in the rendering thread.

Synchronizing contexts in different threads. After sorting out how to create shared contexts and which objects are shared, the last consideration is synchronizing contexts. If there is only a single context, GL commands are processed in the order they are received. This means we can code naturally and write code that creates a resource, then renders with it, as shown in Figure 10.9. For example, if the resource is a texture, we can call Device.**CreateTexture2D**, which under the hood calls glGenTexture, glBind Texture, glTexImage2D, etc., then we can immediately issue a draw command with a shader using the texture. The updates to the texture are seen by the draw command.

When using multiple contexts, each context has its own command stream. It is still guaranteed that a change to an object at time t is completed when a command is issued at time $t+1$, as long as both commands are in the same context. If an object is changed at time t in one context and the CPU is synchronized (perhaps by posting a message to a

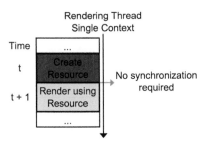

Figure 10.9. No CPU or GL synchronization is required when using a single context.

message queue), it is not guaranteed that a command issued in another context at time $t + 1$ sees the change (see Figure 10.10). For example, if a call to `glTexSubImage2D` is issued in a worker thread and then the thread posts a message to a message queue to notify the rendering thread, the rendering thread cannot issue a draw command with the guarantee that `glTexSubImage2D` is completed, even though the CPU was synchronized and `glTexSubImage2D` was called before the draw command.

Contexts can be synchronized in two ways. The simplest approach is to call `glFinish`, which is abstracted as Device.Finish in `Open Globe.Renderer`, after the resource is created, as shown in Figure 10.11. Although easy to implement, this synchronization approach can cause a significant performance hit because it requires a round-trip to the GL server. It blocks until the effects of all previous GL commands are realized in the GL client, server, and the framebuffer!

Fence sync objects, part of OpenGL 3.2 and also available through the ARB_sync extension [96], provide a more flexible way to synchronize contexts than `glFinish`. Instead of calling `glFinish`, create a fence object using `glFenceSync`. This inserts a fence into the context's command stream.

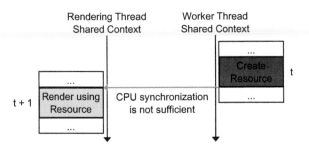

Figure 10.10. CPU synchronization is not sufficient to synchronize two contexts.

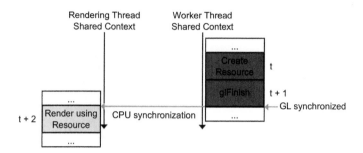

Figure 10.11. Synchronizing multiple contexts with `glFinish`.

Think of a fence as a barrier that separates commands before it and commands after it. To synchronize, call `glWaitSync` or `glClientWaitSync` to wait on the fence, that is, to block until all commands before the fence are completed.

> There are some reports of successfully using the much lighter `glFlush` instead of `glFinish` to synchronize contexts [160]. In fact, we've used it successfully in practice in Insight3D and STK. Unfortunately, this is not guaranteed to work. The OpenGL spec only mentions using `glFinish` or fences to determine command completion: "Completion of a command may be determined either by calling Finish, or by calling FenceSync and executing a WaitSync command on the associated sync object" (pp. 338–339) [152].

○○○○ Patrick Says

Two examples of synchronizing with fences are shown in Figure 10.12. In Figure 10.12(a), a fence is created in the worker thread after creating a resource. Before synchronizing the CPU with the rendering thread, `glFlush` is used to ensure the fence executes. Without flushing the command stream, it is possible that the fence would never execute, and a thread waiting on the fence would deadlock. After synchronizing the CPU, the rendering thread then waits on the fence to synchronize the GL. After waiting, a draw command can be issued with the assurance that resource creation in the other context completed.

A fence can be waited on in a few different ways. A call to `glWaitSync` can be used to block just the GL server until the fence is signaled. Blocking could occur on the CPU or GPU, an implementation detail; `glClientWait Sync` blocks the calling thread until the fence is signaled or a time-out

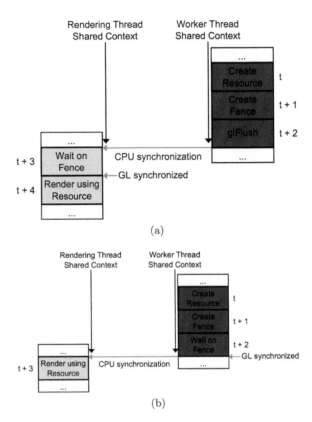

Figure 10.12. (a) GL synchronization by creating a fence in one context and waiting for it in another. (b) Creating a fence, then waiting on it for GL synchronization.

expires. When called with a time-out of zero, `glClientWaitSync` can be used to poll a fence. This is useful for the arrangement shown in Figure 10.12(b).

Here, a fence is created, then immediately waited on in the same context before synchronizing the CPU with the rendering thread. This is nearly identical to calling `glFinish`, as in Figure 10.11. If the client wait is done with a time-out of zero, the worker thread can continue to do other work, and occasionally poll the fence, until the resource creation completes.

Fences are more flexible than `glFinish` because `glFinish` waits for the entire command stream to complete; fences allow waiting on partial completion. The code in `OpenGlobe.Renderer` uses the abstract class Fence to represent fence sync objects. It has a straightforward implementation in `OpenGlobe.Renderer.GL3x`.FenceGL3x.

```
namespace OpenGlobe.Renderer
{
  public enum SynchronizationStatus
  {
    Unsignaled,
    Signaled
  }

  public enum ClientWaitResult
  {
    AlreadySignaled,
    Signaled,
    TimeoutExpired
  }

  public abstract class Fence : Disposable
  {
    public abstract void ServerWait();
    public abstract ClientWaitResult ClientWait();
    public abstract ClientWaitResult ClientWait(
        int timeoutInNanoseconds);
    public abstract SynchronizationStatus Status();
  }
}
```

10.4.4 Multithreaded Drivers

Some OpenGL drivers use multithreading internally to better utilize multicore CPUs [8]. OpenGL calls have some CPU overhead, including managing state and building commands for the GPU. Some calls even have significant CPU overhead, such as compiling and linking shaders. A multithreaded driver moves much of this CPU overhead from the application's rendering thread to a driver thread running on a separate core. When the application makes an OpenGL call, a command is put onto a command queue, which is later fully processed by the driver on a separate thread, and submitted to the GPU. Of course, some OpenGL calls, such as glGet*, require synchronization between the rendering and driver thread.

In the absence of a multithreaded driver, a similar multithreading approach can be taken by the application. The rendering thread can be split into a front-end thread that does culling and state sorting, and ultimately generates draw commands, and a back-end thread that executes these commands using OpenGL or Direct3D calls. This pipeline can increase throughput (i.e., frames per second), but it also increases latency as rendering will be a frame behind. The goal is the same as a multithreaded driver: to move the driver's CPU overhead to a separate core. By doing so with Direct3D 9, Pangerl saw improvements up to 200% in tests and 15% in an actual game [133].

Some care needs to be taken when splitting the rendering thread into front- and back-end threads. In particular, doing so can cause significant memory overhead if data associated with draw commands (e.g., vertex buffers) are double buffered. The increased memory bandwidth and cache pollution can significantly degrade performance. Therefore, this design is not recommended on a single-core CPU. Front- and back-end threads were initially implemented in DOOM III but removed because the memory overhead came with a significant performance penalty on the single-core CPUs that were predominant at the time. Later, Quake 4 was released with two modes: one without a back-end thread and no double buffering for single-core CPUs and another with a back-end thread for multicore CPUs. This allowed a significant amount of work to be offloaded to a separate core on multicore systems without sacrificing any performance on single-core systems. When running on a single core, Kalra and van Waveren observed that about half the time in Quake 4 was spent in the OpenGL driver [80].

10.5   Resources

Herb Sutter's article "The Free Lunch Is Over: A Fundamental Turn Toward Concurrency in Software" does an excellent job of motivating the need to write parallel software [161]. The *CUDA Programming Guide* from NVIDIA discusses SIMT parallelism in their GPUs [122].

Van Waveren's work on texture streaming includes coverage of threaded pipelines, compression, and tactics for streaming from slow storage devices [172,173].

Multithreading with OpenGL is a bit of a black art. Appendix D of the OpenGL 3.3 specification covers sharing objects among contexts [152]. It is valuable reading when using multiple threads with multiple shared contexts. The OpenGL spec also covers sync objects and fences, as does the ARB_sync extension [96]. The Hacks of Life blog contains several detailed posts on their experience with OpenGL multithreading in X-Plane [159]. MSDN covers multithreading with Direct3D 11, including resource creation and generating rendering commands from multiple threads using deferred contexts [114].

The Beyond Programmable Shading SIGGRAPH course notes are an excellent source of information on parallel programming architectures [97–99]. Next-generation game engines are moving to job-scheduling architectures to achieve scalability with an increasing number of cores [6,100,174]. Jobs are small, independent, stateless tasks that are organized into a graph based on their dependencies. An engine then schedules jobs using the graph. Think of this as taking fine-grained threading to the extreme.

Part IV

○○○○

Terrain

○○○○° 11

Terrain Basics

Almost every graphics developer we know has written multiple terrain engines. Of course, there is good reason for this—terrain rendering is a fascinating area of computer graphics and the results are often stunning, beautiful landscapes with rolling hills and mountains. As shown in Figure 11.1, virtual globes have particularly impressive terrains given their use of real-world, high-resolution elevation data and satellite imagery.

Although fun and rewarding, developing a virtual globe-caliber terrain engine can be overwhelming. For this reason, we divide our discussion into four chapters. This first chapter covers the fundamentals: terrain representations, techniques for rendering height maps, computing normals, and a wide array of shading options. In order to understand terrain rendering from the bottom up, this chapter makes the simplifying assumptions that

Figure 11.1. A screen shot of Bing Maps 3D showing real terrain and imagery for Mount Everest and surrounding peaks. © Microsoft Corporation © AND © 2010 DigitalGlobe.

the terrain is small enough to fit into memory and be rendered with brute force, and the terrain is extruded from a plane.

For some applications, these assumptions are perfectly valid. In particular, many games extrude terrain from a ground plane. Applications dealing with large datasets, however, cannot assume that the GPU is fast enough to render with brute force or that the terrain will fit into memory—or even on a single hard disk for that matter! Chapter 12 starts the discussion of massive terrain rendering, then Chapters 13 and 14 look at specific algorithms.

11.1   Terrain Representations

Terrain data come from a wide array of sources. Real-world terrain used in virtual globe and GIS applications is typically created using remote-sensing aircraft and satellites. Artificial terrains for games are commonly created by artists, or even procedurally generated in code. Depending on the source and ultimate use, terrain may be represented in different ways.

11.1.1   Height Maps

Height maps are the most widely used terrain representation. A height map, also called a *height field*, can be thought of as a grayscale image where the intensity of each pixel represents the height at that position. Typically, black indicates the minimum height and white indicates the maximum height.

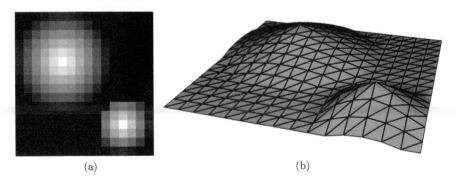

(a)   (b)

Figure 11.2. (a) A 16×16 grayscale image authored in a paint program. Height is indicated by intensity; black indicates minimum height and white indicates maximum height. (b) The image is used as a height map for terrain.

Figure 11.3. Kauai rendered using height-mapped terrain. (Image taken using STK.)

Imagine placing the height map in the ground plane, then extruding each pixel based on its height. The extruded pixels are called *posts*. In this chapter, the ground plane is the xy-plane, and height is extruded along z. It is also common to use the xz-plane with y being "up." Figure 11.2 shows a small 16×16 height map and terrain generated from it.

Height maps are popular because of their simplicity, the large amount of data available as height maps, and the wide array of tools for generating and modifying height maps. Even the gradient fill tool in free paint programs can be used to author simple terrains like the one in Figure 11.2(a).

Height maps are rendered in a wide variety of ways. This chapter will explore three approaches: creating a triangle mesh from a height map, using a height map as a displacement map in a vertex shader, and ray casting a height map in a fragment shader.

Height maps are sometimes called 2.5D. Since only one height per xy-position is stored, a single height map cannot represent terrain features such as vertical cliffs, overhangs, caves, tunnels, and arches. However, we can still produce stunning terrains with height maps (see Figures 11.1 and 11.3). Vertical cliffs can only be approximated by a high-resolution height map with a significant change in height between adjacent posts where the vertical cliff is located. Overhung features, including caves and tunnels, are usually handled using separate models.

11.1.2 Voxels

An alternative terrain representation to height maps that supports truly vertical and overhung features is *voxels*. Intuitively, a voxel is the 3D

(a)   (b)

Figure 11.4. (a) A screen shot of the C4 Engine showing overhangs that can be achieved with voxels. (b) The triangle mesh generated from the voxel representation. (Images courtesy of Eric Lengyel, Terathon Software LLC.)

extension of a pixel. Just as a pixel is a picture element in two dimensions, a voxel is a volume element in three dimensions. Think of a volume as a 3D grid or bitmap, and each cube in the grid, or pixel in the bitmap, is a voxel. In the simplest case, a voxel is binary: a 1 represents solid and a 0 represents empty space. In many real-world cases, a voxel is more than binary, commonly storing density and material.

When terrain is represented using voxels, multiple heights per xy-position are possible. This easily allows vertical cliffs and overhung features, as shown in Figure 11.4.

A downside to voxels is their cubic memory requirements for uniform volumes. A $512 \times 512 \times 512$ volume, with each voxel being 1 byte, requires over 128 MB; whereas a 512×512 height map, with each pixel being one float, requires only 1 MB. Fortunately, several techniques can be used to decrease voxel memory requirements. Fewer bits per voxel can be used; in the extreme case, a single bit per voxel can suffice for simple solid/empty cases. Voxels can also be grouped into larger (e.g., 4×4 or 8×8) cells that can be marked completely solid or empty, therefore avoiding storage of individual voxels for nonboundary cases. Likewise, materials can be assigned per cell instead of per voxel. These two techniques were used for the destructible voxel terrain in Miner Wars [145]. Finally, voxels can be stored using hierarchical data structures, such as sparse octrees, such that branches of the hierarchy containing solid or empty space do not consume memory [129].

Once modeled, voxels may be rendered directly or converted to a triangle mesh for rendering. GPU ray casting is a recent technique for directly rendering voxels [35, 129]. Alternatively, an algorithm such as marching cubes may be used to generate a triangle mesh from a voxel representation [104].

Voxels are becoming an attractive terrain representation in games because of their artistic control. Some game engines are starting to use voxels for terrain editing. Unfortunately for virtual globes, we are unaware of any real terrain data available as voxels.

11.1.3 Implicit Surfaces

There is a whole area of terrain rendering, called *procedural terrain*, that does not rely on real-world or even artist-created terrain. In fact, very little about the terrain is stored. Instead, the terrain surface is described by an implicit function used to procedurally create the terrain at runtime.

For example, Geiss uses a density function for generating procedural terrain on the GPU [59]. Given a point, (x, y, z), the function returns a positive value for solid and a negative value for empty space. The boundary between positive and negative values describes the terrain's surface. For example, to make the terrain surface the plane $z = 0$, the density function is

(a) (b)

(c)

Figure 11.5. Terrain procedurally generated on the GPU using a density function based on noise. (c) A floor is added by increasing the density for positions below a certain z-coordinate. (Images courtesy of NVIDIA Corporation and Ryan Geiss.)

$density = -z$. By combining multiple octaves of noise, each with different frequencies and amplitudes, a wide array of interesting terrains can be generated (see Figure 11.5). Many customizations are possible, such as adjusting the density based on z to create a floor, shelves, or terraces. Similar to voxels, marching cubes can be used to create a triangle mesh for rendering an implicit function.

Procedural generation can produce very stunning terrain with complex overhanging features using a minimal amount of memory. Given the trends in GPU architecture, procedural generation will likely have a bright future in games and simulations. Procedural generation for real terrains is quite challenging to say the least—we are unaware of any endeavors.

11.1.4   Triangulated Irregular Networks

A *triangulated irregular network (TIN)* is basically a triangle mesh. TINs are formed by triangulating a point cloud to create a watertight mesh. Due to the high-resolution TINs available for some geographic regions, TINs are becoming popular in GIS applications.

Point clouds for TINs are commonly retrieved by aircraft using *light detection and ranging (LiDAR)*. LiDAR is a remote-sensing method that samples Earth's surface using timed pulses of laser light. The time between emitting light and receiving its reflection is converted to distance and then to a point in the point cloud based on parameters such as the aircraft and sensor orientation. Since the reflected pulses may include buildings,

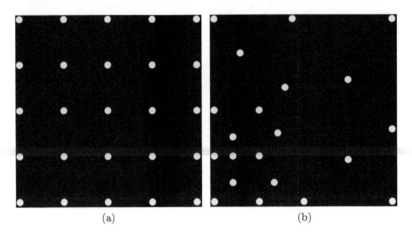

(a)   (b)

Figure 11.6. (a) A standard triangulation of a height map results in a uniform grid. (b) TINs allow for nonuniform sampling. The lower-left part of the triangulation has the highest concentration of triangles.

vegetation, etc., filtering algorithms and manual post-processing are used to create a bare-Earth terrain model.

As shown in Figure 11.6, compared to the uniform structure of height maps, TINs allow for nonuniform sampling; large triangles can cover flat regions and small triangles can represent fine features. Unlike other terrain representations, a TIN is generally ready to render without any additional processing, provided it fits into memory. Given its triangle-mesh nature, a TIN representation allows for vertical and overhung features.

11.1.5 Summary of Terrain Representations

Table 11.1 summarizes the discussed terrain representations. Different representations may be used together or even converted to the same representation for rendering. For example, a height map may be used for the base terrain and voxels may be used when only overhangs are required. A height map may resemble a TIN after it is triangulated and simplified. Likewise, triangle meshes may be derived from voxels or implicit surfaces.

Given their widespread use, we focus on using height maps for terrain rendering.

Height map	Most widely used. Simple to edit, simple to render. A large number of tools and datasets are readily available. Cannot represent vertical cliffs and overhung features.
Voxels	Excellent artistic control, including support for vertical cliffs and overhung features. Not as straightforward to render as height maps or TINs.
Implicit surface	Use very little memory. Used for procedural terrain to generate complex terrains on the fly. Not as straightforward to render as height maps or TINs.
TIN	High-resolution data are becoming available. Its nonuniform grid can make better use of memory than a height map. Easy to render.

Table 11.1. Summary of terrain representations.

11.2 Rendering Height Maps

We will cover three approaches to rendering height-map-based terrain. We'll start with the most obvious: rendering a triangle mesh created from a height map on the CPU. The next two approaches more fully utilize the GPU. One approach uses a height map as a displacement map in a vertex shader and the other casts rays through a height map in a fragment shader.

Throughout this chapter, we use a small 512×512 height map of Puget Sound in Washington state, shown in Figure 11.7.

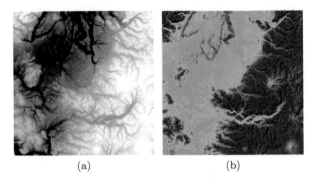

(a)   (b)

Figure 11.7. (a) A low-resolution height map and (b) *color map* (texture) of Puget Sound from the Large Geometric Models Archive at the Georgia Institute of Technology.

In addition to data for heights, height maps typically have associated metadata used to determine how to position the terrain in world space. This chapter uses a class with the following public interface to represent a height map and its metadata:

```
public class TerrainTile
{
    // Constructors ...
    public RectangleD Extent { get; }
    public Vector2I Resolution { get; }
    public float[] Heights { get; }
    public float MinimumHeight { get; }
    public float MaximumHeight { get; }
}
```

The property `Extent` is the height map's world-space boundary in the xy-plane. The property `Resolution` is the x- and y-resolution of the height map, similar to an image's resolution. The property `Heights` includes the actual height-map values in bottom to top row major order. Figure 11.8 shows this layout for a 5×5 height map. This class is named TerrainTile because, as we shall see in Chapter 12, terrain is typically stored in multiple sections, called *tiles*, for efficient culling, level of detail, and paging.

11.2.1 Creating a Triangle Mesh

Given an instance of a TerrainTile, creating a uniform triangle mesh is straightforward. As shown in Figure 11.9, imagine the process by looking straight down at the height map. Create a vertex at each pixel location

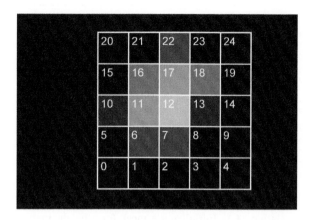

Figure 11.8. Extent.`Heights` stores heights row by row, bottom to top.

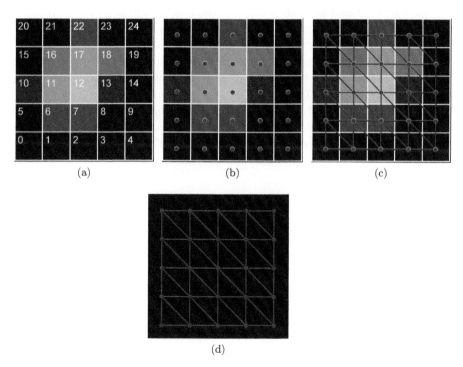

Figure 11.9. (a) A 5×5 height map. (b) Top-down view of vertices. (c) Triangles connecting vertices. (d) Triangle mesh only.

```
Mesh mesh = new Mesh();
mesh.PrimitiveType = PrimitiveType.Triangles;
mesh.FrontFaceWindingOrder = WindingOrder.Counterclockwise;

int numberOfPositions = tile.Resolution.X * tile.Resolution.Y;
VertexAttributeDoubleVector3 positionsAttribute =
    new VertexAttributeDoubleVector3("position",
                                        numberOfPositions);
IList<Vector3D> positions = positionsAttribute.Values;
mesh.Attributes.Add(positionsAttribute);

int numberOfPartitionsX = tile.Resolution.X - 1;
int numberOfPartitionsY = tile.Resolution.Y - 1;
int numberOfIndices =
    (numberOfPartitionsX * numberOfPartitionsY) * 6;
IndicesUnsignedInt indices =
    new IndicesUnsignedInt(numberOfIndices);
mesh.Indices = indices;
```

Listing 11.1. Initializing a mesh for a height map.

using the height of the pixel, then connect surrounding vertices with triangle edges. Once the mesh is created, the height map is no longer required for rendering, but it can be useful for other tasks such as collision detection or reporting the height under the mouse pointer.

An xy-height map yields $x \times y$ vertices and $2(x-1)(y-1)$ triangles. Given a tile, a mesh can be initialized with enough memory for these using the code in Listing 11.1.

The next step is to create one vertex per height-map pixel. A vertex's xy-position is computed by offsetting the tile's lower left position based on the pixel's location in the height map. The vertex's z-coordinate is taken directly from the height map. A simple nested for loop suffices, as shown in Listing 11.2.

```
Vector2D lowerLeft = tile.Extent.LowerLeft;
Vector2D toUpperRight = tile.Extent.UpperRight - lowerLeft;
int heightIndex = 0;
for (int y = 0; y <= numberOfPartitionsY; ++y)
{
  double deltaY = y / (double)numberOfPartitionsY;
  double currentY = lowerLeft.Y + (deltaY * toUpperRight.Y);

  for (int x = 0; x <= numberOfPartitionsX; ++x)
  {
    double deltaX = x / (double)numberOfPartitionsX;
    double currentX = lowerLeft.X + (deltaX * toUpperRight.X);
    positions.Add(new Vector3D(
        currentX, currentY, tile.Heights[heightIndex++]));
  }
}
```

Listing 11.2. Computing vertex positions from a height map.

```
int rowDelta = numberOfPartitionsX + 1;
int i = 0;
for (int y = 0; y < numberOfPartitionsY; ++y)
{
  for (int x = 0; x < numberOfPartitionsX; ++x)
  {
    indices.AddTriangle(new TriangleIndicesUnsignedInt(
        i, i + 1, rowDelta + (i + 1)));
    indices.AddTriangle(new TriangleIndicesUnsignedInt(
        i, rowDelta + (i + 1), rowDelta + i));
    i += 1;
  }
  i += 1;
}
```

Listing 11.3. Computing triangle indices from a height map.

Vertices are stored sequentially row by row, bottom to top, with the same indices as tile.Heights (see Figure 11.9(a)). The final step is to create indices for the actual triangles. For each pixel, except pixels in the top row and right column, create two triangles forming a quad between the pixel, the pixel to the right, the pixel to the upper right, and the pixel above, as done in Listing 11.3.

A complete implementation of creating a triangle mesh from a height map is provided in **OpenGlobe.Scene**.TriangleMeshTerrainTile. Once the triangle mesh is created, rendering is as simple as issuing a draw call. The original height map can be forgotten about if it is not needed for other tasks.

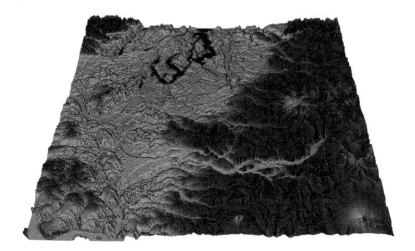

Figure 11.10. The height and color maps of Figure 11.7 rendered with a triangle mesh.

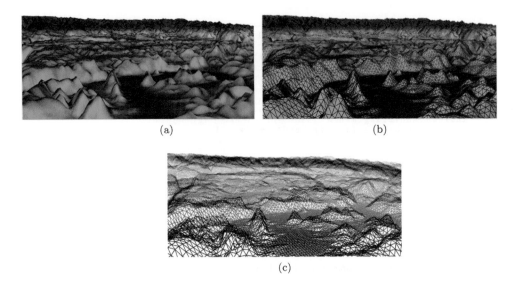

Figure 11.11. (a) Zoomed-in view of terrain. (b) Wireframe overlay. (c) Wireframe only.

An advantage of this uniform layout is, if multiple tiles are rendered, memory can be reduced since each tile can use the same index buffer. To improve rendering performance, indices can be reordered into a cache-coherent layout [50, 148].

Compare Figure 11.10 to the original height map and color map in Figure 11.7. Dark height-map pixels correspond to low heights and bright pixels correspond to higher heights.

Figure 11.11 shows a zoomed-in view of the same terrain, including a wireframe version. The wireframe shows two important points regarding efficiency. First, the mountains in the distance are drawn with an awful lot of triangles. In fact, far fewer triangles could be drawn and the scene would look almost identical. This is not specific to rendering height maps with triangle meshes; the other brute-force height-map-rendering techniques in this chapter have similar problems. Second, the wireframe reveals that flat regions, like the blue area in the foreground, have many unnecessary triangles. This is the result of creating a uniform grid from a height map. In Chapter 12, we discuss level of detail, which can minimize both of these issues.

11.2.2   Vertex-Shader Displacement Mapping

For many years, terrain was rendered using static triangle meshes. When the ability to read from textures was added to vertex shaders, a new GPU

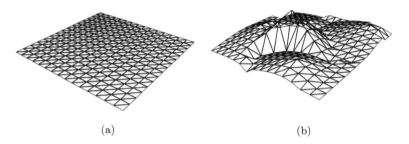

Figure 11.12. (a) Uniform grid in the xy-plane. (b) Terrain is created by displacing grid vertices in the vertex shader based on the height-map texture.

approach became popular [177]. In this approach, a uniform triangle mesh with vertices corresponding to height-map pixels is created as previously described, the difference being that the z-coordinate is not stored in the mesh; only the position in the xy-plane is stored. The height map is written to a one-component floating-point texture, which is sampled in the vertex shader based on the vertex's xy-position. The height map is treated as a *displacement map* applied to the uniform tessellation of the xy-plane, as in Figure 11.12.

At first glance, the advantages of this approach may not be obvious, but there are many:

- *Simplicity.* The implementation is easy, requiring only a one-line vertex shader to do the height map lookup:

```
in vec2 position;
uniform mat4 og_modelViewPerspectiveMatrix;
uniform sampler2DRect u_heightMap;

void main()
{
  gl_Position = og_modelViewPerspectiveMatrix *
      vec4(position, texture(u_heightMap, position).r, 1.0);
}
```

- *Flexibility.* It is easy to implement destructible terrain by modifying a subset of the height-map texture or adding an additional destruction texture. Although not common in virtual globes, destructible terrain is used in games in response to explosions.

- *Low memory usage.* When only a single tile is rendered, displacement mapping uses a similar amount of memory as a uniform static triangle mesh. What about when several tiles are rendered? A height map for each tile is required, but the triangle mesh can be reused by modifying the model matrix to translate and scale it to overlap the tile's world-space extent. This means that additional tiles can be rendered with only the memory costs of a one-component floating-point texture, instead of a three-component vertex buffer.

Of course there are a few downsides to using vertex-shader displacement mapping:

- *Extra texture reads.* An extra texture read is required per vertex. On modern GPUs, these are quite efficient. Even an old NVIDIA GeForce 6800 is capable of 33 million vertex texture reads per second compared to its peak processing of 600 million vertices per second [60].

- *Simplification limitations.* The uniform nature of the height map and triangle mesh allows reusing the same mesh for each tile. But the uniform nature also prevents simplifying flat regions that do not require fine detail.

For many applications, the benefits of vertex-shader displacement mapping far outweigh its downsides, which explains its popularity in game engines such as Frostbite [4].

11.2.3   GPU Ray Casting

The final approach to rendering height-map-based terrain we will consider is GPU ray casting. This algorithm is similar to ray casting an ellipsoid for globe rendering, discussed in Section 4.3. The height map is stored in a one-component floating-point texture. Terrain is rendered by rendering a tile's axis-aligned bounding box (AABB) to invoke a fragment shader that casts rays from the eye through the fragment, shading fragments with rays that intersect the height map and discarding those that miss. The advantages of this approach are the following:

- *Reduced memory usage.* Rendering a tile only requires the tile's height map and AABB. Therefore, more tiles can fit into GPU memory and there is less system bus traffic. This is an important advantage given the widening performance gap between processing power and memory access. Using less memory directly attacks the memory bandwidth bottleneck.

- *Reduced vertex processing.* Since only an AABB is rendered per tile, there is very little vertex shading and triangle setup overhead. On today's GPUs with unified shader architectures, this frees up more GPU resources to process fragments.

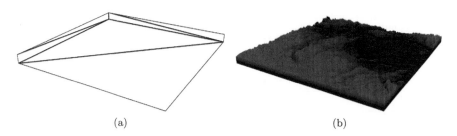

(a)   (b)

Figure 11.13. (a) The back faces of a tile's AABB are rendered to invoke a ray-casting fragment shader. (b) Rays step through the height map to find the visible terrain surface.

The downsides include the following:

- *Complexity.* Ray casting a height map is not as easy to implement as rendering a static mesh or displacement mapping. It is also not as simple as the ellipsoid ray casting we've already seen, which has an analytic solution; ray casting a height map requires stepping the ray through the height map. Fortunately, the complexity is isolated in the fragment shader, making the CPU code straightforward.

- *Increased fragment processing.* Since the ray-casting fragment shader is complex and may require many texture reads, it can require a significant amount of processing.

The question is, does the reduced memory usage and vertex processing offset the complex fragment processing? We'll need to take a closer look at the algorithm to answer this.

Our implementation closely follows the work of Dick et al. [37]. To begin, a triangle mesh for a tile's AABB is created. This AABB is rendered with front-face culling, as in Figure 11.13(a). If back-face culling were used, the terrain would disappear when the viewer enters the AABB because no fragments would be rasterized.

Only the simple vertex shader of Listing 11.4 is required. The world-space position, `position.xyz`, is passed to the fragment shader as `boxExit`. After interpolation, this is the point where a ray cast from the eye through the fragment will exit the AABB. Since only back faces are rendered, we are sure this is the exit point, not the entry point. Computing the ray's entry point is the first task of the fragment shader. Then the fragment shader does the heavy lifting:

- The ray is stepped across height-map pixels, called *texels* here since the height map is stored in a texture, stopping at the first intersection.

```
in vec4 position;
out vec3 boxExit;
uniform mat4 og_modelViewPerspectiveMatrix;

void main()
{
    gl_Position = og_modelViewPerspectiveMatrix * position;
    boxExit = position.xyz;
}
```

Listing 11.4. Vertex shader for GPU ray casting.

```
in vec3 boxExit;
uniform sampler2DRect u_heightMap;
uniform vec3 u_aabbLowerLeft;
uniform vec3 u_aabbUpperRight;
uniform vec3 og_cameraEye;
```

Listing 11.5. GPU ray casting fragment shader inputs.

- If an intersection is found, the intersection point's depth is computed and the fragment is shaded; otherwise, it is discarded.

Except for the ray's exit point, boxExit, the fragment shader takes all of its inputs from uniform variables, as shown in Listing 11.5.

The height map is stored in a rectangle texture, which allows for unnormalized texture coordinates. This simplifies the shader since world space xy-positions more easily line up with texel st-coordinates. The AABB corners are u_aabbLowerLeft and u_aabbUpperRight. Given the ray's exit point and eye position, the ray's direction is simply boxExit − og_cameraEye.

Instead of stepping along the ray starting at the eye, it is more efficient to analytically compute the ray's intersection with the front of the AABB and start stepping the ray from there. Of course, if the viewer is inside the

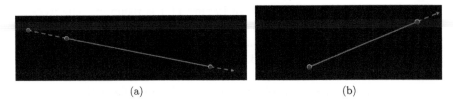

(a)   (b)

Figure 11.14. Side views of a ray and a tile's AABB. Only the segment of the ray that intersects the AABB needs to be considered. (a) Viewer outside the AABB. (b) Viewer inside the AABB.

```
struct Intersection
{
  bool Intersects;
  vec3 IntersectionPoint;
};

bool PointInsideAxisAlignedBoundingBox(vec3 point,
    vec3 lowerLeft, vec3 upperRight)
{ // ... }

Intersection RayIntersectsAABB(vec3 origin, vec3 direction,
    vec3 aabbLowerLeft, vec3 aabbUpperRight)
{ // ... }

void main()
{
  vec3 direction = boxExit - og_cameraEye;

  vec3 boxEntry;
  if (PointInsideAxisAlignedBoundingBox(og_cameraEye,
    u_aabbLowerLeft, u_aabbUpperRight))
  {
    boxEntry = og_cameraEye;
  }
  else
  {
    Intersection i = RayIntersectsAABB(og_cameraEye, direction,
        u_aabbLowerLeft, u_aabbUpperRight);
    boxEntry = i.IntersectionPoint;
  }

  // ...
}
```

Listing 11.6. Computing the ray's entry point.

AABB, then starting at the eye is appropriate. These two cases are shown in Figure 11.14. Since the fragments were rasterized using the AABB, the ray is guaranteed to intersect the AABB, but it is not necessarily true that the ray will intersect the height map. See Listing 11.6 for the portion of the fragment shader used to compute the ray's entry point, boxEntry.

Next, the ray is stepped through the height map in texel space. For simplicity, we assume texel space maps directly to the xy-plane in world space—hence the use of a rectangle texture for the height map. This also assumes the lower-left xy-world position of the tile is $(0, 0)$.

Ray stepping starts at boxEntry and ends when an intersection is found or the end of the height map is reached. To simplify the ray-stepping logic, mirroring is used so the direction is monotonically increasing in x and y as the ray steps over texels. If direction.x and direction.y are nonnegative, nothing is mirrored (see Figure 11.15). Otherwise, if either are negative, the component is negated (e.g., direction.x = −direction.x). To compensate, the corresponding component(s) of the entry point and texture coordinates are mirrored. The relevant fragment-shader snippet

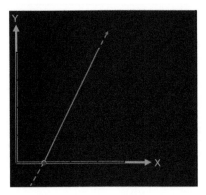

Figure 11.15. When `direction.x` and `direction.y` are nonnegative, no mirroring is required.

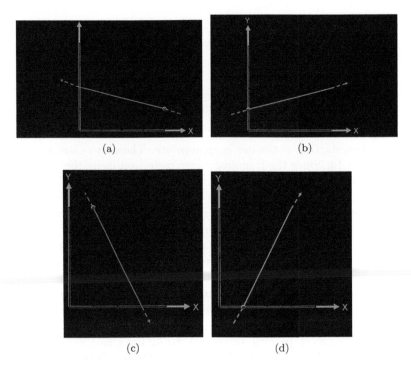

Figure 11.16. In (a), `direction.x` is negative. This is compensated for by mirroring in (b). The same scenario is shown for `direction.y` in (c) and (d).

is shown in Listing 11.7. The adjustments to `direction` and `boxEntry` are shown graphically in Figure 11.16.

After mirroring adjustments, we are ready to step from texel to texel checking for intersection with the height map. The process is similar to line rasterization. Two new variables are used: `texEntry`, the entry point of the ray with the current texel, and `texExit`, the exit point of the ray with the current texel. Initially, `texEntry` is set to the AABB's entry point. As the ray steps from texel to texel, the previous texel exit point becomes the current entry point, as shown in Figure 11.17.

Since mirroring was performed, a ray always enters a texel through the left or bottom edge and exits through the right or top edge. The texel exit point is found using the code in Listing 11.8.

The lower-left corner of the texel is `floorTexEntry`, which makes `floor TexEntry` + `vec2(1.0)` the upper-right corner. This is used to compute `delta`, the distances along the ray to the right and top edges. The smaller

```
void main()
{
  // ...

  vec2 heightMapResolution = vec2(textureSize(u_heightMap));
  bvec2 mirror = lessThan(direction.xy, vec2(0.0));
  vec2 mirrorTextureCoordinates = vec2(0.0);

  if (mirror.x)
  {
    direction.x = -direction.x;
    boxEntry.x = heightMapResolution.x - boxEntry.x;
    mirrorTextureCoordinates.x = heightMapResolution.x - 1.0;
  }

  if (mirror.y)
  {
    direction.y = -direction.y;
    boxEntry.y = heightMapResolution.y - boxEntry.y;
    mirrorTextureCoordinates.y = heightMapResolution.y - 1.0;
  }

  // ...
}
```

Listing 11.7. Mirroring the ray direction.

```
vec2 floorTexEntry = floor(texEntry.xy);
vec2 delta = ((floorTexEntry + vec2(1.0)) - texEntry.xy) /
             direction;
vec3 texExit = texEntry + (min(delta.x, delta.y) * direction);
```

Listing 11.8. Computing `texExit`.

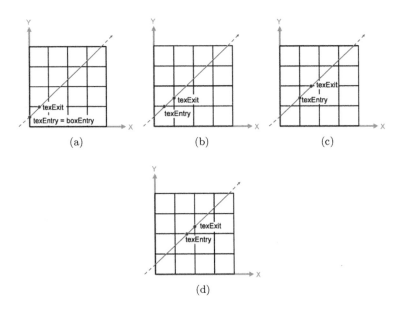

Figure 11.17. `texEntry` and `texExit` for the first four ray steps.

of `delta.x` and `delta.y` is used to offset along the ray, selecting the closer of the two potential exit points, as shown in Figure 11.18.

Now that we know how to step from texel to texel, computing the entry and exit points, the next step is to check for intersection with the height map. If no mirroring was used, the height is looked up using `floorTexEntry` as the texture coordinate. Otherwise, mirrored repeat is emulated as done in Listing 11.9.

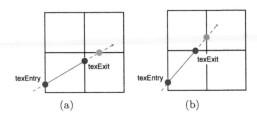

Figure 11.18. The shorter of the two distances along the ray from `texEntry` to the texel's right and top edges is used to find `texExit`. In (a), the right edge is closer. In (b), the top edge is.

```
vec2 MirrorRepeat(vec2 textureCoordinate,
                  vec2 mirrorTextureCoordinates)
{
  return vec2(mirrorTextureCoordinates.x == 0.0
         ? textureCoordinate.x
         : mirrorTextureCoordinates.x - textureCoordinate.x,
      mirrorTextureCoordinates.y == 0.0 ? textureCoordinate.y
         : mirrorTextureCoordinates.y - textureCoordinate.y);
}

// ...

vec2 floorTexEntry = floor(texEntry.xy);
float height = texture(u_heightMap, MirrorRepeat(floorTexEntry,
    mirrorTextureCoordinates)).r;
```

Listing 11.9. Looking up a texel's height.

Given the texel's height, an intersection occurs if the ray goes under the texel. Two intersection cases need to be considered. If the ray is heading upwards (`direction.z >= 0`), the ray intersects the texel if `texEntry.z` < `height`, as shown in Figure 11.19(a). In this case, the intersection point is `texEntry`. A more common view for many virtual globe users is when the ray is heading downwards (`direction.z <= 0`). Here, an intersection occurs if `texExit.z <= height`, as in Figure 11.19(b), with an intersection point of `texEntry` + (`max`)(((`height` − `texEntry.z`)/ `direction.z`, 0.0) ∗ `direction`).

Listing 11.10 brings together the ray-stepping algorithm and intersection test to build a loop that steps across texels until an intersection is found.

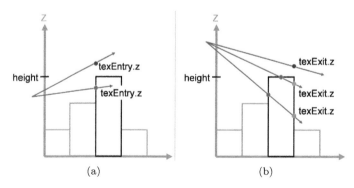

(a)   (b)

Figure 11.19. (a) Upward-heading ray intersecting a texel. (b) Downward-heading ray. Intersects are green and misses are red.

```
void main()
{
  // ...

  vec3 texEntry = boxEntry;
  vec3 intersectionPoint;
  bool foundIntersection = false;

  while (!foundIntersection &&
         all(lessThan(texEntry.xy, heightMapResolution)))
  {
    foundIntersection = StepRay(direction,
        mirrorTextureCoordinates, texEntry, intersectionPoint);
  }

  // ...
}
```

Listing 11.10. Ray-stepping loop.

As stated previously, `texEntry` is initialized with the intersection point of the ray and the AABB. The ray-stepping loop continues until an intersection is found or the end of the tile is reached. The bulk of the work is done in the `StepRay` function in Listing 11.11.

This code should seem familiar. The texel's height is read. The exit point, `texExit`, is computed. Then, the actual intersection test occurs. Note how `texEntry` is set to `texExit` for the next iteration of the loop. Also observe that during the loop, the sign of `direction.z` never changes because the ray does not change direction. Therefore, the (`direction.z >= 0.0`) check can be moved outside of the loop. For that matter, the conditionals in `MirrorRepeat` can be moved as well.

```
bool StepRay(vec3 direction, vec2 mirrorTextureCoordinates,
             inout vec3 texEntry, out vec3 intersectionPoint)
{
  vec2 floorTexEntry = floor(texEntry.xy);
  float height = texture(u_heightMap, MirrorRepeat(
    floorTexEntry, mirrorTextureCoordinates)).r;

  vec2 delta = ((floorTexEntry + vec2(1.0)) - texEntry.xy) /
               direction.xy;
  vec3 texExit = texEntry + (min(delta.x, delta.y) * direction);

  if (delta.x < delta.y)
  {
    texExit.x = floorTexEntry.x + 1.0;
  }
  else
  {
    texExit.y = floorTexEntry.y + 1.0;
  }

  bool foundIntersection = false;
```

```
if (direction.z >= 0.0)
{
    if (texEntry.z <= height)
    {
        foundIntersection = true;
        intersectionPoint = texEntry;
    }
}
else
{
    if (texExit.z <= height)
    {
        foundIntersection = true;
        intersectionPoint = texEntry +
            max((height - texEntry.z) / direction.z, 0.0) *
            direction;
    }
}

texEntry = texExit;
return foundIntersection;
}
```

Listing 11.11. StepRay function.

Since the ray is always going to exit a texel at either the right or top edge, one of the **texExit** components will be exactly one greater than the corresponding **floorTexEntry** component. Although the math for the initial assignment to **texExit** works out this way, Dick et al. warn that roundoff errors can occur, leading to infinite loops. After locking up my computer a few times, I added the explicit assignment suggested by Dick et al. [37].

○○○○ Patrick Says

If an intersection is found, the fragment is shaded and a depth value is computed using the same technique as for ray casting an ellipsoid in Section 4.3. This allows normal rasterized geometry to interact correctly with the ray-casted terrain. The intersection point needs to be adjusted if mirroring was used. If the ray-stepping loop exits without finding an intersection, the fragment is discarded. The end of the fragment shader is shown in Listing 11.12.

```
void main()
{
    // ...

    if (foundIntersection)
    {
        if (mirror.x)
```

```
    {
      intersectionPoint.x =
          heightMapResolution.x - intersectionPoint.x;
    }

    if (mirror.y)
    {
      intersectionPoint.y =
          heightMapResolution.y - intersectionPoint.y;
    }

    fragmentColor = // shade however you like
    gl_FragDepth = ComputeWorldPositionDepth(intersectionPoint);
  }
  else
  {
    discard;
  }
}
```

Listing 11.12. Shading or discarding based on ray intersection.

See **OpenGlobe.Scene.**RayCastedTerrainTile for the completed fragment shader. A zoomed-in screen shot with shading by height is shown in Figure 11.20. Why does the terrain appear blocky?

Our ray-stepping approach is using nearest-neighbor interpolation, so the height is considered constant over an entire texel. In the zoomed view, a projected texel covers several pixels. This produces blockiness because the height map is not of a high enough resolution. Nearest-neighbor interpolation is very efficient, though. For many views, LOD can be used to keep texels less than a pixel.

Figure 11.20. Nearest-neighbor interpolation results in blocky terrain if the height map doesn't have sufficient detail for the view.

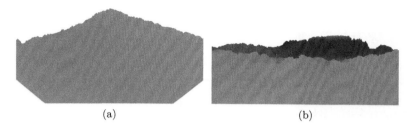

(a)   (b)

Figure 11.21. (a) For downward-facing views, the ray-stepping loop exits quickly. (b) Horizon views require an increasing number of iterations as pixels blend from orange to blue.

Performance. Our brute-force implementation of GPU ray casting has some interesting performance characteristics. Performance is highly dependent on the number of iterations in the ray-stepping loop. More iterations means more texture reads and more instructions executed. The number of iterations is dependent on the view and terrain features. It is instructive to visualize which pixels are the most expensive by shading the terrain based on the number of iterations, as in the following:

```
fragmentColor = mix(vec3(1.0, 0.5, 0.0), vec3(0.0, 0.0, 1.0),
    float(numberOfIterations) / (heightMapResolution.x +
    heightMapResolution.y));
```

With this shading, the color blends between orange, for areas where rays quickly find terrain, and blue, for areas requiring a large number of steps. The two images in Figure 11.21 are rendered with this shading. For nearly top-down views, terrain is quickly found, as shown by the orange in Figure 11.21(a). A more challenging view is looking along the horizon, as in Figure 11.21(b); intersections with peaks near the viewer are found quickly, but reaching terrain in the distance requires a large number of steps, as shown by the pixels blending to blue. A close look at this figure reveals that the top of the distant center peak did not require many steps. For this view, rays intersecting the top of this peak enter the tile's AABB further along the ray than do rays intersecting lower on the peak; therefore, the higher rays step over fewer texels before finding the intersection point. Some of the most expensive pixels aren't even shaded; consider a ray that steps all the way across the height map only to realize that no intersection occurred.

Fortunately, the ray-stepping loop can be optimized quite a bit. It can be reduced to nearly logarithmic time by using a hierarchical data

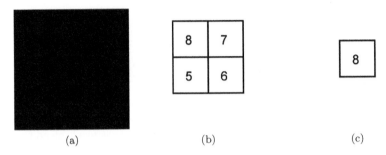

(a)   (b)   (c)

Figure 11.22. A maximum mipmap pyramid. (a) Level zero. (b) Level one.
(c) Level two.

structure called a maximum mipmap pyramid [166]. This is a fully subdivided quadtree where the original height map is the finest level, level
zero, and texels in coarser levels are the maximum height of the 2×2 corresponding heights in the next-finer level, as shown in Figure 11.22. This
data structure can be computed very quickly on the GPU and requires only
an additional one-third of the memory of the original height map.

A maximum mipmap pyramid can be used to implement hierarchical
ray stepping. Instead of stepping texel by texel along the original height
map, the pyramid is used to accelerate stepping by stepping over large parts
of the height map in one loop iteration. Since the ray always intersects the
top 1×1 level, ray stepping starts at the next-highest level (2×2). If the
ray does not intersect any texels at this level, it will not intersect any texels
in the original height map, so the fragment is quickly discarded. If the ray
intersects a texel, the ray is stepped forward to the intersection point and
the pyramid is descended to the next-finer level. Ray stepping continues at
this intersection point. If the ray misses the texel, the ray is stepped to the
texel's exit point. If the ray also left a 2×2 region in this pyramid level,
the pyramid is ascended to the next-coarsest level. Recursion continues
until the end of the height map is reached or an intersection is found in the
finest level.

An alternative ray-stepping acceleration technique is cone step mapping [42]. In a preprocessing step, a cone is computed for each texel. The
cone's apex is centered on top of the texel, with the cone opening upward.
The cone's angle is computed to be as large as possible without intersecting other texels. This gives the guarantee that is used to accelerate ray
stepping: if a ray intersects a cone, it can be stepped all the way across the
cone without intersecting any texels. The downside to cone stepping is the
expensive preprocessing step and memory requirements.

(a)   (b)

Figure 11.23. High-resolution terrain and imagery of Vorarlberg, Austria, rendered using GPU ray casting. (Images courtesy of Christian Dick, Computer Graphics and Visualization Group, Technische Universität München, Germany. Geo data provided by the Landesvermessungsamt Feldkirch, Austria.)

> Run Chapter11TerrainRayCasting, enable shading by number of ray steps, and experiment with different views. Implement a maximum mipmap pyramid or cone step mapping and visually observe the drastic reduction in ray steps.

○○○○ Try This

Implementing optimized GPU ray casting of height maps involves a fair amount of work, even when we are only considering a single tile. Is it worth it? Does the reduction in memory bandwidth and vertex processing outweigh the complex fragment shading? Dick et al. compared GPU ray casting using hierarchical ray stepping with a maximum mipmap pyramid and other occlusion optimizations to their optimized rasterization-based terrain engine [37]. Their results showed that for very high-resolution datasets (see Figure 11.23), ray casting both had a higher frame rate and used less memory than rasterization did; using hierarchical ray stepping resulted in a speedup of five over brute-force ray stepping.

11.2.4 Height Exaggeration

Exaggerating terrain heights is a visual aid that makes subtle height differences more noticeable or flattens areas with significant height differences. Figure 11.24 shows how height exaggeration highlights height differences in a nearly flat terrain. Height exaggeration is a common feature in virtual globe applications and goes by many names; it is called elevation exaggeration in Google Earth and vertical exaggeration in ArcGIS Explorer.

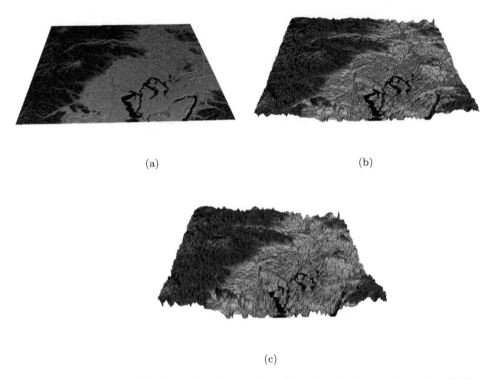

(a)

(b)

(c)

Figure 11.24. (a) A terrain tile spanning 512 units in the xy-plane with height in the range $[0, 1]$ appears flat. (b) Applying a height exaggeration of 30 makes height differences more noticeable. (c) A height exaggeration of 60 overemphasizes height differences.

Implementing height exaggeration is straightforward. The height of a height-map pixel is scaled by the exaggeration value. Values greater than one make height differences more noticeable and values less than one flatten the terrain. For the case of terrain extruded from the xy-plane, each position's z-component is multiplied by the exaggeration value. If normals are procedurally generated, they should be computed using the exaggerated heights.

If the exaggeration is static and will not change, as is the case for terrain formats that include a scale factor, height exaggeration can occur when a tile's triangle mesh is created, or its height map is created in the case of displacement mapping or ray casting. This maps the units for height in the terrain format to your model-space units for height.

```
in vec3 position;

uniform mat4 og_modelViewPerspectiveMatrix;
uniform float u_heightExaggeration;

void main()
{
  vec4 exaggeratedPosition =
      vec4(position.xy, position.z * u_heightExaggeration, 1.0);
  gl_Position =
      og_modelViewPerspectiveMatrix * exaggeratedPosition;
}
```

Listing 11.13. Terrain height exaggeration in a vertex shader.

If the exaggeration is dynamic, perhaps because it can be adjusted by the user via a slider, height exaggeration can occur in a vertex shader, as done in Listing 11.13.

Instead of explicitly multiplying the z-component, height exaggeration can be built into the model matrix by multiplying it by a z-scaling matrix.

Static and dynamic height exaggeration are not mutually exclusive; a terrain format may require scaling the stored heights to determine the true heights, then the user may wish to exaggerate the true heights at runtime.

11.3 Computing Normals

Given the various methods for rendering height maps we've seen thus far, we've solved half of the terrain-rendering battle: finding the visible surface. The other half is shading. The first step we'll take in shading is computing terrain normals. Normals have an obvious use in lighting equations, like the ones presented in Section 4.2.1, and less obvious uses such as shading terrain based on steepness, which we will see in Section 11.4.3.

Traditionally, per-vertex normals for arbitrary triangle meshes are computed by using the cross product to find normals for each triangle sharing a vertex, then averaging these normals. The computed normals are stored with other per-vertex data, such as position. The same algorithm will work for terrain normals. Since we are focusing on terrain derived from height maps, we will also focus on deriving normals from height maps. When normals are stored in a 2D image, similar to a height map, the image is called a *normal map*.

Depending on an application's requirements, terrain normals may be computed at different times:

- *Preprocess.* If an application doesn't have access to programmable GPUs or can't afford to compute normals in a shader, a normal map can be computed as a preprocess step. This may happen offline (e.g., in a game's content-creation pipeline). Major hardware vendors provide normal-map generation tools. NVIDIA has an Adobe Photoshop plug-in to generate normal maps from height maps,[1] and a standalone tool, Melody, to generate normal maps for arbitrary simplified models. AMD provides a standalone tool, Normal Mapper, with C++ source, to generate normals maps from height maps.[2]

Normal maps can also be computed online, after the height map is loaded from disk. This is attractive because it saves on disk space and I/O bandwidth. Rendering, disk I/O, and normal computation can occur on different threads to keep the CPU cores saturated (see Section 12.3). To save memory, a normal map can store two of the three components and compute the third at runtime (e.g., store x and y and compute z), taking advantage of the fact that the normal is of unit length.

- *Procedurally in a shader.* When a shader has access to the height map, which is the case when doing GPU displacement mapping or ray casting, a normal map does not need to be stored. Instead, normals can be procedurally generated based on neighboring heights. Procedural generation saves memory, reduces memory bandwidth, and generally results in less code. In the case of terrain, procedurally generated normals have the added benefit of easily allowing destructible terrain. For example, if a bomb destroys part of the terrain in a game, only the height-map texture needs to be modified; the adjusted normals will be automatically computed.

- *Hybrid.* A hybrid approach can be used to support destructible terrain with normal maps. When the height map is first loaded and every time it is updated, a normal-map texture is generated by a fragment shader that uses the height map to write to an offscreen buffer representing the normal map.

An advantage of using a normal map is that a high-resolution normal map can be used to shade a lower-resolution height map. This improves visual quality without increasing the number of triangles. When a normal map is used in this way, it is called *bump mapping*. The simplicity and memory savings of procedurally generated normals, however, have made it an attractive technique used by many 3D engines.

---

[1]http://developer.nvidia.com/object/photoshop_dds_plugins.html
[2]http://developer.amd.com/gpu/normalmapper/

11.3.1 Forward Difference

Regardless of whether a normal map or procedurally generated normals are used, the same algorithms for computing normals can be applied. They all work by approximating a position's normal using neighboring heights. In the simplest case, the *forward difference* can be used. This is a fancy term for $f(x+h) - f(x)$. Here, $f(x)$ is the function for the height map, and x is the position of the vertex in the xy-plane. If we let $h = (1, 0)$, the forward difference is an approximation of the partial derivative in the x direction. Likewise, if $h = (0, 1)$, it is approximate to the partial derivative in the y direction. As shown in Figure 11.25, taking the cross product of these yields an approximate unnormalized normal. Example code is shown in Listing 11.14.

Given the assumption that terrain is extruded from the xy-plane, Listing 11.14 can be simplified further. Since the forward difference uses samples to the right and above, it also has problems along the top and rightmost rows of the height map because adjacent heights are not available. The *backward difference*, $f(x) - f(x-h)$, can be used in these cases. However, the LOD algorithm usually impacts the solution to this, so we will

```
vec3 ComputeNormalForwardDifference(vec3 position,
                                    sampler2DRect heightMap)
{
  vec3 right = vec3(position.xy + vec2(1.0, 0.0),
      texture(heightMap, position.xy + vec2(1.0, 0.0)).r);
  vec3 top = vec3(position.xy + vec2(0.0, 1.0),
      texture(heightMap, position.xy + vec2(0.0, 1.0)).r);
  return cross(right - position, top - position);
}
```

Listing 11.14. Computing normals for terrain using the forward difference.

Figure 11.25. The forward difference. The approximate normal for a position can be computed by taking the cross product of two vectors created from three height-map samples.

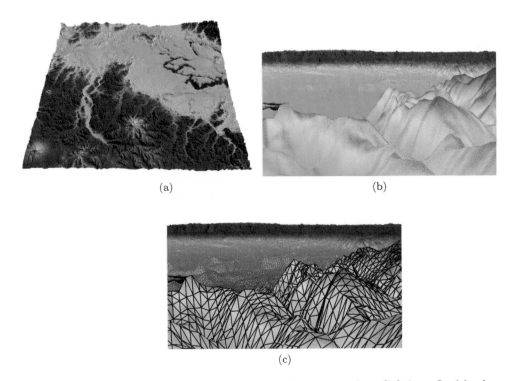

Figure 11.26. Terrain shaded with a color map and no lighting. In (c), the wireframe overlay shows the location of vertices.

defer the full discussion until later chapters. In this chapter, we will not concern ourselves with artifacts along the borders of a terrain tile.

Figure 11.27 shows shaded terrain with normals computed using the forward difference. For comparison, Figure 11.26 shows the same scene without lighting.

Try This ○○○○

> The best way to see the visual differences between the different ways of computing normals is to run Chapter11TerrainShading and rotate through each algorithm. There are options to overlay the terrain's wireframe and normals that will help you notice differences.

Using forward or backward difference has the benefit of being very efficient; it requires only three height-map samples. If this is done in a

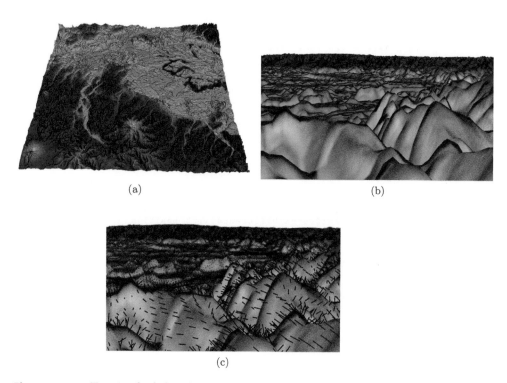

(a)

(b)

(c)

Figure 11.27. Terrain shaded with normals computed using the forward difference. The positional light is attached to the camera. This technique is fast, requiring the fewest texture reads and least amount of computation, but produces less accurate results.

displacement-mapping vertex shader, one of the samples is required anyway, so only two extra texture reads and a small amount of computation is required to derive a normal. The downside is that the normal isn't very accurate since it only depends on two additional heights. This is tolerable for low frequency terrains but can be very inaccurate for high-frequency terrains with steep features.

11.3.2 Central Difference

The *central difference*, $f(x + \frac{1}{2}h) - f(x - \frac{1}{2}h)$, can also be used to compute normals. Using the four adjacent heights to a position to compute its normal provides a nice balance between performance and accuracy. This is a good choice for many applications that procedurally generate normals in a shader. The principle is the same as using the forward difference:

```
vec3 ComputeNormalCentralDifference(vec3 position,
                                    sampler2DRect heightMap)
{
  vec3 left = vec3(position - vec2(1.0, 0.0),
      texture(heightMap, position - vec2(1.0, 0.0)).r);
  vec3 right = vec3(position + vec2(1.0, 0.0),
      texture(heightMap, position + vec2(1.0, 0.0)).r);
  vec3 bottom = vec3(position - vec2(0.0, 1.0),
      texture(heightMap, position - vec2(0.0, 1.0)).r);
  vec3 top = vec3(position + vec2(0.0, 1.0),
      texture(heightMap, position + vec2(0.0, 1.0)).r);
  return cross(right - left, top - bottom);
}
```

Listing 11.15. Computing terrain normals using the central difference.

```
vec3 ComputeNormalCentralDifference(vec3 position,
                                    sampler2DRect heightMap)
{
  float leftHeight =
      texture(heightMap, position.xy - vec2(1.0, 0.0)).r;
  float rightHeight =
      texture(heightMap, position.xy + vec2(1.0, 0.0)).r;
  float bottomHeight =
      texture(heightMap, position.xy - vec2(0.0, 1.0)).r;
  float topHeight =
      texture(heightMap, position.xy + vec2(0.0, 1.0)).r;
  return vec3(leftHeight - rightHeight,
          bottomHeight - topHeight, 2.0);
}
```

Listing 11.16. An optimized version of computing terrain normals using the central difference.

take the cross product of two vectors representing partial derivatives. The vector in the x direction is based on the left and right heights, and the vector in the y direction is based on the top and bottom heights. Using four samples in this way is sometimes called a *star* or *cross filter*.

A GLSL function for computing unnormalized normals using the central difference is shown in Listing 11.15. Compare this to the optimized version of the same function in Listing 11.16 [154]. The optimized version assumes terrain is extruded from the xy-plane and posts are one unit apart. Figure 11.28 shows the visual results.

11.3.3　Sobel Filter

A popular technique for deriving normals from a height map is to use a *Sobel filter* [156]. This is used by AMD's NormalMapper tool; AMD's RenderMonkey provides GLSL and HLSL code examples.[3] The Sobel filter

---

[3]http://developer.amd.com/gpu/rendermonkey/

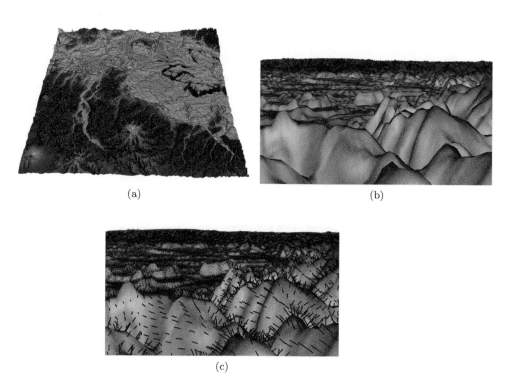

(a)   (b)

(c)

Figure 11.28. Terrain shaded with normals computed from four samples provides a reasonable trade-off between performance and accuracy.

is an edge-detection filter used in image processing with separate kernels to detect horizontal and vertical edges. The horizontal kernel is shown in Equation (11.1) and the vertical kernel is shown in Equation (11.2):

$$\begin{pmatrix} -1 & -2 & -1 \\ 0 & 0 & 0 \\ 1 & 2 & 1 \end{pmatrix}, \tag{11.1}$$

$$\begin{pmatrix} -1 & 0 & 1 \\ -2 & 0 & 2 \\ -1 & 0 & 1 \end{pmatrix}. \tag{11.2}$$

In image processing, a horizontal or vertical edge is detected by "centering" the corresponding kernel over a pixel and summing the product of the kernel with the pixel and its eight neighbors component-wise. If the sum

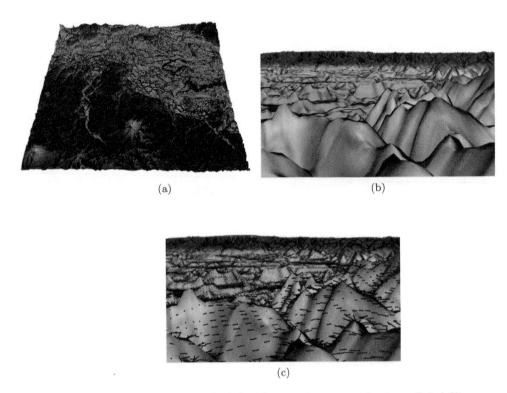

Figure 11.29. Terrain shaded with normals computed using a Sobel filter.

```
vec3 ComputeNormalSobelFilter(vec3 position,
                              sampler2DRect heightMap)
{
    float upperLeft =
        texture(heightMap, position.xy + vec2(-1.0,  1.0)).r;
    float upperCenter =
        texture(heightMap, position.xy + vec2(0.0,  1.0)).r;
    float upperRight =
        texture(heightMap, position.xy + vec2(1.0,  1.0)).r;
    float left =
        texture(heightMap, position.xy + vec2(-1.0,  0.0)).r;
    float right =
        texture(heightMap, position.xy + vec2(1.0,  0.0)).r;
    float lowerLeft =
        texture(heightMap, position.xy + vec2(-1.0,  -1.0)).r;
    float lowerCenter =
        texture(heightMap, position.xy + vec2(0.0,  -1.0)).r;
    float lowerRight =
        texture(heightMap, position.xy + vec2(1.0,  -1.0)).r;

    float x = upperRight + (2.0 * right) + lowerRight -
            upperLeft - (2.0 * left) - lowerLeft;
    float y = lowerLeft + (2.0 * lowerCenter) + lowerRight -
```

```
              upperLeft - (2.0 * upperCenter) - upperRight;

    return vec3(-x, y, 1.0);
}
```

Listing 11.17. Computing terrain normals from a height map using a Sobel filter.

is above a threshold, an edge is detected. Since the kernels approximate the partial derivatives in x and y of the image's intensity, a Sobel filter can also be used to compute terrain normals, as shown in Listing 11.17. Like all functions shown thus far, the returned normal is not normalized. The constant 1.0 is used for the normal's z-component; this can be adjusted to smooth out or sharpen the shading.

Figure 11.29 shows terrain shaded with normals computed with a Sobel filter. The edge-detection nature of the Sobel filter is apparent in the zoomed-out view. Increasing the normal's z-component would smooth out the shading and make edges less apparent.

11.3.4   Summary of Normal Computations

For many applications, normals are used primarily for lighting. In this case, the accuracy of the normal is not that important as long as the result looks right. In other cases, the normal may be used for analytic visualization, such as shading based on steepness or compass direction. In these cases, the accuracy of the normal may be more important than its speed of computation. When a normal map is computed offline, the speed of computation is typically even less important.

Figures 11.30, 11.31, and 11.32 show side-by-side comparisons of the three normal computations described in this chapter. In all cases a single positional light is located at the camera.

11.4   Shading

A wide array of creative terrain-shading algorithms are used in virtual globes, GIS applications, and games. Virtual globes tend to focus on shading with high-resolution satellite imagery. GIS applications commonly shade terrain to highlight its features, such as valleys or steep areas. Perhaps games have the ultimate challenge as their terrain shading needs to portray a realistic image of an artificial, potentially destructible landscape and enable artists to easily author such scenes.

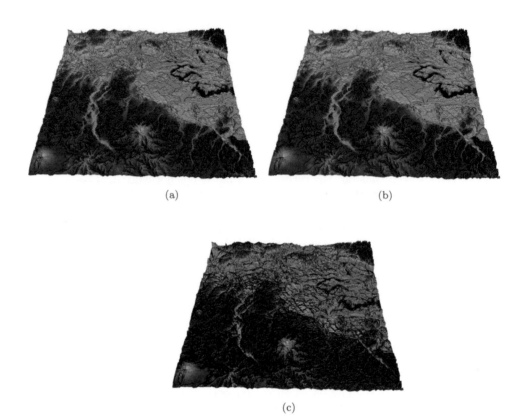

Figure 11.30. In a zoomed-out view, the (a) forward difference and (b) central difference are similar. The central difference produces slightly smoother shading. The (c) Sobel filter results in the sharpest shading, highlighting features that are less apparent in (a) and (b).

11.4.1 Color Maps and Texture Coordinates

A common way to shade terrain is to use a color map like the one shown earlier in Figure 11.7(b). A color map is a texture containing colors used for shading. It is exactly what we think of when we think of a texture map. Since we may store other things in textures when rendering terrain (e.g., normals for lighting or alpha values for blending multiple textures), we explicitly state we are storing colors by calling the texture a color map. A color map could be real satellite imagery or artist-created grass, dirt, stone, etc.

When a single color map is draped over a terrain tile, the fragment shader can be as simple as a single texture read, usually modulated with

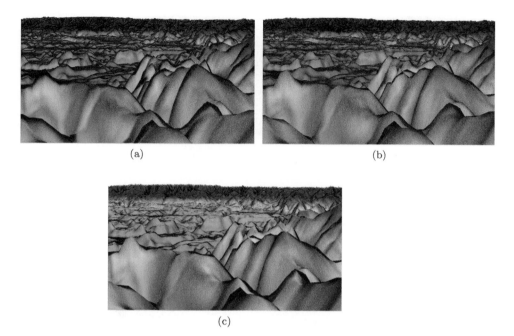

(a)

(b)

(c)

Figure 11.31. In a zoomed-in view, the (a) forward difference and (b) central difference are again pretty similar. The differences are most noticeable on the left side in the close peaks. In this view, the (c) Sobel filter produces smoother shading (see Figure 11.32).

lighting intensity. If a color map is only modulated with the diffuse-lighting component, it is called a *diffuse map*. Likewise, it is called a *specular map* if it is only modulated with the specular component.

Texture coordinates used to read a color map can be procedurally generated in the vertex shader. A texture coordinate can be computed from the vertex's xy-world-space coordinates using a simple linear mapping, as done in Listing 11.18.

When `u_textureCoordinateScale` is set to $\left(\frac{1}{x \text{ tile resolution}}, \frac{1}{y \text{ tile resolution}} \right)$, the texture coordinates span from $(0,0)$ in the bottom left corner of the tile to $(1,1)$ in the upper-right corner.

When color maps have repeating patterns (e.g., grass or dirt), it is useful to use repeat or mirrored-repeat texture filtering and set `u_texture CoordinateScale` to $\left(\frac{n}{x \text{ tile resolution}}, \frac{m}{y \text{ tile resolution}} \right)$ to repeat the color map

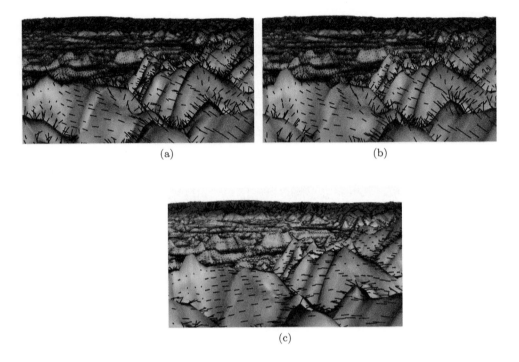

Figure 11.32. The same zoomed-in view as in Figure 11.31 with normals drawn at vertices. The (c) Sobel filter's normals do not point in the z direction as much as the other methods and, therefore, produce smoother, lighter shading because of the light source's position at the camera.

```
in vec2 position;
out vec2 textureCoordinate;

uniform vec2 u_textureCoordinateScale;
uniform vec3 u_aabbLowerLeft;

void main()
{
  // ...
  textureCoordinate =
      (position - u_aabbLowerLeft.xy) * u_textureCoordinateScale;
}
```

Listing 11.18. Generating texture coordinates in a vertex shader.

along the tile n times in the x direction and m times in the y direction. This can improve visual quality by creating a more even texel to pixel ratio without increasing the size of the actual color map. To make the repetition less noticeable, the repeat can be made view dependent by blending between two sets of texture coordinates based on the distance to the camera [145].

It is also common to generate more than one set of texture coordinates to look up different textures. For example, the texture coordinate used to read a base color map may span from $(0, 0)$ to $(1, 1)$ while a color map with fine detail may be repeated across a tile using texture coordinates spanning from $(0, 0)$ to $(4, 4)$. In fact, this technique is common enough that the fine-detail color map is called a *detail map*.

11.4.2 Detail Maps

Due to limited texture resolution, terrain near the viewer can appear blurry. For example, in virtual globes, sparsely populated areas tend to have lower-resolution satellite imagery than densely populated areas. When the user is zoomed in close to the low-resolution imagery, one texel maps to many pixels, making the terrain look blurry. For applications that can tolerate synthesizing detail, a detail map can be used to reduce blurring. A detail map is a texture containing high-frequency details, including dips and bumps, commonly authored using a noise function. To avoid blurring near the viewer, a color map is blended with a tiled detail map to increase the appearance of detail.

11.4.3 Procedural Shading

Procedural-shading techniques compute a fragment's color at runtime using properties of the terrain and typically very little memory. Procedural techniques create visually rich shading with little memory or shading that highlights terrain features such as steepness. Either way, we are amazed at how so little code can generate such visually pleasing and useful shading!

> The figures in this section provide a good sense of procedural shading, but to really get acquainted, run Chapter11TerrainShading. In addition to rotating through the algorithms, enable and disable lighting to get a sense of how each shading technique highlights the terrain's features.

○○○○ **Try This**

Height-based shading. Many terrain-shading algorithms are based on the terrain's height. A vertex shader typically passes the height, which is interpolated during rasterization, to the fragment shader. Perhaps the simplest approach is to map the minimum height to one color and the maximum height to another color and linearly interpolate between the two based on the fragment's height. The fragment shader in Listing 11.19 does this by assigning black to the minimum height and green to the maximum height.

The result is shown in Figure 11.33(c). Compared to no shading at all (see Figure 11.33(a)), shading by height provides some sense of terrain features. Compared to just lighting (see Figure 11.33(b)), shading by height provides a better sense of height for a particular location; it is easy to determine that dark regions are low and bright regions are high—such observations are not as easy to make with just lighting.

Figure 11.33(d) shows lighting combined with shading by height, achieving the best of both worlds. These are combined by multiplying the height-based color with the light intensity. Combining the two also avoids a weak-

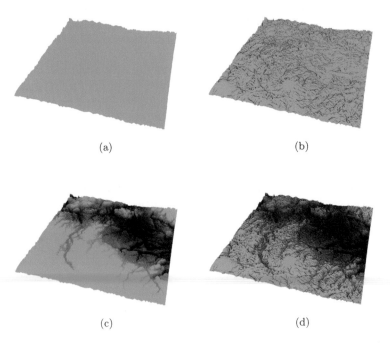

(a)

(b)

(c)

(d)

Figure 11.33. (a) Without shading, terrain features are unrecognizable. (b) With lighting, terrain features become apparent. (c) Shading by height provides a sense of terrain features even without lighting. (d) Combining lighting and shading by height conveys both terrain features and height.

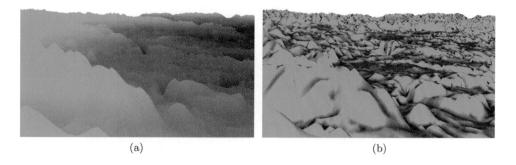

(a)   (b)

Figure 11.34. (a) For horizon views, shading by height alone does not give a good sense of terrain features. (b) Lighting brings out terrain features for all views, especially horizon views.

ness of shading by height alone—horizon views. As shown in Figure 11.34, just shading by height makes it hard to distinguish peaks in the background and foreground that are at similar heights.

Allowing the user to determine the approximate terrain height at a quick glance is useful for scientific visualization and GIS applications. Shading by height is one way to achieve this. Another approach, which can be combined with shading by height, is to render contour lines of constant height, as shown in Figure 11.35.

Height contours can be procedurally generated in a fragment shader using the same technique used to render a latitude-longitude grid on a globe, as described in Section 4.2.4. Listing 11.20 shows a fragment-shader snippet that shades contours red and terrain green.

The interval between contours is used to determine the distance from the fragment to the closest contour. If the distance is small enough, the

```
in float height;
out vec3 fragmentColor;

uniform float u_minimumHeight;
uniform float u_maximumHeight;

void main()
{
    fragmentColor = vec3((height - u_minimumHeight) /
                         (u_maximumHeight - u_minimumHeight),
                         0.0, 0.0);
}
```

Listing 11.19. Shading by height in a fragment shader.

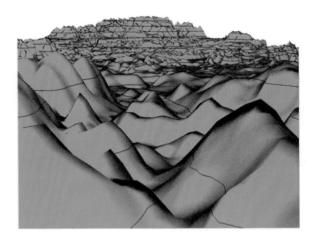

Figure 11.35. Contour lines procedurally generated in a fragment shader allow the user to quickly gauge approximate terrain height and curvature.

fragment is shaded appropriately. Screen-space partial derivative functions, dFdx and dFdy, are used to maintain a constant pixel width.

Try This ○○○○

> Similar to latitude-longitude grids, a variety of shading options exist for height contours. Make contours a different color or width based on height.

In addition to providing an approximate sense of height, the curvature of height contours can highlight terrain features, as shown in Figure 11.36.

Thus far, our shading techniques have used just height to determine a fragment's color. A wide array of creative techniques become possible when additional data in the form of textures are also used. One such technique

```
float distanceToContour = mod(height, u_contourInterval);
float dx = abs(dFdx(height));
float dy = abs(dFdy(height));
float dF = max(dx, dy) * u_lineWidth;
fragmentColor = mix(vec3(0.0, intensity, 0.0),
                    vec3(intensity, 0.0, 0.0),
                    (distanceToContour < dF));
```

Listing 11.20. Creating height contours.

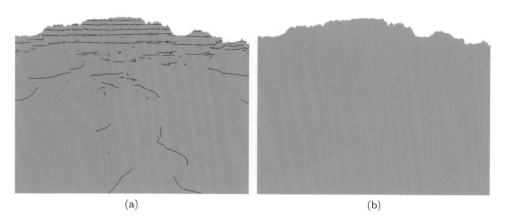

                 (a)   (b)

Figure 11.36. (a) As seen by the curvature in the foreground, height contours provide some sense of terrain features even without shading. (b) Without shading, terrain features are unrecognizable.

uses a fragment's height to look into a 1D texture, or a one-texel-wide 2D texture, called a *color ramp*, to determine the fragment's color. A color ramp, such as the one in Figure 11.37, can easily be authored in a paint program. Figure 11.38 shows terrain shaded with this color ramp, which ramps from water to grass to dirt to snow.

    Using a color-ramp texture is more general and easier to implement than trying to blend between several colors in a fragment shader. It requires only a single line of code, as demonstrated in Listing 11.21. It is very efficient

Figure 11.37. Color ramp used for height-based shading.

Figure 11.38. Terrain shaded with the color ramp in Figure 11.37.

in terms of speed and memory, costing only a single read from a usually small texture.

Color ramps are useful in virtual globes when satisfactory satellite imagery is not available or as a memory-efficient alternative to satellite imagery for certain areas. Height-based color ramps do not need to represent materials such as grass and dirt. They can represent height intervals, giving the user another method of approximating height in addition to height contours and height-based shading. This can be useful in aircraft simulation and in water level what-if scenarios (e.g., if the water level rose two meters, what areas would flood?).

Using a color ramp can add significant visual richness with very little performance and memory costs. Its major limitation is that the same color is used for all fragments at a given height. This means that materials like grass and stones have to be described by a single color. In many cases, it is desirable to use color maps, like the ones in Figure 11.39, for materials.

A simple way to do this is to extend the color-ramp approach. Instead of storing colors in the ramp, store an alpha value that is used to blend two color maps. An example blend ramp is shown in Figure 11.40. Linearly interpolating between the two color maps based on the alpha in the blend ramp is usually sufficient (i.e., color = $((1 - \text{alpha}) * \text{grassColorMap}) + (\text{alpha} * \text{stoneColorMap})$).

```
vec2 coord = vec2(0.5, (height - u_minimumHeight) /
                  (u_maximumHeight - u_minimumHeight));
fragmentColor = intensity * texture(u_colorRamp, coord).rgb;
```

Listing 11.21. Using a height-based color ramp for terrain shading.

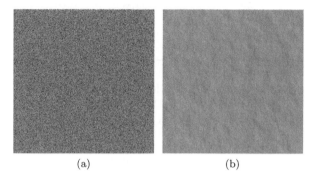

(a)   (b)

Figure 11.39. (a) Grass and (b) stone color maps that are blended together based on the blend ramp in Figure 11.40. When the blend ramp is black, grass is used. When the blend ramp is white, stone is used.

Listing 11.22 shows how this is implemented. A texel is read from both the grass and stone color maps. The texture coordinates for the color maps are not dependent on the blend ramp or each other; therefore, color maps can be repeated inside terrain tiles independent of each other. Next, the fragment's height is used to look up the alpha value in the blend ramp. This alpha value is used to linearly interpolate between the two texels with

```
float normalizedHeight = (height - u_minimumHeight) /
                         (u_maximumHeight - u_minimumHeight);
fragmentColor = intensity * mix(
    texture(u_grass, repeatTextureCoordinate).rgb,
    texture(u_stone, repeatTextureCoordinate).rgb,
    texture(u_blendRamp, vec2(0.5, normalizedHeight)).r);
```

Listing 11.22. Blending two color maps by height.

Figure 11.40. Blend ramp used for height-based shading.

mix. To make the blend between color maps more realistic, a noise function can be used.

Try This ○○○○

A blend ramp is likely to have a lot of 0s and 1s, making one of the color map texture reads in Listing 11.22 unnecessary. Try optimizing the shader in Chapter11TerrainShading using dynamic branches to avoid the unnecessary texture read, similar to the night-lights fragment shader in Listing 4.15 in Section 4.2.5. Does it improve performance? Why or why not?

Figure 11.41 shows the result of using a blend ramp to blend between grass and stone based on height. A blend ramp allows flexible nonlinear control of blending, which is evident as most heights are grass and only the highest locations have stone.

More than two color maps can be used with a blend ramp between each. Consider a scene with water, sand, and grass color maps and a blend ramp, $ramp0$, between water and sand, and a blend ramp, $ramp1$, between sand and grass. First the water and sand are blended: $color0 = ((1 - ramp0.alpha) * waterColorMap) + (ramp0.alpha * sandColorMap)$. This color is then blended with grass to compute the final color: $color = ((1 - ramp1.alpha) * color0) + (ramp1.alpha * grassColorMap)$. A more

(a)   (b)

Figure 11.41. (a) The result of using the blend ramp in Figure 11.40 to blend grass and stone color maps. (b) The blend ramp is applied to terrain as a color ramp.

memory efficient, but slightly harder to author, approach is to use a single blend ramp, where the integer part of the alpha value determines what color maps to blend between and the fractional part is the alpha value used for blending. In this example, $0.x$ would indicate to blend between water and sand, and $1.x$ would indicate to blend between sand and grass.

In some cases, blending can occur offline, generating a single color map, so only a single texture read in the fragment shader is required at runtime. This approach lacks the flexibility and memory efficiency of independently repeating small color maps across a terrain tile. Since the terrain's height is used to create the offline color map, the color map cannot be repeated across the tile.

Slope-based shading. All the terrain-shading techniques that use height can also use slope. In games, this is useful to make flat areas grassy and steep areas rocky. In GIS applications, slope-based shading allows the user to quickly approximate steepness.

We define a slope of zero to indicate a steep, $90°$ face and a slope of one to indicate flat land. This allows us to formulate the terrain's slope as the cosine of the angle between the terrain's normal and the ground plane's normal, which is the dot product of the two:

$$\text{slope} = \hat{\mathbf{n}}_{\text{terrain}} \cdot \hat{\mathbf{n}}_{\text{groundplane}}.$$

When the ground plane is the xy-plane, the slope is just the z-component of the terrain's normal. Visualizing slope, as in Figure 11.42, requires a very short fragment shader, $\texttt{fragmentColor} = \text{vec3}(\texttt{normal.z})$, assuming $\texttt{normal}$ is normalized.

Figure 11.42. Terrain shaded by slope. Flat areas are white, steeper areas fade to black. Slope is computed as the dot product of the terrain's normal and the ground plane's normal.

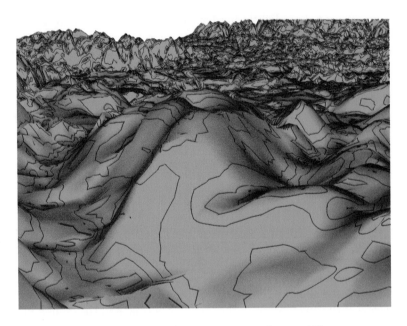

Figure 11.43. Slope contours spaced every 15°.

Contour lines of constant slope are useful to identify areas of similar steepness and areas where steepness changes (see Figure 11.43). These slope contours are rendered in the same manner as height contours. The potential point of confusion is that our definition of slope is not an angle. It is the cosine of the angle between the terrain's normal and the ground plane's normal. Therefore, to render contour lines every x degrees, acos must be used to determine the angle of the slope for the fragment. See the first line in Listing 11.23. Compare this listing to the code for rendering height contours in Listing 11.20.

Slope-based color ramps are useful in GIS applications. A user planning new roads may want to easily determine steep areas so they can be avoided or the required switchbacks can be added. Likewise, a military user plan-

```
float slopeAngle = acos(normal.z);
float distanceToContour = mod(slopeAngle, u_contourInterval);
float dx = abs(dFdx(slopeAngle));
float dy = abs(dFdy(slopeAngle));
float dF = max(dx, dy) * u_lineWidth;
fragmentColor = mix(vec3(0.0, intensity, 0.0),
                    vec3(intensity, 0.0, 0.0),
                    (distanceToContour < dF));
```

Listing 11.23. Creating slope contours.

Figure 11.44. Color ramp used for slope-based shading.

ning a mission may want to identify areas that are too steep to walk along. The color ramp in Figure 11.44 was colored to warn of steep areas. A slope of 60° or greater is red and a slope between 30° and 60° is orange. Given the fragment shader has the cos of the slope, the color ramp was authored such that the bottom half is red ($\cos 60° = 0.5$), the next area is orange (0.5 to $\cos 30° \approx 0.866$), and the top section, representing the flattest areas, is green. If the color ramp were authored linearly in degrees, an acos would be required in the shader. Since exact intervals are desired, the color ramp doesn't have smooth transitions.

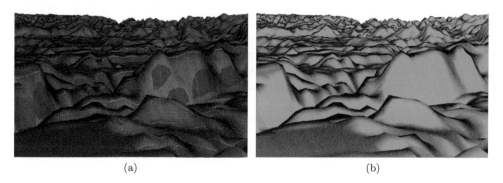

(a) (b)

Figure 11.45. Slope-based shading using the color ramp from Figure 11.44. (a) Slope-based color ramp. (b) Just lighting. It is much easier to approximate the steepness of the terrain in (a) than in (b).

(a)   (b)

Figure 11.46. (a) A fragment's slope is used to look up an alpha value used to blend grass and stone such that flat areas are grassy and steep areas are rocky. (b) The blend ramp is applied to terrain as a color map.

Figure 11.45 shows the result of slope-based color-ramp shading. The slope is used to read the color ramp in the same way the height is used to read a color ramp in Listing 11.21.

Try This ○○○○

> Run Chapter11TerrainShading and select "Color Ramp By Slope." Use the up and down arrow keys to change the height exaggeration, which in turn adjusts the slope. Color-ramp shading makes it easy to see areas of different steepness. Try adjusting height exaggeration without it to see the difference.

Slope-based blend ramps are useful for games and simulations. Just like height can be used to look up an alpha value to blend two color maps, slope can also be used. The code is identical to the height-based approach shown in Listing 11.22 except that slope is used. Perhaps the most common use, shown in Figure 11.46, is to make flat areas grassy and steep areas rocky as they are unlikely to grow vegetation. Slope-based blend ramps create quite a bit of visual complexity using very little memory. Slope-based shading is also useful for other techniques, such as snow accumulation, as is done in the Frostbite engine [4].

Comparison of height- and slope-based shading. Table 11.2 shows the major height-based and slope-based shading techniques. These techniques can be combined. For example, a user may want to see height contours on slope-based color-ramp-shaded terrain. Likewise, height- and slope-based shading can be combined with other shading techniques like detail maps.

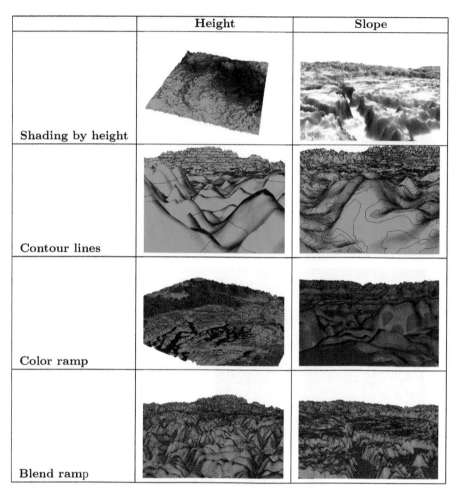

	Height	Slope
Shading by height		
Contour lines		
Color ramp		
Blend ramp		

Table 11.2. Comparison of procedural height-based and slope-based shading techniques.

Besides shading based on height and slope, it is also common to shade based on the terrain's normal. In GIS applications, a normal-based color ramp may be used to help the user determine which compass direction terrain faces. In games, normal-based blend ramps may be used to render denser vegetation on terrain faces that capture more sun. Modify Chapter11TerrainShading to use one of these techniques.

○○○○ Try This

Silhouette-edge rendering. A silhouette edge with respect to the viewer is an edge in which one triangle is front-facing and the other is back-facing. Loosely speaking, it can be thought of as the outline of an object (including interior outlines), as shown in Figures 11.47(a) and 11.47(b). Silhouette-edge rendering is an area of non-photorealistic rendering (NPR). It is commonly combined with toon shading to provide a cartoon-like appearance. It can also be used in terrain rendering to emphasize peaks, slopes, and other features.

A simple technique for rendering silhouette edges is surface angle silhouetting. A silhouette is detected in a fragment shader by checking the dot product of the surface normal and the vector to the viewer. When the dot product is zero, the vectors are perpendicular, and thus an edge is detected. In practice, the dot product is checked to be near zero, as in

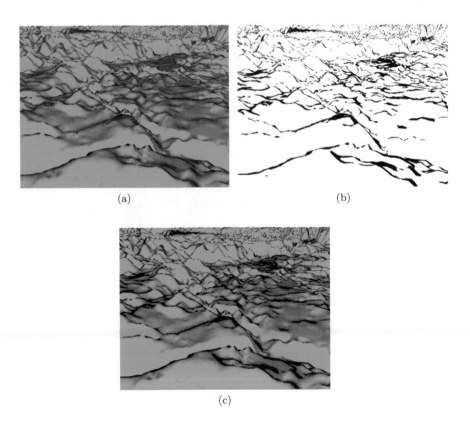

(a)   (b)

(c)

Figure 11.47. (a) Terrain shaded with a solid color. (b) The same terrain with only its silhouette edges rendered. (c) Shading combined with silhouette edges.

```
if (abs(dot(normal, positionToEye)) < delta)
{
    fragmentColor = vec3(0.0);
    return;
}
```

Listing 11.24. Fragment-shader snippet for surface angle silhouetting.

Listing 11.24, or used to index into a 1D texture, similar to using height or slope to index into a color ramp.

Surface angle silhouetting only requires a single rendering pass, but it fails to capture some silhouettes, while smearing black on flat areas (see areas in the distance in Figure 11.47(c)). This is one of many silhouette-edge-rendering techniques. Other techniques include silhouette-edge detection in a geometry shader, edge detection via image processing, and multipass rendering. A multipass wireframe approach has been evaluated for use with terrain [16]. Akenine-Moller et al. include an excellent survey of NPR, including silhouette-edge rendering [3].

Blend maps. Procedural shading is an elegant approach to shading terrain. Before programmable fragment shaders, techniques like slope-based color ramps required offline computation. Now, a small shader will do the trick. Unfortunately, some terrain-shading techniques cannot utilize procedural shading. A prominent example in virtual globes is rendering high-resolution satellite imagery. The fragment shader is usually not much more than a texture read and lighting, and perhaps some atmospheric effects. The real challenge is in managing the enormous amount of imagery. It is also difficult to utilize procedural shading when complete control over blending is desired. In this case, blend maps can be used.

Height- and slope-based blend ramps use terrain features to select an alpha value for blending. There are times when a user wants complete control over where and how color maps are blended. In the case of a game, an area of terrain may have been impacted by an explosion, or a vehicle coming to a screeching halt may have left brake marks behind. Cases like these call for a blend map as shown in Figure 11.48. With a blend map, an alpha value in a 2D alpha texture is used to blend between two color maps.

```
fragmentColor = intensity * mix(
    texture(u_grass, repeatTextureCoordinate).rgb,
    texture(u_stone, repeatTextureCoordinate).rgb,
    texture(u_blendMask, textureCoordinate).r);
```

Listing 11.25. Using a blend map.

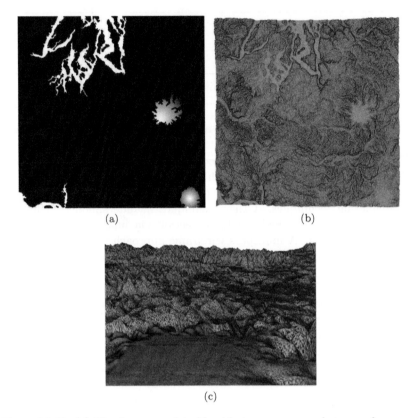

Figure 11.48. (a) Blend map used to blend between grass and stone color maps. (b) Top-down view of terrain rendered with the blend map. Black areas in the blend map correspond to grass and white areas correspond to stone. (c) A horizon view of the same terrain showing blending between grass and stone.

As shown in Listing 11.25, the code for a blend map is nearly identical to using a blend ramp. A blend map requires more memory than procedural techniques. To reduce memory consumption, the blend map does not need to be high resolution or cover an entire terrain tile, just the parts that require blending. It is also possible to use a sparse quadtree texture representation to significantly reduce memory usage [5].

For more information on blending color maps, see Jenks's article [77] and the early work on texture splatting by Bloom [17], which uses multipass rendering to composite textures in the framebuffer via alpha blending.

Texture stretching. When texture coordinates are computed based on a simple linear mapping of the vertex's xy-world-space position to texel space,

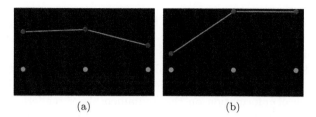

(a)   (b)

Figure 11.49. (a) When adjacent posts have similar heights, little texture stretching occurs because texel space is nearly evenly assigned to world space. (b) When adjacent posts have significant height differences, texture stretching occurs. Compare the world-space difference between the first two posts and the last two posts. Stretching will occur between the first two posts because not enough texel space is used.

texture-stretching artifacts occur on steep slopes. As shown in Figure 11.49, stretching occurs because uniformly distributing texture coordinates results in an equal amount of texels between each post, even if the height difference, and thus world-space distance, between adjacent posts is large. The same amount of texel space is assigned to long, steep areas as to shorter, flat areas, causing the texture to stretch on steep slopes.

There are many ways to minimize stretching. If detail mapping is used, stretching may not even be noticed. Since detail maps are tiled more frequently than color maps, there is less mismatch between a detail map's texel space and world space. The result of blending the detail map and color map tends to create enough detail on all but the steepest slopes.

Alternative solutions to stretching include assigning texture coordinates based on the steepness of adjacent posts, although this can shift the distortion to flatter areas. *Triplanar texturing* can reduce texture stretching by using the terrain's normal to blend between textures from one of three planes: the xy-plane, the xz-plane, and the yz-plane [58]. Of course, triplanar texturing comes with additional runtime overhead. Two-sided triplanar texturing is used in the C4 Engine to allow texturing of terrain features such as steep cliffs and cave interiors [165]. Finally, a precomputed *indirection map* can minimize stretching artifacts at the cost of a precomputation step that iteratively relaxes a spring network and uses one texture map and texture read at runtime [112].

11.5   Resources

As their popularity increases, a wealth of information is becoming available on non-height-map terrain representations. For voxels, Rosa's article on destructible volumetric terrain in Miner Wars is a recent description of how voxels were used in a game [145]. Geiss's article on GPU-based procedural terrain is an excellent example of modeling terrain with an implicit function [59]. It is freely available online, along with all of *GPU Gems 3*.

Although most of the terrain-rendering literature focuses on LOD, there is still plenty of information on the fundamentals of rendering individual tiles. Vistnes's article is a great source for a complete implementation of displacement mapping in a vertex shader [177]. For ray-casting terrain, the work of Dick et al. includes a detailed analysis [38]. The DirectX SDK includes an implementation of GPU ray casting using cone step mapping [115].

Kris Nicholson's master's thesis provides a nice overview of GPU-based procedural terrain texturing [119]. Perhaps the ultimate resource to continue exploring terrain shading is Johan Andersson's description of terrain rendering in Frostbite [5].

○○○○○ 12

Massive-Terrain
Rendering

Virtual globes visualize massive quantities of terrain and imagery. Imagine a single rectangular image that covers the entire world with a sufficient resolution such that each square meter is represented by a pixel. The circumference of Earth at the equator and around the poles is roughly 40 million meters, so such an image would contain almost a quadrillion (1×10^{15}) pixels. If each pixel is a 24-bit color, it would require over 2 million gigabytes of storage—approximately 2 petabytes! Lossy compression reduces this substantially, but nowhere near enough to fit into local storage, never mind on main or GPU memory, on today's or tomorrow's computers. Consider that popular virtual globe applications offer imagery at resolution higher than one meter per pixel in some areas of the globe, and it quickly becomes obvious that such a naïve approach is unworkable.

Terrain and imagery datasets of this size must be managed with specialized techniques, which are an active area of research. The basic idea, of course, is to use a limited storage and processing budget where it provides the most benefit. As a simple example, many applications do not need detailed imagery for the approximately 70% of Earth covered by oceans; it makes little sense to provide one-meter resolution imagery there. So our terrain- and imagery-rendering technique must be able to cope with data with varying levels of detail in different areas. In addition, when high-resolution data are available for a wide area, more triangles should be used to render nearby features and sharp peaks, and more texels should be used where they map to more pixels on the screen. This basic goal has been pursued from a number of angles over the years. With the explosive growth in GPU performance in recent years, the emphasis has shifted from minimizing the number of triangles drawn, usually by doing substantial computations on the CPU, to maximizing the GPU's triangle throughout.

We consider the problem of rendering planet-sized terrains with the following characteristics:

- They consist of far too many triangles to render with just the brute-force approaches introduced in Chapter 11.

- They are much larger than available system memory.

The first characteristic motivates the use of terrain LOD. We are most concerned with using LOD techniques to reduce the complexity of the geometry being rendered; other LOD techniques include reducing shading costs. In addition, we use culling techniques to eliminate triangles in parts of the terrain that are not visible.

The second characteristic motivates the use of out-of-core rendering algorithms. In out-of-core rendering, only a small subset of a dataset is kept in system memory. The rest resides in secondary storage, such as a local hard disk or on a network server. Based on view parameters, new portions of the dataset are brought into system memory, and old portions are removed, ideally without stuttering rendering.

Beautifully rendering immense terrain and imagery datasets using proven algorithms is pretty easy if you're a natural at spatial reasoning, have never made an off-by-one coding error, and scoff at those who consider themselves "big picture" people because you yourself live for the details. For the rest of us, terrain and imagery rendering takes some patience and attention to detail. It is immensely rewarding, though, combining diverse areas of computer science and computer graphics to bring a world to life on your computer screen.

Presenting all of the current research in terrain rendering could fill several books. Instead, this chapter presents a high-level overview of the most important concepts, techniques, and strategies for rendering massive terrains, with an emphasis on pointing you toward useful resources from which you can learn more about any given area.

In Chapters 13 and 14, we dive into two specific terrain algorithms: *geometry clipmapping* and *chunked LOD*. These two algorithms, which take quite different approaches to rendering massive terrains, serve to illustrate many of the concepts in this chapter.

We hope that you will come away from these chapters with lots of ideas for how massive-terrain rendering can be implemented in your specific application. We also hope that you will acquire a solid foundation for understanding and evaluating the latest terrain-rendering research in the years to come.

12.1 Level of Detail

Terrain LOD is typically managed using algorithms that are tuned to the unique characteristics of terrain. This is especially true when the terrain is represented as a height map; the regular structure allows techniques that are not applicable to arbitrary models. Even so, it is helpful to consider terrain LOD among the larger discipline of LOD algorithms.

LOD algorithms reduce an object's complexity when it contributes less to the scene. For example, an object in the distance may be rendered with less geometry and lower resolution textures than the same object if it were close to the viewer. Figure 12.1 shows the same view of Yosemite Valley, El Capitan, and Half Dome at different geometric levels of detail.

LOD algorithms consist of three major parts [3]:

- *Generation* creates different versions of a model. A simpler model usually uses fewer triangles to approximate the shape of the original model. Simpler models can also be rendered with less-complex shaders, smaller textures, fewer passes, etc.

- *Selection* chooses the version of the model to render based on some criteria, such as distance to the object, its bounding volume's estimated pixel size, or estimated number of nonoccluded pixels.

- *Switching* changes from one version of a model to another. A primary goal is to avoid *popping*: a noticeable, abrupt switch from one LOD to another.

(a) (b)

Figure 12.1. The same view of Yosemite Valley, El Capitan, and Half Dome at (a) low detail and (b) high detail. The differences are most noticeable in the shapes of the peaks in the distance. Image USDA Farm Service Agency, Image ©2010 DigitalGlobe. (Figures taken using Google Earth.)

Furthermore, we can group LOD algorithms into three broad categories: *discrete*, *continuous*, and *hierarchical*.

12.1.1   Discrete Level of Detail

Discrete LOD is perhaps the simplest LOD approach. Several independent versions of a model with different levels of detail are created. The models may be created manually by an artist or automatically by a polygonal simplification algorithm such as vertex clustering [146].

Applied to terrain, discrete LOD would imply that the entire terrain dataset has several discrete levels of detail and that one of them is selected for rendering at each frame. This is unsuitable for rendering terrain in virtual globes because terrain is usually both "near" and "far" at the same time.

The portion of terrain that is right in front of the viewer is nearby and requires a high level of detail for accurate rendering. If this high level of detail is used for the entire terrain, the hills in the distance will be rendered with far too many triangles. On the other hand, if we select the LOD based on the distant hills, the nearby terrain will have insufficient detail.

12.1.2   Continuous Level of Detail

In continuous LOD (CLOD), a model is represented in such a way that the detail used to display it can be precisely selected. Typically, the model is represented as a base mesh plus a sequence of transformations that make the mesh more or less detailed as each is applied. Thus, each successive version of the mesh differs from the previous one by only a few triangles.

At runtime, a precise level of detail for the model is created by selecting and applying the desired mesh transformations. For example, the mesh might be encoded as a series of *edge collapses*, each of which simplifies the mesh by removing two triangles. The opposite operation, called a *vertex split*, adds detail by creating two triangles. The two operations are shown in Figure 12.2.

CLOD is appealing because it allows a mesh to be selected that has a minimal number of triangles for a required visual fidelity given the viewpoint or other simplification criteria. In days gone by, CLOD was the best way to interactively render terrain. Many historically popular terrain-rendering algorithms use a CLOD approach, including Lindstrom et al.'s CLOD for height fields [102], Duchaineau et al.'s real-time optimally adapting mesh (ROAM) [41], and Hoppe's view-dependent progressive meshes [74]. Luebke et al. have excellent coverage of these techniques [107].

Today, however, these have largely fallen out of favor for use as runtime rendering techniques. CLOD generally requires traversing a CLOD data

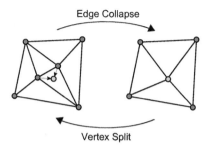

Figure 12.2. For an edge interior to a mesh, edge collapse removes two triangles and vertex split creates two triangles.

structure on the CPU and touching each vertex or edge in the current LOD. On older generations of hardware, this trade-off made a lot of sense; triangle throughput was quite low, so it was important to make every triangle count. In addition, it was worthwhile to spend extra time refining a mesh on the CPU in order to have the GPU process fewer triangles.

Today's GPUs have truly impressive triangle throughput and are, in fact, significantly faster than CPUs for many tasks. It is no longer a worthwhile trade-off to spend, for example, 50% more time on the CPU in order to reduce the triangle count by 50%. For that reason, CLOD-based terrain-rendering algorithms are inappropriate for use on today's hardware.

Today, these CLOD techniques, if they're used at all, are instead used to preprocess terrain into view-independent blocks for use with hierarchical LOD algorithms such as chunked LOD. These blocks are static; the CPU does not modify them at runtime, so CPU time is minimized. The GPU can easily handle the additional triangles that it is required to render as a result.

A special form of CLOD is known as infinite level of detail. In an infinite LOD scheme, we start with a surface that is defined by a mathematical function (e.g., an implicit surface). Thus, there is no limit to the number of triangles we can use to tessellate the surface. We saw an example of this in Section 4.3, where the implicit surface of an ellipsoid was used to render a pixel-perfect representation of a globe without tessellation.

Some terrain engines, such as the one in Outerra,[1] use fractal algorithms to procedurally generate fine terrain details (see Figure 12.3). This is a form of infinite LOD. Representing an entire real-world terrain as an implicit surface, however, is not feasible now or in the foreseeable future. For that reason, infinite LOD has only limited applications to terrain rendering in virtual globes.

---

[1]http://www.outerra.com

(a)   (b)

Figure 12.3. Fractal detail can turn basic terrain into a beautiful landscape.
(a) Original 76 m terrain data. (b) With fractal detail. (Images courtesy of
Brano Kemen, Outerra.)

12.1.3 Hierarchical Level of Detail

Instead of reducing triangle counts using CLOD, today's terrain-rendering
algorithms focus on two things:

- Reducing the amount of processing by the CPU.

- Reducing the quantity of data sent over the system bus to the GPU.

The LOD algorithms that best achieve these goals generally fall into
the category of hierarchical LOD (HLOD) algorithms.

HLOD algorithms operate on *chunks* of triangles, sometimes called
patches or *tiles*, to approximate the view-dependent simplification achieved
by CLOD. It some ways, HLOD is a hybrid of discrete LOD and CLOD. The
model is partitioned and stored in a multiresolution spatial data structure,
such as an octree or quadtree (shown in Figure 12.4), with a drastically
simplified version of the model at the root of the tree. A node contains one
chunk of triangles. Each child node contains a subset of its parent, where
each subset is more detailed than its parent but is spatially smaller. The
union of all nodes at any level of the tree is a version of the full model. The
node at level 0 (i.e., the root) is the most simplified version. The union of
the nodes at maximum depth represents the model at full resolution.

If a given node has sufficient detail for the scene, it is rendered. Otherwise, the node is refined, meaning that its children are considered for
rendering instead. This process continues recursively until the entire scene
is rendered at an appropriate level of detail. Erikson et al. describe the
major strategies for HLOD rendering [48].

HLOD algorithms are appropriate for modern GPUs because they help
achieve both of the reductions identified at the beginning of this section.

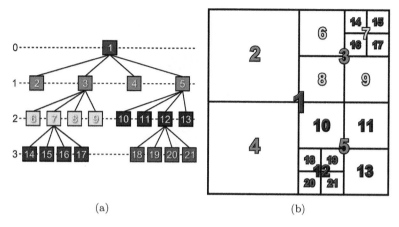

(a)   (b)

Figure 12.4. In HLOD algorithms, a model is partitioned and stored in a tree. The root contains a drastically simplified version of the model. Each child node contains a more detailed version of a subset of its parent.

In HLOD algorithms, the CPU only needs to consider each chunk rather than considering individual triangles, as is required in CLOD algorithms. In this way, the amount of processing that needs to be done by the CPU is greatly reduced.

HLOD algorithms can also reduce the quantity of data sent to the GPU over the system bus. At first glance, this is somewhat counterintuitive. After all, rendering with HLOD rather than CLOD generally means more triangles in the scene for the same visual fidelity, and triangles are, of course, data that need to be sent over the system bus.

HLOD, however, unlike CLOD, does not require that new data be sent to the GPU every time the viewer position changes. Instead, chunks are cached on the GPU using static vertex buffers that are applicable to a range of views. HLOD sends a smaller number of larger updates to the GPU, while CLOD sends a larger number of smaller updates.

Another strength of HLOD is that it integrates naturally with out-of-core rendering (see Section 12.3). The nodes in the spatial data structure are a convenient unit for loading data into memory, and the spatial ordering is useful for load ordering, replacement, and prefetching. In addition, the hierarchical organization offers an easy way to optimize culling, including hardware occlusion queries (see Section 12.4.4).

12.1.4 Screen-Space Error

No matter the LOD algorithm we use, we must choose which of several possible LODs to use for a given object in a given scene. Typically, the

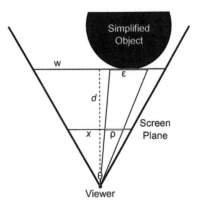

Figure 12.5. The screen-space error, ρ, of an object is estimated from the distance between the object and the viewer, the parameters of the view, and the geometric error, ϵ, of the object.

goal is to render with the simplest LOD possible while still rendering a scene that looks good. But how do we determine whether an LOD will provide a scene that looks good?

A useful objective measure of quality is the number of pixels of difference, or screen-space error, that would result by rendering a lower-detail version of an object rather than a higher-detail version. Computing this precisely is usually challenging, but it can be estimated effectively. By estimating it conservatively, we can arrive at a guaranteed error bound; that is, we can be sure that the screen-space error introduced by using a lower-detail version of a model is less than or equal to a computed value [107].

In Figure 12.5, we are considering the LOD to use for an object a distance d from the viewer in the view direction, where the view frustum has a width of w. In addition, the display has a resolution of x pixels and a field of view angle θ. A simplified version of the object has a geometric error ϵ; that is, each vertex in the full-detail object diverges from the closest corresponding point on the reduced-detail model by no more than ϵ units. What is the screen-space error, ρ, that would result if we were to render this simplified version of the object?

From the figure, we can see that ρ and ϵ are proportional and can solve for ρ:

$$\frac{\epsilon}{w} = \frac{\rho}{x},$$
$$\rho = \frac{\epsilon x}{w}.$$

The view-frustum width, w, at distance d is easily determined and substituted into our equation for ρ:

$$w = 2d \tan \frac{\theta}{2},$$

$$\rho = \frac{\epsilon x}{2d \tan \frac{\theta}{2}}. \qquad (12.1)$$

Technically, this equation is only accurate for objects in the center of the viewport. For an object at the sides, it slightly underestimates the true screen-space error. This is generally considered acceptable, however, because other quantities are chosen conservatively. For example, the distance from the viewer to the object is actually larger than d when the object is not in the center of the viewport. In addition, the equation assumes that the greatest geometric error occurs at the point on the object that is closest to the viewer.

○○○○ Kevin Says

For a bounding sphere centered at c and with radius r, the distance d to the closest point of the sphere in the direction of the view, $\mathbf{v}$, is given by

$$d = (c - \text{viewer}) \cdot \mathbf{v} - r.$$

By comparing the computed screen-space error for an LOD against the desired maximum screen-space error, we can determine if the LOD is accurate enough for our needs. If not, we refine.

12.1.5 Artifacts

While the LOD techniques used to render terrain are quite varied, there's a surprising amount of commonality in the artifacts that show up in the process.

Cracking. *Cracking* is an artifact that occurs where two different levels of detail meet. As shown in Figure 12.6, cracking occurs because a vertex in a higher-detail region does not lie on the corresponding edge of a lower-detail region. The resulting mesh is not watertight.

The most straightforward solution to cracking is to drop vertical skirts from the outside edges of each LOD. The major problem with skirts is that they introduce short vertical cliffs in the terrain surface that lead to texture stretching. In addition, care must be taken in computing the normals of the skirt vertices so that they aren't visible as mysterious dark or light

Figure 12.6. Cracking occurs when an edge shared by two adjacent LODs is divided by an additional vertex in one LOD but not the other.

lines around an LOD region. Chunked LOD uses skirts to avoid cracking, as will be discussed in Section 14.3.

Another possibility is to introduce extra vertices around the perimeter of the lower LOD region to match the adjacent higher LODs. This is effective when only a small number of different LODs are available or when there are reasonable bounds on the different LODs that are allowed to be adjacent to each other. In the worst case, the coarsest LOD would require an incredible number of vertices at its perimeter to account for the possibility that it is surrounded by regions of the finest LOD. Even in the best cases, however, this approach requires extra vertices in coarse LODs.

A similar approach is to force the heights of the vertices in the finer LOD to lie on the edges of the coarser LOD. The geometry-clipmapping terrain LOD algorithm (see Chapter 13) uses this technique effectively. A danger, however, is that this technique leads to a new problem: *T-junctions.*

T-junctions. T-junctions are similar to cracking, but more insidious. Whereas cracking occurs when a vertex in a higher-detail region does not lie on the corresponding edge of a lower-detail region, T-junctions occur because the high-detail vertex *does* lie on the low-detail edge, forming a T shape. Small differences in floating-point rounding during rasterization of the adjacent triangles lead to very tiny pinholes in the terrain surface. These pinholes are distracting because the background is visible through them.

Ideally, these T-junctions are eliminated by subdividing the triangle in the coarser mesh so that it, too, has a vertex at the same location as the vertex in the finer mesh. If the T-junctions were introduced in the first place in an attempt to eliminate cracking, however, this is a less than satisfactory solution.

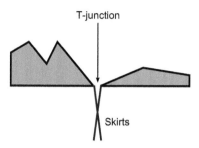

Figure 12.7. A side view of an exaggerated T-junction between two adjacent LODs. Skirts can hide T-junctions if they are angled slightly outward.

Another possibility is to fill the T-junctions with degenerate triangles. Even though these degenerate triangles mathematically have no area and thus should produce no fragments, the same rounding errors that cause the tiny T-junction holes to appear in the first place also cause a few fragments to be produced from the degenerate triangles, and those fragments fill the holes.

A final possibility, which is effective when cracks are filled with skirts, is to make the skirts of adjacent LODs overlap slightly, as shown in Figure 12.7.

Popping. As the viewer moves, the level of detail of various objects in the scene is adjusted. When the LOD of a given object is abruptly changed from a coarser one to a finer one, or vice versa, the user may notice a "pop" as vertices and edges change position.

This may be acceptable. To a large extent, virtual globe users tend to be more accepting of popping artifacts than, say, people playing a game. A virtual globe is like a web browser. Users instinctively understand that a web browser combines a whole lot of content loaded from remote servers. No one complains when a web browser shows the text of a page first and then "pops" in the images once they have been downloaded from the web server. This is much better than the alternative: showing nothing until all of the content is available.

Similarly, virtual globe users are not surprised that data are often streamed incrementally from a remote server and, therefore, are also not surprised when it suddenly pops into existence. As virtual globe developers, we can take advantage of this user expectation even in situations where it is not strictly necessary, such as in the transition between two LODs, both cached on the GPU. In many cases, however, popping can be prevented.

One way to prevent popping is to follow what Bloom refers to as the "mantra of LOD": a level of detail should only switch when that switch will be imperceptible to the user [18]. Depending on the specific LOD algorithm in use and the capabilities of the hardware, this may or may not be a reasonable goal.

Another possibility is to blend between different levels of detail instead of switching them abruptly. The specifics of how this blending is done are tied closely to the terrain-rendering algorithm, so we cover two specific examples in Sections 13.4 and 14.3.

12.2   Preprocessing

Rendering a planet-sized terrain dataset at interactive frame rates requires that the terrain dataset be preprocessed. As much as we wish it were not, this is an inescapable fact. Whatever format is used to store the terrain data in secondary storage, such as a disk or network server, must allow lower-detail versions of the terrain dataset to be obtained efficiently.

As described in Chapter 11, terrain data in virtual globes are most commonly represented as a height map. Consider a height map with 1 trillion posts covering the entire Earth. This would give us approximately 40 m between posts at the equator, which is relatively modest by virtual globe standards. If this height map is stored as a giant image, it will have 1 million texels on each side.

Now consider a view of Earth from orbit such that the entire Earth is visible. How can we render such a scene? It's unnecessary, even if it were possible, to render all of the half a trillion visible posts. After all, half a trillion posts is orders of magnitude more posts than there are pixels on even a high-resolution display.

Terrain-rendering algorithms strive to be *output sensitive*. That is, the runtime should be dependent on the number of pixels shaded, not on the size or complexity of the dataset.

Perhaps we'd like to just fill a 1,024 × 1,024 texture with the most relevant posts and render it using the vertex-shader displacement-mapping technique described in Section 11.2.2. How do we obtain such a texture from our giant height map? Figure 12.8 illustrates what is required.

First we would read one post. Then we would seek past about 1,000 posts before reading another post. This process would repeat until we had scanned through nearly the entire file. Seeking through a file of this size and reading just a couple of bytes at a time would take a substantial amount of time, even from local storage.

Worse, reading just one post out of every thousand posts would result in aliasing artifacts. A much better approach would be to find the average or maximum of the 1,000 × 1,000 "skipped" post heights to produce one

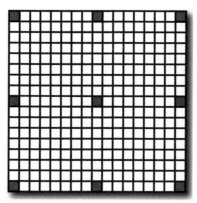

Figure 12.8. Rendering a low-detail representation of a height map when only a high-detail representation is available requires scanning past extraneous posts (white) to read the relevant ones (red). With a large height map, this is unworkable. Instead, we preprocess the height map.

height in the target low-resolution height map. Thus, rendering the lowest-detail view of the terrain requires that we read the entire high-detail terrain dataset. Clearly this is unworkable for real-time rendering.

Instead, we must preprocess the terrain dataset so that the low-detail representation is available quickly as a contiguous unit.

12.2.1   Height Maps to Mipmaps to Clipmaps

Thinking of a height map as an image, it is natural to address the problem described above by creating a *mipmap* from the image.

Mipmapping is a technique commonly used with textures. A mipmap consists of a number of mip levels, as shown in Table 12.1, each of which is conceptually a separate image. Mip level 0 contains the entire high-resolution image. Mip level 1 has half the texels of level 0 in each direction while still covering the entire extent of the texture. Mip level 2 has half again the texels of level 1. The highest mip level consists of just one pixel.

Many online sources of worldwide terrain and imagery present imagery as a giant mipmap. The Esri World Imagery map service,[2] for example, has 20 mip levels. The highest mip level covers the entire world with a single 256×256 image. The lowest mip level covers select areas with less than a meter between texels. If the entire world were instead covered by a single image with the highest resolution, it would contain quadrillions of pixels.

---

[2]http://server.arcgisonline.com/ArcGIS/rest/services/World_Imagery/MapServer

0	16,384 × 16,384	8	64 × 64
1	8,192 × 8,192	9	32 × 32
2	4,096 × 4,096	10	16 × 16
3	2,048 × 2,048	11	8 × 8
4	1,024 × 1,024	12	4 × 4
5	512 × 512	13	2 × 2
6	256 × 256	14	1 × 1
7	128 × 128		

Table 12.1. The mipmap levels and associated resolutions for an example 16,384×
16,384 texture. Creating a mipmap from a terrain height map allows faster access
to low-resolution representations of the height map.

Consider rendering the 16,384×16,384 mipmapped texture in Table 12.1
to a full-screen quad on a 1,024×1,024 pixel display. In this case, mip level 4
(1,024 × 1,024) will be selected. As we zoom out, higher mip levels will be
selected, consisting of fewer texels from the original texture. As we zoom
in, a lower mip level will be selected (consisting of more texels), but only
a subset of it will be visible.

So far, this is just a simplified explanation of mipmapping. However,
there is an important insight in this that led to the development of a technique called clipmapping: the visible subset of the lower mip levels is limited by the resolution of the display. In fact, in our example, no more than
2,048 texels, or twice the resolution of the display, will ever be required
from any given mip level. For this reason, it is not necessary to keep the
entire 16,384 × 16,384 texture in video memory. We can simply compute
the region of each mip level that can possibly be selected for a given scene,
and "clip" the mipmap to that region.

Clipmapping is just such a method of clipping a mipmap to the subset that is needed to render the current scene [164]. A clipmap can be
graphically depicted as a stack of levels forming an inverted pyramid, as in
Figure 12.9. The topmost levels in the stack—the most detailed levels—are
clipped to the same number of texels, which is determined from the display
resolution.

In contrast with the usual convention for mip levels, clipmap levels are
numbered from 0, the coarsest, least-detailed level, to $L-1$, the finest, most-
detailed level. This convention is convenient for virtual globes because the
number of clipmap levels often varies from region to region.

Virtual globes such as Google Earth use techniques like clipmapping
to allow you to seamlessly navigate through enormous quantities of imagery [12]. In fact, it is not only unnecessary for the entire worldwide
texture to fit into video memory—which is good because the texture is
several thousands of gigabytes in size—but it also does not need to reside

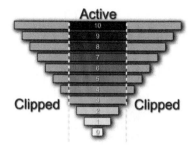

Figure 12.9. A clipmap is graphically depicted as a stack of levels forming an inverted pyramid. The topmost levels in the stack are all clipped to the same number of texels.

in system memory or even on a local disk. Virtual globes stream the clipped mipmap from a web server on demand.

While clipmapping was originally envisioned by Tanner et al. as a texture-mapping technique [164], a similar technique can be applied to terrain elevation data in a height map, which will be discussed in detail in Chapter 13.

12.2.2 Tiling

Mipmapping is a convenient way of preprocessing both height map and imagery data in order to render them at interactive frame rates. The alert reader may have noticed, however, that mipmapping alone does not completely solve the "seek a lot, read a little" problem that was our original motivation for creating mipmaps from the height-map data.

The problem, shown in Figure 12.10, occurs at the most detailed mip levels.

While it might make sense to render the entirety of a low-detail mip level, rendering all of a high-detail mip level is just as impractical as rendering the entire original height map. If each mip level is a single file, we still need to do a lot of seeking in order to read the subset of the level that is necessary to render the current scene.

For that reason, it is common to break each mip level into tiles. A tile has a specific number of texels, such as 256×256, but the area covered by the tile decreases with finer mip levels. For example, the coarsest mip level might cover the entire world $((-180°, -90°)$ to $(180°, 90°)$ in longitude and latitude, respectively) with a single 256×256 tile. The next level has four 256×256 tiles with extents $(-180°, -90°)$ to $(0°, 0°)$, $(0°, -90°)$ to $(180°, 0°)$, $(-180°, 0°)$ to $(0°, 90°)$, and $(0°, 0°)$ to $(180°, 90°)$. Each successive level doubles the number of tiles in each direction and halves the extent

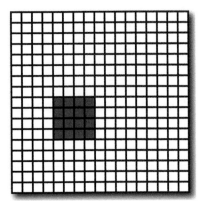

Figure 12.10. If an entire mip level is stored in a single file, rendering a subset of the mip level still requires scanning past clipped texels (white) to read the unclipped ones (red). This problem can be addressed by separating mip levels into tiles.

covered by each tile, as shown in Figure 12.11. Any given tile can be found quickly given its file name, well-known offset in a file, or, for the common case where the terrain or imagery is hosted on a server, its URL.

NASA, Esri, and other organizations arrange their terrain and imagery data in this manner and offer their enormous datasets through publicly accessible web servers. Though they differ in certain details, such as the size of each tile, the number of mip levels, the resolution of data available, and the scheme used for locating individual tiles, it's generally possible to adapt a given virtual globe rendering engine to render any of these datasets.

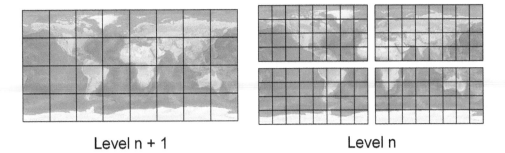

Figure 12.11. A mipmapped image is broken into tiles, all of which have the same number of texels. Each successive mip level has twice the number of tiles in each direction, and each tile covers one-quarter of the area.

Of course, there is some efficiency to be gained if the tiling scheme closely aligns with the rendering engine's expectations.

12.2.3 Mesh Simplification

We consider mipmapping and tiling to be the minimum amount of preprocessing necessary to render planet-sized height maps at interactive frame rates. However, some terrain-rendering algorithms require or benefit from additional preprocessing.

Mesh simplification is used to eliminate redundancy in an input terrain. For example, a height map for a flat region, such as the Great Plains of the United States, will have many more vertices than are necessary to faithfully represent the terrain.

There are many techniques and algorithms that can be used to simplify meshes, including *vertex decimation* [150], *quadric error metrics* [55], and *vertex clustering* [146]. These algorithms vary in their performance, their complexity, and the visual fidelity of the resulting mesh. In addition, some algorithms are only applicable to certain types of models. Height-map-based terrain, however, with its manifold topology, can be simplified with any of these simplification algorithms. These algorithms and more are covered by Luebke et al. [107].

In Section 14.5, we describe in detail one effective technique for simplifying a height-map terrain for use with the chunked LOD terrain-rendering algorithm.

12.3 Out-of-Core Rendering

Virtual globes invariably render terrain using out-of-core (OOC) algorithms. This means that the algorithms keep a subset of the terrain dataset in memory at any given time, while the rest of the dataset resides in secondary storage, such as a local hard disk or a cluster of network servers. Considering that virtual globe terrain datasets are often measured in terabytes, it comes as no surprise that the entire dataset cannot be in memory all at once.

Even with a relatively small terrain dataset, or on a machine with an unusually generous quantity of RAM, loading an entire terrain dataset into memory is impractical. An application that can't render its first scene until multiple gigabytes of data are loaded from a hard disk will have a long start up time. While a virtual globe might be a useful tool to visualize the progression of geologic time, it's best if we aren't tempted to use geologic time scales to measure its start up time.

The subset of data in memory at any given time is referred to as the *working set*, and in many cases, it is a very small fraction of the total dataset. The goal of OOC rendering is to seamlessly bring new data into the working set as they are needed.

To that end, we need a policy for deciding which items to bring into the working set and in what order, which we call a *load-ordering policy*. Because the working set is not infinite in size, we eventually need to decide which items to remove from the working set in order to make room for new items; this is known as a *replacement policy*. Ideally, we also include an effective *prefetching* strategy so that items are brought into the working set just before the user notices that they're missing.

OOC rendering is not restricted to bringing data into system memory, however. In practice, the OOC data management system is responsible for moving data within a hierarchy of caches.

12.3.1   Cache Hierarchies

Effective OOC rendering of massive-terrain datasets requires a hierarchy of caches. A cache hierarchy consists of multiple types of memory along a continuum. At one end of the continuum is small, fast memory. At the other end, memory is much larger but also much slower. The multiple types of memory are used together to create the illusion of large, fast memory by taking advantage of coherence between memory accesses.

Consider an example cache architecture for a virtual globe application, as shown in Figure 12.12.

At the lowest level, all terrain and imagery data are stored on a network server, or perhaps a cluster of servers. The data are measured in tens or hundreds of terabytes, so accessing the entire dataset is an extremely time-consuming process. In addition, network latency and the need to serve multiple users simultaneously severely limits the speed at which the client application can retrieve data from the server.

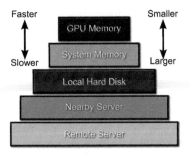

Figure 12.12. Virtual globe applications store terrain and imagery data in a hierarchy of caches. Caches at the bottom are the largest and slowest, while caches at the top are fast and relatively small.

Perhaps this network server is far away or can only be accessed over a slow network. In that case, it's useful to include another server in the hierarchy of caches. This server is accessible over a faster network, perhaps a LAN, and serves fewer users simultaneously. It caches a large subset of the data on the primary network server, but not all of it.

Next, the data are cached on the local hard disk of the computer that is running the virtual globe client application. This cache is rather large, perhaps tens of gigabytes, but much smaller than the tens or hundreds of terabytes of data stored on the network servers.

A subset of the data available on disk is stored in the process memory of the virtual globe client application. This level of the cache hierarchy might be measured in hundreds of megabytes or less, a substantial reduction again from the amount of data stored on disk.

Finally, the smallest subset of the dataset is stored in GPU memory, where it is available extremely quickly for rendering a frame.

A cache miss at any of these levels necessitates going to the next-lower cache in the hierarchy. As we go down the hierarchy, the cost of a miss gets higher, and the time until we actually have the data we need increases. For example, if a chunk of terrain data exists in system memory but is not yet in GPU memory, we can realistically copy it to the GPU within the current render frame. Of course, we still don't want to do that too often, though, or our frame rate will plummet.

On the other hand, if we need to go all the way to a remote network server to obtain the chunk of data, we may not actually be able to use it to render for several seconds. In the meantime, we must continue rendering using the best data we have available. This is an important point. In conventional computer architecture, a cache miss may result in a stall; we sit there and wait until the data are available. Although modern CPU architectures can frequently turn a stall into a context switch and allow the CPU to continue doing useful work while waiting for the data, the instruction using that data cannot execute until the data are returned. In terrain rendering, however, a cache miss does not result in a stall. Instead, it results in decreased visual quality as we use a lower-detail representation or simply omit the missing data.

The best way to decouple cache population from rendering is to use multiple threads. One or more worker threads download a new chunk of terrain data from a network server, load it into memory, and create the GPU buffer, all while the rendering thread renders potentially hundreds of frames. An alternative architecture where the rendering thread tries to do just a bit of this work each frame would be much more complicated.

In addition, a multithreaded architecture enables effective use of today's multicore CPUs. This is especially important if the worker thread does more than just shuffle data around. For example, it might decompress data from its stored format, recompress it for the GPU, compute normals, etc. Job systems take this to an extreme and use several threads executing a large number of tasks related via a dependency graph [6].

Chapter 10 describes how a multithreaded architecture can be used to prepare resources, and it is just as applicable to terrain as it is to vector data.

12.3.2 Load-Ordering Policies

In the event of a cache miss during rendering, the item goes to the *request queue*. Perhaps several such misses occur while rendering a particular frame, and in the next frame, the viewer has moved, so new cache misses occur. How do we decide in what order to load the various items that have been requested?

Loading the cache items in the order they're requested is probably not a great strategy. When the viewer is moving quickly and generating a lot of cache misses, the worker thread is simply unable to keep up. By the time an item is loaded, the viewer may have moved such that the item is not even visible anymore.

A better strategy is to first load the item that was requested most recently and work backward in time from there. Typically, "most recently" means the most recent render frame. This way, the items that were needed for rendering in the last frame are loaded first, which makes sense because those same items are likely to be needed again in the next frame.

In some ways, the request queue, shown in Figure 12.13, acts more like a stack than a queue. A new item is pushed onto the head as a result of a cache miss, and the loading thread pops an item off the head to select the next item to load. One difference, however, is that the request queue does not operate in a strictly last-in, first-out (LIFO) order. Instead, existing

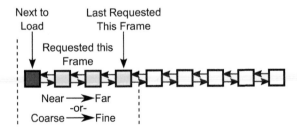

Figure 12.13. The request queue is a priority queue where the priority is determined by the frame number in which the item was requested and optionally by another metric, such as the distance of the viewer to the item. In many cases, the request queue can be implemented as a doubly linked list.

items in the queue are moved to the head each time they are requested by the rendering thread. In fact, the request queue is a *priority queue* in which the priority is determined by the frame number in which the item was last requested.

The priority of a request is determined by more than just the frame number in which it was last requested, however. For example, one useful intraframe ordering gives the items closest to the viewer higher priority in order to maximize the detail near the viewer. Alternatively, the items with the largest world-space extent, which are usually the items with the coarsest detail, can be loaded first in order to maximize the average detail across the scene.

Priority queues are usually implemented using tree-like data structures that provide $O(\log n)$ performance for inserting new items and for removing the highest-priority item. In the case of the request queue, however, we can implement the priority-based queuing using a double-linked list instead, gaining both performance and simplicity in the process. This is possible because the priority in the queue is monotonically increasing. New items always have the highest priority at the time they're added. An existing item that is already in the queue is also bumped to being the highest-priority item if it is requested again.

This property of the request queue makes it easy to implement using a doubly linked list. Newly requested items are added to the head of the list. If a requested item is already in the list, it is removed from its current position in the list and reinserted at the head. The loading thread, for its part, removes its next item to load from the head of the list. All of these operations happen in $O(1)$ time.

This request queue, as designed so far, will load first the items that were requested most recently. What may not be obvious, however, is how it can be used to achieve the intraframe load ordering mentioned previously. Among all the items in a single frame, how can we ensure that the items closest to the viewer are loaded first? Or, how can we ensure that the items with the largest extent are loaded first?

Fortunately, the natural structure of most terrain LOD algorithms makes it relatively easy to sort terrain items by these criteria. In quadtree-based algorithms, for example, it is easy to traverse child nodes in near-to-far order (see Section 12.4.5). Similarly, a breadth-first traversal of a quadtree results in visiting first the nodes with the largest extents.

These traversals result in orders that are exactly opposite to what we require, however. In our simplified priority queue, the priority of an item is determined from the order in which the item was added to the queue. So, in order to give closer items higher priority, we need to request those items last in the render frame. We do have one trick up our sleeves, however. Within a frame, we can request items in the order we'd like them loaded, instead of in the opposite order, by simply keeping track of the *last* item requested in the frame and adding the new item after it. At the start of each render frame, we reset the reference to the *last* so that the first item requested is placed at the head of the list. This is the main reason we use a linked list instead of an array: the linked list allows us to insert items in the middle of the collection in constant time.

It is often useful to limit the number of items that can be waiting in the request queue in order to prevent the queue from growing indefinitely. For example, if the queue contains over 200 items, an old item at the tail of the list is removed each time a new one is added at the head.

12.3.3   Replacement Policies

When a cache is full and we need to unload an existing item in order to make room for a new one, we need to decide which item to unload. This is called a replacement policy because we are selecting which item to replace in the cache.

Replacement policies are important because a cache without a replacement policy is just another name for a memory leak. Data are added to the cache as needed and according to the load-ordering policy. Without a replacement policy, however, data are never removed from the cache, and the size of the cache grows indefinitely.

As you might imagine, the load-ordering policy and the replacement policy are closely related. It makes sense that the item we're loading right now should be last to be unloaded among the items currently loaded. In a particularly egregious violation of this principle, an item might be loaded, only to be immediately replaced with the next item loaded. In the next frame, the first item is loaded again, only to be unloaded again shortly thereafter. This effect, called *ping-ponging* or *cache thrashing*, is a waste of resources and should, of course, be avoided whenever possible.

It's not strictly true, though, that items should be replaced in the order in which they're loaded. Changes in viewer position affect the relative priority of items in the cache. For example, viewer movement to the other side of the Earth makes the cache items on the first side more eligible for replacement.

The canonical replacement policy for any cache is the least-recently used (LRU) replacement policy. The idea is simple and effective: the next item to be replaced is the one that was least recently needed for anything. Using this replacement policy requires us to keep track of the last time each cache item was accessed. Cache-item replacement is controlled by a *replacement queue*, shown in Figure 12.14, which is a restricted priority queue implemented as a linked list, much like the one we used for the request queue.

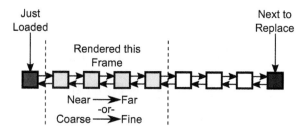

Figure 12.14. Newly loaded items are added to the head of the replacement queue, and the next item to replace is selected from the end. The items that were used to render the most recent frame are located near the head of the list and are sorted in one of two ways.

Once an item is loaded, it is added to the head of the replacement queue. When an item needs to be unloaded in order to make room for a new item, the item to unload is selected from the tail of the replacement queue. Putting newly loaded items at the head of the replacement queue minimizes cache thrashing because only items rendered in subsequent frames will be less eligible for replacement than the item just loaded. Each time a loaded item is used for rendering, it is moved to the head of the replacement queue, which delays its replacement.

It's important, however, that a new item is only loaded if it was last used more recently than the LRU item in the replacement queue. Otherwise, items near the tail of the request queue that were requested a very long time ago can replace much more useful items that were used relatively recently.

Over the course of several frames, the more recently used items migrate toward the head of the list and are less likely to be replaced. The less recently used items lag behind at the end of the list and are more likely to be replaced. If the same traversal strategy is used to update item positions in the replacement queue as is used to update item positions in the request queue, the items near the head of the replacement queue will be sorted near to far or coarse to fine, matching the priority with which the items would be loaded. At the head of the queue are the items that were just loaded since the last render frame.

12.3.4 Prefetching

As the viewer moves, different subsets of terrain and imagery data are required to render the scene most effectively. When required data are not yet available, a lower-resolution version of the terrain is rendered. For the best user experience, we should try to minimize the frequency with which this happens.

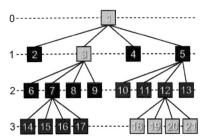

Figure 12.15. In an HLOD algorithm, an effective prefetching strategy is to prefetch the child nodes (red) of the nodes currently being rendered (blue).

In the normal course of events, we notice a missing piece of terrain or imagery data just at the moment we'd like to render it, so we request that it be loaded. At this point, we've already lost, though. The data will not be available for several frames. This is reactive; we notice that data are missing and react by requesting that they be loaded.

Prefetching, on the other hand, is proactive. We try to predict the subset of data that is likely to be needed in the near future and load it before it is needed. If we do this well, and perhaps with a little bit of luck, the data will already be loaded before we actually need it during rendering.

Prefetching for terrain is often dependent on the specific LOD algorithm used, but there are some common themes.

Ideally, the same metric is used for refinement and prefetching. Thus, items that are closer to becoming visible as a result of refinement are more likely to be prefetched. A simple but useful technique is to prefetch the data that are closest to the current position of the viewer, up to the limit of a specified cache budget. This maximizes detail near the viewer. Varadhan and Manocha describe a priority-based prefetching algorithm based on predicting when objects are going to switch LODs [175].

For HLOD algorithms, a simple and effective prefetching strategy is to prefetch the child nodes of nodes currently being rendered. Since the child nodes represent a more detailed representation of the current node, prefetching them is preparing for the eventuality that the viewer moves closer to the node and thus more detail is required. This is shown in Figure 12.15.

Similarly, it may be worthwhile to prefetch the parent node of a rendered node. This is optimizing for the case where the user zooms out or moves away from the node. While these coarser data are somewhat less likely to be missed by the user, rendering without it can cause aliasing artifacts and low frame rates because faraway detail is rendered with a high level of

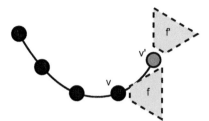

Figure 12.16. In the current frame, the viewer is at position v. The position of the next frame, v', is predicted, and the predicted view frustum, f', is used for prefetching.

detail. Also, the coarser data in parent nodes occupy far less memory than the fine data in child nodes because there are fewer parent nodes.

Another common technique for prefetching is to determine the data items that are needed, and load them based on an inflated view frustum. This is especially useful for LOD algorithms that are dependent on the direction of the viewer in addition to its position. It minimizes the chances of visual artifacts when rotating the camera in place.

A particularly interesting prefetching technique is to extrapolate the viewer's next position from his or her current position and linear and angular speeds and prefetch the data that will be needed for the new view. In this scheme, shown in Figure 12.16, a background thread analyzes the predicted next view frustum and issues requests to prefetch the necessary data. Correa et al. describe such an approach [29].

Depending on the speed of the viewer, the quantity of memory dedicated to prefetched data, and the accuracy with which the viewer's motion can be predicted, prefetching may completely eliminate cache misses during rendering. More commonly, however, especially in virtual globes, the maximum speed of viewer motion significantly exceeds the rate at which new data can be loaded. The goal is to improve the user experience, particularly when navigating within a relatively small area.

12.3.5 Compression

Compression reduces the size of geometry and texture data and is valuable at all levels of the cache hierarchy. Smaller data are retrieved from a remote server more quickly, they're faster to read from disk, and they take less time to send to the GPU over the system bus. Size is speed, especially when it comes to I/O. In addition, compressed data take less space in system and GPU memory, effectively allowing more data to fit in the cache or leaving more memory available for other purposes.

Instead of reducing memory usage and increasing performance, compression can also be used to increase visual quality within the same performance and storage budget. For example, a 512×512 DXT compressed RGBA texture consumes the same amount of memory as an uncompressed 256×256 RGBA texture and provides better image quality. DXT decompression is implemented in hardware on modern GPUs.

One simple form of compression is to store information implicitly rather than explicitly. A great example of this is the representation of terrain as a height map, as described in Section 11.1.1. A height map can be thought of as a compressed mesh where heights are stored explicitly and the horizontal coordinates are implicitly defined by the position of the height in the map. As a result, a height map occupies approximately one-third of the space required for an equivalent mesh. Of course, height maps are not appropriate for representing arbitrary meshes.

A related technique is mesh simplification, described in Section 12.2.3. Mesh simplification can be thought of as a form of compression because it reduces the number of vertices in a mesh without overly impacting its appearance. In fact, mesh simplification is a form of lossy compression because the original mesh cannot be reconstructed perfectly from its simplified representation. However, the simplified mesh is close enough to the original, and the space savings are substantial enough, that the lossiness is considered acceptable.

These two techniques compress by eliminating unnecessary information, for example, the horizontal coordinates in a height map or unimportant vertices in a simplified mesh. Most well-known compression algorithms, however, instead operate by eliminating redundancy or efficiently encoding patterns in the data. Such algorithms include everything from the lossless Deflate algorithm commonly used to compress ZIP and PNG files to lossy algorithms such as DXT and JPEG. Recent GPUs make it now possible to send compressed geometry to the GPU and decompress it in a geometry shader [101].

Section 10.3.3 describes an architecture for multithreaded decompression and recompression of textures and other resources.

12.4   Culling

In many scenes, only a small portion of the triangles composing the scene are actually visible. The rest are invisible, either because they are hidden behind other triangles or because they are outside of the field of view. For example, when looking at a location in Europe, it's unnecessary to render the Rocky Mountains in the western United States. Also, when zoomed in close to a mountain, it is unnecessary to render the foothills hidden behind it. Culling reduces the amount of detail that needs to be rendered by eliminating these triangles that don't contribute to the scene.

12.4.1 Back-Face Culling

Perhaps the simplest form of culling is back-face culling, in which triangles that are facing away from the viewer are not rasterized (see Figure 12.17). Back-face culling can be used when the viewer is outside of a closed opaque object, such as a cube, without resulting in missing triangles. If a viewer were to enter the cube, for example, back-face culling would make the cube disappear; instead, front-face culling should be used when the viewer enters.

In virtual globes, back-face culling is also used for polygons rendered on the globe (see Section 8.2.6) and terrain. Missing back-facing triangles will be evident if a viewer were to go underneath terrain, but most virtual globes use collision detection to avoid this.

Back-face culling is simple to implement: a call to `glEnable` and `glCullFace` in OpenGL is all that is required to configure the rasterization pipeline for back-face culling. The only other requirement is that the vertices making up each triangle be specified in a consistent order, clockwise or counter-clockwise, so that the rasterization pipeline can identify which side of a triangle is considered to be the front.

Back-face culling happens late in the rasterization pipeline, however, so its benefits are limited. It is most valuable in applications like games with expensive fragment shaders. In applications that typically use simple fragment shaders, like virtual globes, the performance benefits are modest.

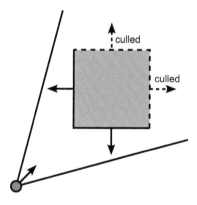

Figure 12.17. A top-down view showing two of four faces culled by back-face culling. Back-face culling is simple to implement and offers modest performance benefits.

12.4.2   View-Frustum Culling

View-frustum culling is a simple and effective culling technique that is used in virtually all terrain-rendering algorithms. Before rendering an object, the object is tested against the six planes of the view frustum. The test may indicate that the object is entirely inside or outside the frustum or that it intersects with one of the planes. If the object is inside or intersects, it is rendered; otherwise, it is discarded. Since the test happens on the CPU, the object is not processed at all by the GPU if it is not visible.

The object's true geometry is almost never tested against the view frustum. Instead, the test is done much more efficiently by fitting a bounding volume, such as a bounding sphere or AABB, around the object and testing that against the view frustum.

In terrain rendering, the number of objects to test against the view frustum can quickly become prohibitive. For example, a planet-sized terrain divided into average-sized chunks can result in millions of chunks to test against the view frustum. To optimize view-frustum culling, objects are organized into spatial data structures such as quadtrees. The bounding volumes of nodes in the spatial data structure are organized such that, if a node is not visible, neither are its children. This allows large parts of the scene to be culled quickly, minimizing CPU overhead. Furthermore, if a node is completely within the view frustum, so are its children, so the children do not need to be checked against the view frustum.

View-frustum culling is of limited value for terrain rendering when the viewer is far away from the globe and looking toward it. When the view encompasses the entire globe, the entire terrain is inside the frustum and nothing is culled. Level of detail, however, becomes important; at such an altitude, few, if any, terrain details are visible. When zoomed in, however, view-frustum culling is much more effective, potentially eliminating a large percentage of the terrain chunks that would otherwise be rendered.

12.4.3   Horizon Culling

Despite the effectiveness of view-frustum culling, objects inside the view frustum are not necessarily visible. In particular, objects are not visible if they're hidden behind other objects. Culling out objects that are occluded by other objects is known as *occlusion culling*.

When it comes to rendering terrain on a globe, one very big and important occluder is worth special consideration: the Earth itself. We'd like a way to quickly determine that a region of terrain is below the horizon, as shown in Figure 12.18, so that we don't waste any time rendering it. When the viewer is far away from the planet, horizon culling obviates the need to render about half of the planet. When the viewer is close to the

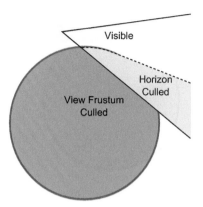

Figure 12.18. Horizon culling eliminates objects that are invisible because they are below the horizon of a globe's surface.

planet's surface, horizon culling really shines, potentially eliminating even more of the planet from consideration. While back-face culling eliminates the fragments on the back side of the planet, horizon culling allows us to eliminate the geometry as well.

Ohlarik presents a fast way to perform horizon culling given a bounding sphere for the potentially occluded object [125]. In Figure 12.19, the sphere centered at o is the bounding sphere of an object, perhaps a terrain tile, that is potentially occluded. The sphere centered at e is the object doing the occluding, such as the Earth. The point v is the position of the viewer.

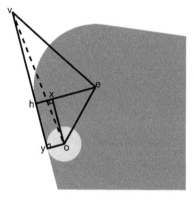

Figure 12.19. Occlusion of a smaller sphere by a larger one can be determined as a function of the distances between the spheres, their radii, and the distance from each to the viewer.

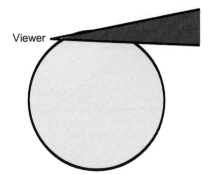

Figure 12.20. An occlusion sphere (yellow) that is larger than the Earth occludes the red region, but the Earth itself does not. Instead, the occlusion sphere must fit inside the Earth.

Anything in the tan region is occluded by the Earth from the viewpoint of the viewer.

If we were to wrap the Earth in a bounding sphere, we would use a sphere that completely encloses the Earth. In this case, though, we're using the sphere for occlusion, so we must use a sphere that fits entirely within the Earth's actual shape. If the occlusion sphere poked through the actual Earth surface anywhere, the portion above the surface could falsely occlude objects, as shown in Figure 12.20.

In Figure 12.19, the object o is invisible, but it is just on the edge of being visible. If it were to rotate clockwise around e, while remaining the same distance from it, it would move closer to v and also become visible. Given the distances between the two spheres; their radii, r_e and r_o; and the distance from the viewer to the center of the Earth, we can compute the smallest distance from the viewer to the object, $\|\mathbf{vo}\| = \|v - o\|$, for which the object is occluded by the Earth. If the distance to the viewer is smaller, at least part of the bounding sphere is not occluded.

First, observe that by the Pythagorean theorem

$$\|\mathbf{vo}\|^2 = (\|\mathbf{vh}\| + \|\mathbf{hy}\|)^2 + r_o^2. \tag{12.2}$$

Also, since $\mathbf{vh}$ and $\mathbf{eh}$ are perpendicular at point h, the Pythagorean theorem can be used again to compute $\|\mathbf{vh}\|$ and $\|\mathbf{hy}\|$:

$$\|\mathbf{vh}\| = \sqrt{\|\mathbf{ve}\|^2 - r_e^2},$$

$$\|\mathbf{hy}\| = \sqrt{\|\mathbf{eo}\|^2 - (r_e - r_o)^2}.$$

Note that $\|\mathbf{vh}\|$ is only a function of Earth, not of the object that is being tested for occlusion by Earth. Thus, when testing many objects for occlusion, we can compute that quantity once and reuse it for all objects. Substituting $\|\mathbf{hy}\|$ into Equation (12.2), we get

$$\|\mathbf{vo}\|^2 = \left(\|\mathbf{vh}\| + \sqrt{\|\mathbf{eo}\|^2 - (r_e - r_o)^2} \right)^2 + r_o^2.$$

If $\|\mathbf{vo}\|^2$ is less than the square of the actual distance between the viewer and the center of the object's bounding sphere, the object is occluded. Otherwise, it is at least partially visible.

For terrain tiles, a bounding sphere that encompasses all of the vertices in the tile is not a very tight fit, which results in the horizon test determining that the tile is visible even when it is not. Ohlarik also presents a useful technique for choosing a very small bounding sphere—a single point, in fact—that can be used with this horizon-culling algorithm to cull terrain tiles more accurately [128].

Many virtual globe engines perform horizon and view-frustum culling hierarchically. Terrain chunks and other objects are arranged in a hierarchy of bounding volumes so that entire subtrees of objects can be tested for culling quickly. If a parent bounding volume is entirely below the horizon or outside the view frustum, all of its children are as well, so they do not need to be tested individually. If the parent bounding volume is entirely above the horizon and inside the view frustum, so are its children. Only if a parent bounding volume lies across the horizon or view-frustum boundary do its children need to be tested.

12.4.4   Hardware Occlusion Queries

When rendering terrain in a scene where the viewer is near the terrain surface and looking out toward the horizon, nearby terrain features are likely to occlude more distant terrain features. In an extreme case, imagine a viewer near the floor of the Grand Canyon. Terrain beyond the canyon walls is invisible to the viewer because it is occluded by the canyon walls themselves. Similarly, buildings or a dense forest might occlude much of the terrain in a scene.

Ideally, we would like to be able to take this occlusion into account in order to avoid rendering the invisible terrain. With large numbers of irregularly shaped occluders like trees and terrain features, however, determining occlusion is an extremely CPU-intensive process. Most applications that perform occlusion culling on the CPU make simplifying assumptions about the scene and the occluders. For example, in a city scene, we might assume that all buildings are rectangles extruded perpendicular to the ground. Horizon culling is a useful technique for the special case where the occludee is a large ellipsoid-shaped planet like Earth.

Figure 12.21. Neither of the two rectangles in this scene occlude the red circle, but the combination of both of them makes the red circle invisible to the viewer. This is called occluder fusion.

The problem is especially difficult when a region of terrain is only occluded by the fusion of multiple occluders, as shown in Figure 12.21. One tree, for example, is unlikely to occlude any reasonably sized terrain region, but an entire forest is another matter altogether. For general scenes like these, and in particular for the case of terrain occluding other terrain, the best approach to occlusion culling is to use the GPU to determine which objects are occluded.

When we render a scene, many of the fragments resulting from our rasterized triangles are discarded because they are clipped; that is, the triangles fall outside the view frustum. Other fragments are discarded because they fail a depth or stencil test or are explicitly discarded by the fragment shader. The rest pass all the way through the rasterization pipeline and modify the framebuffer. Hardware occlusion queries (HOQs) enable us to ask the GPU how many fragments were written to the framebuffer over the course of one or more draw calls. If no fragments were written while rendering an object, that object is occluded and does not need to be drawn in subsequent frames from the same viewpoint.

HOQs use the rasterization power of the GPU to determine if an object is visible. Typically, HOQs are performed by testing a bounding volume for occlusion rather than testing a detailed version of the object because an object's bounding volume usually has much less geometry than the object. In addition, HOQs are performed with color and depth writes disabled, allowing today's GPUs to use a higher-performance rendering path.

Because HOQs involve a roundtrip from the CPU, through the entire GPU pipeline, and back to the CPU, their naïve use creates two problems: *CPU stalls* and *GPU starvation*. If the CPU issues a query and then immediately waits for the result, the CPU is stalled; it does no useful work during the potentially lengthy period that the GPU is executing the query.

Figure 12.22. Improper use of HOQs stalls the CPU as it waits on the result of a query. The stalled CPU cannot issue commands to the GPU, which starves the GPU.

While the CPU is stalled, it does not issue new commands to the GPU, and thus starves the GPU. This is shown in Figure 12.22.

> It can be useful to perform HOQs using multiple bounding volumes for an object instead of just one. This improves the tightness of the fit of the bounding volume around the object so it is more likely to be found to be occluded, while only minimally increasing the vertex costs.

○○○○ Kevin Says

Effective occlusion algorithms based on HOQs exploit temporal coherence between frames to improve the parallelism between the CPU and GPU [153]. The basic idea is to issue a query in one frame but not check the result of the query until a later frame. Items that were visible in the earlier frame are assumed to be visible in the later one as well.

In addition to temporal coherence, effective occlusion algorithms take advantage of the spatial coherence of visibility within a frame [181]. A spatial data structure such as a quadtree, octree, or kd-tree is used to cull large occluded segments of a scene with minimal overhead.

HOQs are useful in scenes with high depth complexity, that is, scenes where many fragments compete for the same pixel. In urban walkthroughs, for example, large buildings near the viewer occlude most of the scene, so only a small subset of objects actually needs to be rendered. For terrain rendering, HOQs are potentially valuable with ground-level views. When the viewer position is high above the terrain, however, the terrain's depth complexity is low and occlusion culling offers little benefit.

12.4.5   Rendering Front to Back

Another simple culling technique is to render the scene from front to back; triangles closest to the viewer are rendered first, and triangles rendered thereafter are increasingly far from the viewer.

This functions as occlusion culling because of the characteristics of the GPU's depth buffer. Only fragments are culled, however, not triangles. A fragment that fails the depth test is not written to the framebuffer. By rendering front to back, more fragments fail the depth test, so less memory bandwidth is spent writing to the framebuffer.

More importantly, however, today's GPUs often do the depth test *before* invoking the fragment shader, an optimization called *early-z* [123,136]. In this case, the fragment shader is not invoked if the fragment fails the depth test, which is desirable because there is no point in shading a fragment that never becomes a pixel. When complex fragment shaders are used, this offers substantial performance improvements.

Early-z is a fine-grained per-fragment test. Today's GPUs also implement *z-cull*, also called *hierarchical z*, which is a coarse-grained check that quickly tests a tile (e.g., an 8×8 block of fragments). These optimizations are enabled by default, but can be disabled in certain circumstances, most notably when a fragment shader outputs depth or uses discard as done for GPU ray casting in Section 4.3.

Obviously, to get the most out of early-z and z-cull, we should use these fragment-shader features sparingly. We can take it a step further: the more fragments that fail the depth test, the more effective these optimizations become. In many scenes, several fragments with the same screen-space location fight to win the depth test, and only one fragment becomes a pixel. The number of fragments per pixel is called the *depth complexity*. Since one fragment occludes the other fragments for a particular pixel (ignoring translucency), we must render triangles yielding the front-most fragments first so later fragments are not shaded or written to the framebuffer.

This is done by rendering in ascending order based on the distance to the viewer. Sorting individual triangles on the CPU is typically impractical, and precisely sorting entire objects may even be too expensive. Instead, we must balance the amount of time spent on the CPU optimizing for the GPU. A bucket sort can provide a coarse front-to-back ordering. Certain data structures also lend themselves to efficient front-to-back traversal.

Approximate front-to-back sorting is straightforward using quadtrees and octrees (see Figure 12.23). A node's children do not need to be explicitly sorted. Instead, the traversal order can be looked up in a table based on the viewer's position only; orientation isn't even needed. For a quadtree, only four unique child-traversal orders are required for front-to-back sorting, as shown in Table 12.2. The quadrant of the viewer relative

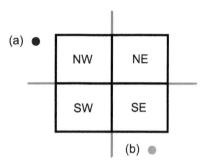

Figure 12.23. The front-to-back traversal order of a quadtree is easily determined from the quadrant of the viewer. For viewer location (a), the traversal is NW, SW or NE, SW or NE, SE. For (b), it is SE, SW or NE, SW or NE, NW.

to the node is used to look up the traversal order. First, the child in the same quadrant as the viewer is rendered, followed by the two children adjacent to that child, and finally the child adjacent to the two children. The order of the second and third children does not matter; neither occludes the other, but they both potentially occlude the final child.

Another technique for taking advantage of early-z and z-cull is to render the scene in two passes. In the first pass, the scene is rendered to the depth buffer only (see Figure 12.24(a)). This pass is efficient because very simple fragment shaders can be used since no shading is required. In addition, GPUs also support an optimization called *double speed z-only* that improves rendering performance when color writes are disabled. This depth pass can also benefit from a coarse front-to-back sorting to reduce the number of depth buffer writes.

The second pass renders the entire scene again to perform the actual shading (see Figure 12.24(b)). This pass does not need to write to the depth

Viewer Quadrant	First	Second	Third	Fourth
Southwest	Southwest	Northwest or Southeast	Northwest or Southeast	Northeast
Southeast	Southeast	Southwest or Northeast	Southwest or Northeast	Northwest
Northwest	Northwest	Southwest or Northeast	Southwest or Northeast	Southeast
Northeast	Northeast	Northwest Southeast	Northwest Southeast	Southwest

Table 12.2. A quadtree can be rendered in front-to-back order by choosing one of four traversal orders based on the quadrant of the viewer.

(a)   (b)

Figure 12.24. Rendering a (a) depth-only pass before (b) shading takes advantage of GPU depth buffer optimizations and effectively reduces the shading depth complexity to one.

buffer. Instead it reads from the depth buffer using a less than or equal to depth test. Only one fragment per pixel is shaded, effectively reducing the depth complexity to one. This shading pass can also be sorted by renderer state, reducing the tension between sorting by distance and sorting by state (see Section 3.3.6).

The downside is the scene's geometry is transformed twice. This can be mitigated by only rendering major occluders during the initial depth pass. Deferred shading is another rendering technique, which has become quite popular, that shades only nonoccluded fragments [142].

Early-z and z-cull, and occlusion culling in general, are particularly effective for horizon views like that of Figure 12.24(b) because the depth complexity is high. Other techniques can be used to reduce the depth complexity, such as defining a maximum view distance and using fog to gracefully fade out rendering over an interval at this distance. For some applications this is appropriate, but for many virtual globe use cases (e.g., simulations), users want the full view distance.

12.5   Resources

The Virtual Terrain Project[3] is an excellent resource for the aspiring (or even experienced) terrain-engine author. In particular, the "LOD Papers" page has an extensive catalog of published terrain LOD algorithms.

Level of Detail for 3D Graphics has thorough coverage of a number of topics relevant to massive-terrain rendering, including mesh simplification, error metrics, continuous LOD, and the perceptual aspects of LOD [107]. Because of advances in GPU hardware, however, the specific advice for terrain rendering is not as useful as it was when the book was originally

---

[3]http://www.vterrain.org/

published. Also, *Real-Time Rendering* has an extensive survey of culling techniques [3].

Erikson et al. cover many of the important techniques for rendering scenes using hierarchical LOD [48]. Gobbetti et al. survey useful techniques for visualizing massive models like terrain, including culling, cache-coherent layouts, data management, and simplification [63].

Many authors cover out-of-core rendering and prefetching, including Correa et al. and Varadhan and Manocha [29,175]. Szofran offers a fascinating look at how terrain is managed in Microsoft Flight Simulator [163].

The blogs of folks working on virtual globes and terrain rendering are excellent sources of terrain-rendering tidbits, as well as inspiring screen shots. Some of our favorites include Outerra[4] and X-Plane.[5] Visit our website at http://www.virtualglobebook.com for more.

---

[4]http://outerra.blogspot.com/
[5]http://www.x-plane.com/blog/

$\bigcirc\bigcirc\bigcirc\circ$ 13

Geometry Clipmapping

Geometry clipmapping is a terrain LOD technique in which a series of nested, regular grids of terrain geometry are cached on the GPU. Each of the grids is centered around the viewer and is incrementally updated with new data as the viewer moves. The grid levels form concentric squares, as shown in Figure 13.1.

Geometry clipmapping renders rasterized elevation data in the form of a height map. Prior to rendering, the height map is prefiltered into a mipmap pyramid, as described in Section 12.2. Each concentric square, called a clipmap level, corresponds to a level in the mipmap. Clipmap levels each have a texture in GPU memory that holds the subset of posts in the corresponding mip level that is closest to the viewer.

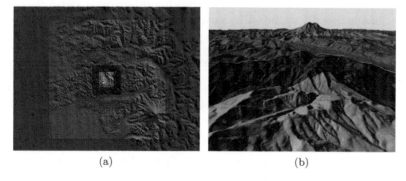

(a) (b)

Figure 13.1. (a) A geometry clipmap is rendered as a set of nested rings centered around the viewer. Each successive ring radiating away from the viewer has half the resolution of the ring before it while covering four times the area. (b) The clipmap from the viewer's perspective.

403

The innermost clipmap level closest to the viewer corresponds to the most detailed mip level of the height map, and successive levels radiating away from the viewer are increasingly less-detailed mip levels. Each clipmap level is square, and all levels have exactly the same number of posts, while covering four times the area of the next-finer level. All of the levels, except the innermost one, are rendered as hollow rings encircling the next, more-detailed level. At the finest level, the entire grid is rendered.

The clipmap levels are aligned with world space, meaning that the horizontal direction in the grid always corresponds to the x direction in world space, and the vertical direction in the grid corresponds to the y direction in world space.

Each level is rendered using the vertex-shader displacement-mapping technique described in Section 11.2.2. The grid (x, y) coordinate is input to the vertex shader, and the height at the grid point is read from the level's height-map texture using vertex texture fetch (VTF). The vertices in the grids are precisely aligned with posts in the associated mip level of the height map, so interpolation between texels is not required.

As the viewer moves, the clipmap levels are updated such that the clipmap pyramid remains centered on the viewer. Two-dimensional wraparound texture addressing, also known as *toroidal addressing*, is used to eliminate the need to rewrite the entire height-map texture each time the viewer moves. Instead, only the newly visible posts are copied into the clipmap.

Geometry clipmapping has some nice features:

- *Simplicity.* The LOD and rendering algorithm are straightforward to implement.

- *Effective use of the GPU.* The algorithm makes good use of the GPU and is friendly to the CPU, leaving it largely available for other tasks.

- *Visual continuity.* Transitions between levels of detail are seamless and are easily implemented in shader programs.

- *Consistent rendering rate.* The number of triangles used to render a region is independent of the roughness or other characteristics of the terrain. Thus, the frame rate remains relatively steady.

- *Graceful degradation.* The rendering load can be reduced, when necessary, by reducing the size of each clipmap level or by disabling finer clipmap levels for a fast-moving viewer.

- *Minimal preprocessing.* Terrain is efficiently rendered from a simple mipmapped height map. Expensive or difficult to implement preprocessing steps are not required.

- *Compression and synthesis.* The hierarchy of nested, regular grids naturally lends itself to efficient compression and synthesis of terrain data.

On the other hand, geometry clipmapping has a few disadvantages:

- *More triangles.* Geometry clipmapping uses more triangles to attain a similar visual quality than do other terrain algorithms, such as the chunked LOD algorithm described in Chapter 14. This is because of the use of completely regular grids. Other terrain algorithms use irregular meshes that can reduce detail in flat areas of the terrain while maintaining high resolution in the bumpy sections. Effectively, geometry clipmapping assumes a worst-case terrain that is highly detailed throughout and optimizes for that case.

- *Modern GPU required.* Efficient implementation of geometry clipmapping requires a GPU that can quickly sample a texture from the vertex shader. While mainstream GPUs have had this capability for several years now, it may be a concern if older hardware must be supported.

- *Loose screen-space error guarantees.* Geometry clipmapping aims to produce triangles that are approximately the same size in screen space at each level. However, if the terrain has steep slopes, the triangles can be stretched vertically to arbitrary size. Thus, the error in screen space is a function of the terrain data itself and cannot be directly controlled by the algorithm. For applications that need a very precise representation of the terrain, this may be unacceptable.

- *It's patented.* US patent number 7436405, held by Microsoft, covers some aspects of this technique.

In some sense, geometry clipmapping continues the inevitable march toward GPU-centric terrain rendering. It requires the GPU to process more triangles than prior techniques. In exchange, it reduces CPU time. Perhaps surprisingly, memory usage and bus bandwidth are also reduced because the regular grid structure allows terrain data to be represented much more compactly.

This chapter begins with a detailed explanation of how to implement geometry clipmapping to render terrain extruded from a flat, horizontal plane. Then, we discuss how this technique can be used in a virtual globe application where terrain is extruded instead from an ellipsoid or sphere.

13.1   The Clipmap Pyramid

The entire clipmap pyramid, shown in Figure 13.2, consists of L levels, where L is based primarily on the number of mip levels of terrain data available. In contrast to the usual convention for mip levels, clipmap levels are numbered from 0, the coarsest, least detailed level, to $L-1$, the finest, most detailed level. This convention is convenient for virtual globes because the number of clipmap levels often varies from region to region. The NASA World Wind `mergedElevations` terrain dataset, for example, has 12 mip levels, numbered 0 through 11. It has 10 m resolution terrain data for most of the United States, 900 m resolution for the oceans, and 90 m resolution data for most of the rest of the world. Other World Wind terrain datasets include detail at 1 m resolution for Denmark and 30 m resolution for the rest of the world.

Each of the L clipmap levels has an $n \times n$ height-map texture associated with it. There is some flexibility in the value of n, subject to a couple of constraints.

First, n must be an odd number. As shown in Figure 13.3, this allows the posts in a coarser level to be coincident with those in the next-finer level at the boundary where the two meet. This is vital to ensure that there are no cracking artifacts between clipmap levels.

Second, n should be chosen with some consideration for texture sizes that are efficient for the GPU, and that generally means sizes that are nearly powers of two. The texture size is rounded up to the next power of two, and the extra row and column are unused. Asirvatham and Losasso both use $n = 255$, but values such as 511 and 1,023 are reasonable as well [9, 105]. Clipmap sizes that are not near powers of two can be used as well without impacting the algorithm, and they need not be rounded up to the next power of two, but it may lead to less efficient use of the GPU.

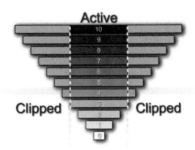

Figure 13.2. A geometry clipmap is graphically depicted as a stack of levels forming an inverted pyramid. The topmost levels in the stack are all clipped to the same number of posts.

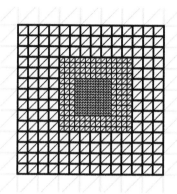

Figure 13.3. Vertices at the boundary between adjacent clipmap levels must be coincident to prevent cracks between the levels. This requires that clipmap levels have an odd number of posts.

The value n can be tweaked to trade rendering quality for performance on slower hardware.

The L levels of the pyramid are arranged in concentric squares, as shown in Figure 13.1. Level 0 covers the largest region of the world. Level 1 covers half of the world extent in each direction, or one-fourth of the area. Level 2 covers half again of the world extent in each direction, and so on. Each level is approximately centered around the viewer, subject to the overriding constraint that the posts of a finer clipmap level must be coincident with the posts of the next coarser-level at the clipmap level boundary.

Given a desired longitude and latitude on which to center the clipmap pyramid, the post indices of the southwest and northeast corners—the extent—of the finest, most detailed clipmap level are computed in Listing 13.1.

```
double centerLongitude = /* ... */;
double centerLatitude = /* ... */;

Level level = _clipmapLevels[_clipmapLevels.Length - 1];
double longitudeIndex =
    level.Terrain.LongitudeToIndex(centerLongitude);
double latitudeIndex =
    level.Terrain.LatitudeToIndex(centerLatitude);

int west = (int)(longitudeIndex - _clipmapPosts / 2);
if ((west % 2) != 0) ++west;
int south = (int)(latitudeIndex - _clipmapPosts / 2);
if ((south % 2) != 0) ++south;

level.NextExtent.West = west;
```

```
level.NextExtent.East = west + _clipmapPosts - 1;
level.NextExtent.South = south;
level.NextExtent.North = south + _clipmapPosts - 1;
```

Listing 13.1. The extent of the finest clipmap level within the terrain level is computed such that it is approximately centered around a given location.

```
int levelOffset = (_clipmapPosts + 1) / 4 - 1;

Level currentLevel = _clipmapLevels[i];
Level finerLevel = _clipmapLevels[i + 1];

level.NextExtent.West =
    finerLevel.NextExtent.West / 2 - levelOffset;
level.OffsetStripOnEast = (level.NextExtent.West % 2) == 0;
if (!level.OffsetStripOnEast) --level.NextExtent.West;
level.NextExtent.East =
    level.NextExtent.West + _clipmapPosts - 1;

level.NextExtent.South =
    finerLevel.NextExtent.South / 2 - levelOffset;
level.OffsetStripOnNorth = (level.NextExtent.South % 2) == 0;
if (!level.OffsetStripOnNorth) --level.NextExtent.South;
level.NextExtent.North =
    level.NextExtent.South + _clipmapPosts - 1;
```

Listing 13.2. The extent of a coarser clipmap level is computed from the extent of the finer level that it encloses.

Then, the origins of the remaining, coarser levels are computed in Listing 13.2.

Since each clipmap level's height-map texture has the exact same $n \times n$ size, and we now know the extent of the level in terms of terrain post indices, we are ready to fill the textures with post data.

Clipmap levels are incrementally filled with post data as the viewer moves, and this process is described in detail in Section 13.5. In addition, normals are incrementally computed from the post data and stored in a normal map for the level. For now, let's assume that the levels are already filled and move on to the matter of how to render them.

13.2   Vertex Buffers

With the height data stored in textures, the only per-vertex data we need to pass to the vertex shader is the (x, y) position of each grid point. Furthermore, these 2D vertex data are, up to a scale and translation, identical between clipmap levels. This is huge—it means that we can use a small number of static vertex and index buffers to render terrain for the entire scene. We create them once, hand them off to the GPU, and never have to update them again. This is a major reason why the geometry-clipmap algorithm makes such efficient use of modern GPU hardware, and why it is so friendly to the CPU.

At a minimum, we need two vertex buffers. One vertex buffer contains vertices for an entire clipmap-level grid and is used for the finest-detail level that is rendered. The other vertex buffer is a ring with a hollow space in the middle for the next-finer level and is used for all clipmap levels except the finest. However, Asirvatham and Hoppe recommend breaking up the vertex buffer into a number of smaller patches, as shown in Figure 13.4 [9]. The major advantage of this approach, which is the one we use in OpenGlobe, is that it allows portions of a level to be culled by testing them against the view frustum. Asirvatham and Hoppe report that this results in a reduction of rendering load by a factor of about two to three. Another advantage is that less total vertex data are required, saving a bit of memory.

> Modify OpenGlobe to test each patch against the view frustum before rendering it. By how much does this improve performance?

○○○○ **Try This**

Each clipmap level is broken up into the following patches:

- *Fill.* Most of the area covered by a clipmap level is filled by instances of a single $m \times m$ vertex buffer, where $m = \frac{(n+1)}{4}$. In the case of a clipmap size of $n = 255$, $m = 64$, so our fill patch has 64×64 vertices. We fill the outer ring of a non-finest clipmap level with 12 instances of this block, shown in yellow in Figure 13.4. For the finest level, we fill the interior of the ring with four additional instances. Adjacent blocks are overlapped at their edges to avoid gaps between blocks.

- *Horizontal and vertical fixups.* This leaves gaps two quads wide at the top and bottom of the ring and two quads high at the left and right of the ring. We fill these using vertex buffers of size $3 \times m$ and $m \times 3$, respectively, shown in blue in Figure 13.4. As before, these ring fixup buffers overlap the fill patches at their edges to avoid gaps. These fixup patches are also used to fill the gaps in the interior of the finest-detail ring.

- *Horizontal and vertical offset strips.* There is a one-quad gap along two inside edges of the outer ring, depending on how the next-finer clipmap level is situated within the ring. We render an offset strip on the south or north and west or east, as appropriate, shown in green

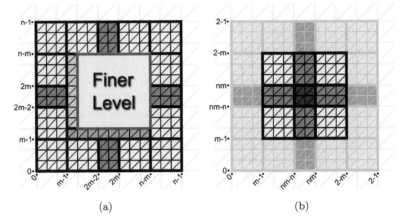

Figure 13.4. (a) All clipmap levels except the finest are rendered using multiple instances of six distinct, static vertex buffers. (b) The finest clipmap level is filled using a different configuration of three of the same static vertex buffers, plus one additional one. The labels on the axes indicate the indices of the posts within the level.

in Figure 13.4. The location of the offset strips is determined by the values of the `OffsetStripOnNorth` and `OffsetStripOnEast` properties that were computed in Listing 13.2.

- *Center.* The very center of the finest clipmap level is filled with a 3×3 patch, shown in red.

- *Degenerate triangles.* Finally, a ring of degenerate triangles is rendered where two different clipmap levels meet, shown as a purple border around the finer level in Figure 13.4. Even though these degenerate triangles technically have no area and thus should produce no fragments, we render them in order to avoid T-junctions in the clipmap mesh. As described in Section 12.1.5, T-junctions are places where an edge shared by two adjacent triangles is divided by an additional vertex when rendering one triangle but not when rendering the other. These T-junctions can show up as very small cracks between clipmap levels, so we fill them with degenerate triangles.

We use OpenGlobe's RectangleTessellator.`Compute` method to easily compute a vertex and index buffer for each of these patches.

Note that this particular breakdown of the clipmap level into patches imposes a requirement on the size, n, of the clipmap levels, in addition to the requirement that it must be an odd number: $n + 1$ must be evenly divisible by four. Otherwise, the patches will not line up properly.

13.3   Vertex and Fragment Shaders

With the vertex buffers for the various patches ready to go, we now turn our attention to the shaders. A basic vertex shader is shown in Listing 13.3.

The only vertex attribute input to the vertex shader is `position`, which is a 2D vertex from one of the patches described above. The `position` is relative to the origin of the patch; in other words, the southwest vertex in the patch has position $(0,0)$ and, for a "fill" patch, the northeast vertex is $(m-1, m-1)$.

```
in vec2 position;

out vec2 fsFineUv;
out vec3 fsPositionToLight;

uniform mat4 og_modelViewPerspectiveMatrix;
uniform vec3 og_sunPosition;
uniform vec2 u_patchOriginInClippedLevel;
uniform vec2 u_levelScaleFactor;
uniform vec2 u_levelZeroWorldScaleFactor;
uniform vec2 u_levelOffsetFromWorldOrigin;
uniform vec2 u_fineTextureOrigin;
uniform float u_heightExaggeration;
uniform float u_oneOverClipmapSize;
uniform sampler2D u_heightMap;

float SampleHeight(vec2 levelPos)
{
  fsFineUv = (levelPos + u_fineTextureOrigin) *
             u_oneOverClipmapSize;
  return texture(u_heightMap, fsFineUv).r *
         u_heightExaggeration;
}

void main()
{
  vec2 levelPos = position + u_patchOriginInClippedLevel;

  float height = SampleHeight(levelPos);
  vec2 worldPos = (levelPos * u_levelScaleFactor *
                  u_levelZeroWorldScaleFactor) +
                  u_levelOffsetFromWorldOrigin;
  vec3 displacedPosition = vec3(worldPos, height);

  fsPositionToLight = og_sunPosition - displacedPosition;

  gl_Position = og_modelViewPerspectiveMatrix *
                vec4(displacedPosition, 1.0);
}
```

Listing 13.3. A simple vertex shader for clipmap terrain rendering.

The first task of the shader is to translate the integer-valued components of the incoming vertex by the origin of the patch, u_patchOriginIn ClippedLevel. This gives us a new set of integer-valued coordinates that are expressed relative to the origin of the clipped level.

Next, the height of the vertex is sampled from the height-map texture. The texture coordinates are computed by adding the uniform u_fineTex tureOrigin to the coordinates of the vertex within the level. Initially, u_fineTextureOrigin is set to $(0.5, 0.5)$ in order to make sure the vertex's height is sampled from the center of the corresponding height-map texel. As the viewer moves, the texture origin moves as well. This will be discussed in more detail in Section 13.5.

We now scale the vertex position up to world space by multiplying by u_levelScaleFactor and then by u_levelZeroWorldScaleFactor; u_lev elScaleFactor is set by the CPU to 2^{-L}, where L is the zero-based index of the clipmap level currently being rendered. At level 0 (the coarsest), u_levelScaleFactor is 1. At level 1 it is 0.5, at level 2 it is 0.25, and so on. This reflects the fact that the distance between posts halves with each successive clipmap level. The object u_levelZeroWorldScaleFactor is the real-world space between posts at the coarsest clipmap level, level 0. Multiplying these two values together gives us the real-world space between posts at the current clipmap level.

Kevin Says ○○○○

Keeping u_levelScaleFactor and u_levelZeroWorldScaleFactor separate, rather than multiplying them together on the CPU and passing one uniform, ensures that coincident posts in adjacent levels actually end up coincident. Small integers like our post indices can be represented perfectly by a float, and they remain perfectly accurate when multiplied by powers of two like u_levelScaleFactor. Posts on the interlevel boundary exist in both levels. In the coarser level, levelPos is half the size it is in the finer level, but u_levelScaleFactor is twice as large. When multiplied with perfect accuracy in each level, the same answer is obtained both times, and u_levelZeroWorldScaleFactor is constant across all levels, so the posts really are coincident. However, the product of u_levelScaleFactor and u_levelZeroWorldScaleFactor is not necessarily representable with perfect accuracy. Errors due to rounding the two different values in the two levels can cause the common posts to not be coincident, creating cracks.

To this product, we add u_levelOffsetFromWorldOrigin, the world coordinates of the southwest corner of the clipped level, to get the 2D world

```
in vec2 fsFineUv;
in vec3 fsPositionToLight;

out vec3 fragmentColor;

uniform vec4 og_diffuseSpecularAmbientShininess;
uniform sampler2D u_normalMap;

vec3 SampleNormal()
{
  return normalize(texture(u_normalMap, fsFineUv).rgb);
}

void main()
{
  vec3 normal = SampleNormal();
  vec3 positionToLight = normalize(fsPositionToLight);

  float diffuse = og_diffuseSpecularAmbientShininess.x *
                  max(dot(positionToLight, normal), 0.0);
  float intensity = diffuse +
                    og_diffuseSpecularAmbientShininess.z;
  fragmentColor = vec3(0.0, intensity, 0.0);
}
```

Listing 13.4. The corresponding fragment shader for clipmap terrain rendering.

coordinates of the vertex. The final, displaced, world position of the vertex is simply this (x, y) position with the sampled height as the z-coordinate.

The only remaining steps for the vertex shader are to compute the vector to the sun for use by lighting calculations in the fragment shader and to transform the displaced position into clip coordinates using the model-view-perspective matrix.

(a)   (b)

Figure 13.5. (a) Without blending, cracks (light blue) are clearly visible between clipmap levels. (b) Blending between clipmap levels eliminates the cracks.

The corresponding fragment shader is shown in Listing 13.4.

This simple fragment shader samples the level's normal map using the same texture coordinates that were used to sample the height map in the vertex shader. It then shades the fragment with a simplified version of the Phong-lighting computation explained in Section 4.2.1.

At this point, we can render nice scenes like the one shown in Figure 13.5(a). The cracks between clipmap levels are a blight on an otherwise picturesque landscape, though, so let's fix them.

13.4   Blending

Cracks occur because the vertices on the boundary between two clipmap levels are displaced twice; once when rendering the finer clipmap level and once when rendering the coarser one. The two heights read from the two different height maps are unlikely to be the same. An elegant solution to this problem is to introduce a "transition region" near the outer boundary of each clipmap level, as shown in Figure 13.6. Within the transition region, the vertex height is a blend of the heights obtained from the two height maps.

Losasso and Hoppe recommend that the width, w, of the transition region be $\frac{n}{10}$ [105]. In other words, vertices one-tenth of the way into the clipmap level will begin blending with the coarser level surrounding it. Specifically, if the distance from the viewer to the vertex along either

Figure 13.6. A transition region around the perimeter of each clipmap level smoothly blends the geometry with the next-coarser level.

dimension is greater than

$$\delta = \frac{n-1}{2} - w - 1,$$

then the vertex is within the transition region. Furthermore, we want vertices at the outer perimeter of this region to be displaced exclusively by the coarser height map. We compute a blend parameter, $\alpha = \max(\alpha_x, \alpha_y)$, where

$$\alpha_x = \text{clamp}\left(\frac{|x - v_x| - \delta}{w}, 0, 1\right), \qquad (13.1)$$

and similarly for α_y. The position (v_x, v_y) is the position of the viewer expressed in the $[0, n-1]$ coordinate system of the clipped level.

We can easily modify the `SampleHeight` vertex-shader function in Listing 13.3 to compute α and blend the two texture samples. The updated function, plus the new required uniforms, are shown in Listing 13.5.

This function obtains the height of the vertex from the current, finer height map just as before. However, it also obtains a height from the next-coarser height map. The texture coordinates in the coarse height

```
out vec2 fsCoarseUv;
out float fsAlpha;

uniform vec2 u_fineLevelOriginInCoarse;
uniform vec2 u_viewPosInClippedLevel;
uniform vec2 u_unblendedRegionSize;
uniform vec2 u_oneOverBlendedRegionSize;
uniform sampler2D u_fineHeightMap;
uniform sampler2D u_coarserHeightMap;

float SampleHeight(vec2 levelPos)
{
  fsFineUv = (levelPos + u_fineTextureOrigin) *
             u_oneOverClipmapSize;
  fsCoarseUv = (levelPos * 0.5 + u_fineLevelOriginInCoarse) *
             u_oneOverClipmapSize;

  vec2 alpha = clamp((abs(levelPos - u_viewPosInClippedLevel) -
                     u_unblendedRegionSize) *
                     u_oneOverBlendedRegionSize, 0, 1);
  fsAlpha = max(alpha.x, alpha.y);

  float fineHeight = texture(u_fineHeightMap, fsFineUv).r;
  float coarseHeight =
     texture(u_coarserHeightMap, fsCoarseUv).r;
  return mix(fineHeight, coarseHeight, fsAlpha) *
         u_heightExaggeration;
}
```

Listing 13.5. The `SampleHeight` vertex-shader function is modified to sample heights from two adjacent levels and blend them.

map are obtained by halving the `levelPos` coordinates and then adding `u_fineLevelOriginInCoarse`, which is the origin of the fine clipmap level expressed in the coordinates of the coarse clipmap level.

The blend parameter `alpha` is computed according to Equation (13.1); `u_viewPosInClippedLevel` is (v_x, v_y), `u_unblendedRegionSize` is δ, and `u_oneOverBlendedRegionSize` is $\frac{1}{w}$. The blend parameter is used to linearly blend, using the GLSL mix function, between the `coarseHeight` at the outer perimeter of the transition region and the `fineHeight` at the inner perimeter of the transition region. Outside the transition region, `alpha` is zero so the `fineHeight` is used exclusively.

Note that we output both the coarse texture coordinates and the `alpha` value to the fragment shader.

Try This ○○○○

> Reduce the number of required texture lookups by packing more information into a single texture. For example, the coarse height can be stored as the whole-number portion of the float, and the difference between the coarse and fine heights, divided by a suitable scale factor, can be stored as the fractional portion of the float. A simpler approach is to use a 16-bit floating-point texture instead of a 32-bit one.

In addition to blending the heights, we also need to blend the normals. The updated `SampleNormal` fragment-shader function is shown in Listing 13.6.

This blending technique does a nice job of hiding transitions between different levels of detail, as shown in Figure 13.5(b).

Kevin Says ○○○○

> When I first implemented the clipmap-based terrain rendering in Open-Globe, I thought I was clever. Instead of blending the coarse and fine normals, as shown here, I computed the normal in the vertex shader by finite differencing the blended heights, as shown in Section 11.3.1. This sounds like almost the same thing, but it's not. It created clear discontinuities at the level boundaries that looked like advancing waves as the viewer moved around the world.

Cracking between different levels of detail is a common problem with many terrain-rendering algorithms. The specifics of the clipmap-based rendering approach allow this problem to be dealt with in an extremely elegant

```
in vec2 fsCoarseUv;
in float fsAlpha;

uniform sampler2D u_fineNormalMap;
uniform sampler2D u_coarserNormalMap;

vec3 SampleNormal()
{
  vec3 fineNormal =
      normalize(texture(u_fineNormalMap, fsFineUv).rgb);
  vec3 coarseNormal =
      normalize(texture(u_coarserNormalMap, fsCoarseUv).rgb);
  return normalize(mix(fineNormal, coarseNormal, fsAlpha));
}
```

Listing 13.6. The `SampleNormal` fragment-shader function is modified to sample normals from two adjacent levels and blend them according to the `alpha` value computed by the vertex shader.

way. Unlike the skirts or flanges that are commonly used in other terrain-rendering algorithms, blending produces a mesh without discontinuities at the boundaries between levels of detail that can lead to texture stretching. Significantly, this extends to the surface normals used for lighting.

13.5 Clipmap Update

As the viewer moves, the clipmap levels move as well, remaining approximately centered around the viewer. The process of updating a clipmap level as a result of viewer motion is shown in Figure 13.7.

We keep a series of terrain height-map tiles cached in textures on the GPU. As explained in Section 12.2.2, terrain datasets are typically broken up into tiles so that a particular region can be accessed more quickly. When rendering such a dataset, these terrain tiles are convenient units in which to send terrain data to the GPU, particularly if they already have a power-of-two size. In the unlikely event that the terrain data source is not already tiled, it should be logically broken up into tiles anyway in order to allow data to be uploaded to the GPU in convenient tile-sized chunks.

Clipmap level updates take place almost entirely on the GPU by copying newly visible post data from one or more of these tiles to the appropriate place in the clipmap level's height-map texture.

The copy operation is performed by rendering a quad textured with the tile's height-map texture to a framebuffer with the clipmap level's height-map texture as a color attachment. A simple fragment shader does the actual copying. This strategy allows us to send the tile's height-map data over the system bus once, at least as long as the tile remains cached (see

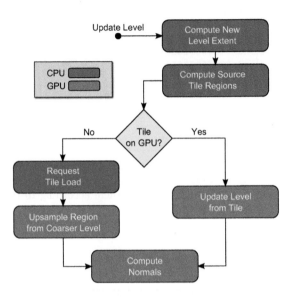

Figure 13.7. The overall process of updating a clipmap level in response to viewer motion. The colors of the boxes indicate whether the operation primarily takes place on the CPU or GPU.

Section 13.5.7), and then do clipmap updates with little additional bus traffic.

Similarly, the normal map for each level is updated by rendering a quad textured with the level's height map to a framebuffer with the level's normal map attached and computing the normal in the fragment shader.

Kevin Says ○○○○

> In OpenGlobe, we use a single-channel, floating-point texture to store the heights, and a three-channel, floating-point texture to store the normals. For some applications, using 4 bytes per height and 12 bytes per normal will be overkill, in which case a more compact representation can be used instead.

If a tile is not yet resident on the GPU, it is queued for load by a worker thread, and the corresponding region in the level is filled by upsampling heights from the next-coarser level.

13.5.1   Toroidal Addressing

We use TextureWrap.**Repeat**, also known as 2D wraparound or toroidal addressing, to incrementally update clipmap levels. The texture can be thought of as occupying a torus, where the texture wraps around on itself in both directions.

Consider the case where the viewer moves toward the east, as shown in Figure 13.8. In this case, in order to remain centered around the viewer, the complete set of height data for the level must shift toward the left. The westernmost columns of height data fall off the edge of the level height texture, and the easternmost columns are filled with posts from the height map that were not previously visible.

Because every post shifts position when the viewer moves, it may appear at first that we need to rewrite the entire texture every time. Fortunately, this is not the case.

Instead, we toroidally address the height texture using a sampler configured to use TextureWrap.**Repeat**. Then, when the viewer moves, the origin of the level within the texture moves as well. In our example above, the origin is initially at $(0,0)$. After the viewer moves one post to the east, the origin is changed to $(1,0)$. This means that column 1, the second column, is now the westernmost column of posts in this clipmap level. By wraparound texture addressing, column 0, the first column, is the easternmost column of posts.

This is immensely convenient. It means that the region in the texture occupied by the *old* westernmost posts, which are no longer necessary when

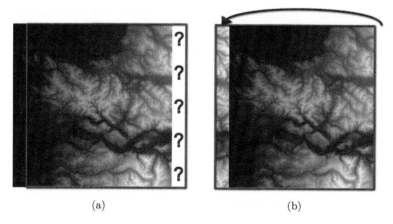

(a)   (b)

Figure 13.8. (a) As the viewer moves, clipmap levels are incrementally updated with new data. (b) Toroidal addressing eliminates the need to write the entire height-map texture. Instead, new posts replace old ones.

the clipmap level moves to the east, is the same region that is now needed for the *new* easternmost posts. Instead of rewriting the entire clipmap level, we only need to incrementally populate it with the newly visible posts, overwriting old posts that are no longer visible.

The origin of the level within the texture is passed to the vertex shader shown in Listing 13.3 as `u_fineTextureOrigin`, plus 0.5 in each direction so that heights are sampled from the centers of height texels.

13.5.2 The Update Region

Section 13.1 described how to compute the extent of a clipmap level from the viewer position. The displacement of this extent since the last render frame defines the region that needs to be updated. It also determines the origin of the clipmap level in the level's height map.

The code to compute the origin, `level.OriginInTextures`, is shown in Listing 13.7. The object `CurrentExtent` is the extent, in integer-valued terrain coordinates, of the clipmap level on the last render frame; `NextExtent` is the extent computed from the new viewer position. At the end of the update process, the `NextExtent` becomes the `CurrentExtent`.

The `OriginInTextures` is simply displaced by the change in the southwest corner of the clipmap level's extent since the last frame. If the origin falls off one side of the texture, it wraps around to the other side.

The region that needs to be updated is the region that is new to the clipmap level extent this render frame and is computed as shown in Listing 13.8.

If the viewer moved to the east (`delta X > 0`), the region that needs to be updated starts just after the easternmost post in the previous clipmap-

```
int deltaX = level.NextExtent.West - level.CurrentExtent.West;
int deltaY = level.NextExtent.South - level.CurrentExtent.South;

// if deltaX and deltaY are 0, no update is necessary

int newOriginX =
    (level.OriginInTextures.X + deltaX) % _clipmapPosts;
if (newOriginX < 0)
  newOriginX += _clipmapPosts;
int newOriginY =
    (level.OriginInTextures.Y + deltaY) % _clipmapPosts;
if (newOriginY < 0)
  newOriginY += _clipmapPosts;

level.OriginInTextures = new Vector2I(newOriginX, newOriginY);
```

Listing 13.7. The new origin of the clipmap level within the level's height-map texture is computed from the motion of the level's extent since the last frame.

```
int minLongitude = deltaX > 0 ? level.CurrentExtent.East + 1
                              : level.NextExtent.West;
int maxLongitude = deltaX > 0 ? level.NextExtent.East
                              : level.CurrentExtent.West - 1;
int minLatitude = deltaY > 0 ? level.CurrentExtent.North + 1
                             : level.NextExtent.South;
int maxLatitude = deltaY > 0 ? level.NextExtent.North
                             : level.CurrentExtent.South - 1;

int width = maxLongitude - minLongitude + 1;
int height = maxLatitude - minLatitude + 1;

if (height > 0)
{
  ClipmapUpdate horizontalUpdate = new ClipmapUpdate(
      level,
      level.NextExtent.West,
      minLatitude,
      level.NextExtent.East,
      maxLatitude);
  _updater.Update(context, horizontalUpdate);
}

if (width > 0)
{
  ClipmapUpdate verticalUpdate = new ClipmapUpdate(
      level,
      minLongitude,
      level.NextExtent.South,
      maxLongitude,
      level.NextExtent.North);
  _updater.Update(context, verticalUpdate);
}
```

Listing 13.8. The regions to update are computed from the motion of the level's extent since the last frame.

level extent and it continues to the easternmost post in the next extent of the same clipmap level. If the viewer moved to the west (deltaX < 0), the region starts at the next westernmost post and continues just shy of the previous westernmost post. The minimum and maximum latitudes that need to be updated are computed similarly from the deltaY, where deltaY < 0 is movement to the south and deltaY > 0 is movement to the north.

If the viewer moved in only one direction, east/west or north/south, the update region is a simple rectangle. If the viewer moved in both directions since the last render frame, the texels that need to be updated form an L shape, as shown in Figure 13.9, and we update the L using two rectangular updates.

Just like CurrentExtent and NextExtent, the update regions express a range of posts in terms of *terrain coordinates*. Terrain coordinates are the integer-valued coordinates of the post within a particular level of the terrain dataset, numbered sequentially starting from the southwest corner

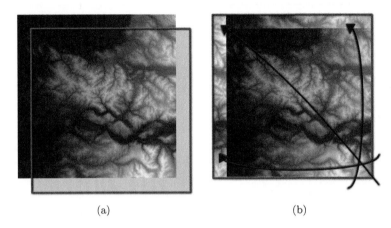

(a)   (b)

Figure 13.9. (a) When the viewer moves diagonally over the course of a single frame, the region that needs to be updated with new posts forms an L shape. (b) Wraparound addressing in both directions avoids the need to copy existing posts.

of the world. This is in contrast with *texture coordinates*, which are the coordinates of the post in the height-map texture for a particular clipmap level. Texture coordinates take into account the current `OriginInText ures` of the clipmap level, as well as the use of toroidal texture addressing.

For a given set of terrain coordinates, it is straightforward to determine the corresponding texture coordinates:

$$\text{texture} = (\text{textureOrigin} + (\text{terrain} - \text{levelOrigin}))\,\%\text{clipmapSize},$$

where *terrain* is the terrain coordinates of the post; *levelOrigin* is the coordinates of the southwest corner of the clipmap level expressed in terrain coordinates; *textureOrigin* is `OriginInTextures`, the location within the texture that corresponds to the *levelOrigin*; and *clipmapSize* is the length of one side of the clipmap height texture.

Of course, if the terrain coordinates of a post are outside the extent of the clipmap level, the post's texture coordinates are undefined.

13.5.3   Updating Heights

We use GPU rasterization to update a level's height-map texture. We attach the level's height-map texture as the color attachment of a framebuffer and then render a quad. This presents a problem, however.

Due to the `OriginInTextures`, it is likely that the clipmap updates determined in the previous section wrap around the height-map texture. In other words, the northeasternmost post in the update region will have texture coordinates less than the texture coordinates of the southwesternmost post. Effectively, it appears that we need to use toroidal addressing when writing to the framebuffer. This is a problem because framebuffers don't support toroidal addressing.

We work around this problem by using OpenGlobe's ClipmapUp dater.`SplitUpdateToAvoidWrapping` method to split an update into between one and four simple, rectangular, nonwrapping updates.

We're not done splitting the update yet, however. We also need to split the update based on the distinct tiles that are the sources of the height data. In OpenGlobe, this additional splitting is performed by Ras terTerrainLevel.`GetTilesInExtent`. Given the extent of an update in terrain coordinates, it returns a list of regions of tiles that compose the update region. The union of all the tile regions is equivalent to the update region.

Finally, we're ready to instruct the GPU to do the height update. The height-update vertex shader is shown in Listing 13.9.

The input `position` vertex attribute comes from a simple unit quad and has one of four possible values: $(0.0, 0.0)$, $(1.0, 0.0)$, $(0.0, 1.0)$, and $(1.0, 1.0)$.

The uniform `u_updateSize` is the width and height of the region to be updated. Multiplying it by the `position` scales the unit quad up to the actual dimensions of the region.

The uniform `u_destinationOffset` is the offset in the clipmap level's height texture to which to write the update. Adding it to the scaled position locates the update within the destination texture.

```
in vec2 position;
out vec2 fsTextureCoordinates;

uniform mat4 og_viewportOrthographicMatrix;
uniform vec2 u_destinationOffset;
uniform vec2 u_updateSize;
uniform vec2 u_sourceOrigin;

void main()
{
  vec2 scaledPosition = position * u_updateSize;
  gl_Position = og_viewportOrthographicMatrix *
                vec4(scaledPosition + u_destinationOffset,
                     0.0, 1.0);
  fsTextureCoordinates = scaledPosition + u_sourceOrigin;
}
```

Listing 13.9. The vertex shader used to update heights in a clipmap.

```
in vec2 fsTextureCoordinates;
out float heightOutput;
uniform sampler2DRect u_tileHeightMap;

void main()
{
  heightOutput =
      texture(u_tileHeightMap, fsTextureCoordinates).r;
}
```

Listing 13.10. The fragment shader used to update heights in a clipmap.

The uniform `u_sourceOrigin` is the origin of the update in the source tile's height-map texture. Adding it to the scaled position locates the source of the update data.

Kevin Says ○○○○

> Don't forget to set the context's viewport to cover the entire height-map texture, from $(0, 0)$ to $(\text{width}, \text{height})$, and then restore it after rendering. Also note that, in this case, we don't need to offset `u_destinationOffset` or `u_sourceOrigin` by 0.5 in order to line up texels. The fragment shader will be invoked based on the centers of fragments, which are already aligned with the centers of tile height-map texels.

The corresponding fragment shader, shown in Listing 13.10, is as simple as can be. It simply writes the height sampled from the source texture to the color attachment associated with `heightOutput`. In other words, it copies a single value from one texture to another.

13.5.4 Updating Normals

After a level's height-map texture is completely updated, we're ready to update the normal map. The approach is basically the same: we render a quad to a framebuffer with the normal map configured as its color attachment. We compute the normals in the fragment shader using central differencing (see Section 11.3.2). The source texture, the one used to texture the quad, is the level's height map.

The vertex shader for updating normals, shown in Listing 13.11, is very similar to the one used to update heights. Unlike the height-update shader, however, this shader only has one origin, which is used to determine both the destination vertex coordinates and the source texture coordinates because normals are computed from the heights around the same

```
in vec2 position;
out vec2 fsTextureCoordinates;

uniform mat4 og_viewportOrthographicMatrix;
uniform vec2 u_updateSize;
uniform vec2 u_origin;
uniform vec2 u_oneOverHeightMapSize;

void main()
{
  vec2 coordinates = position * u_updateSize + u_origin;
  gl_Position = og_viewportOrthographicMatrix *
                vec4(coordinates, 0.0, 1.0);
  fsTextureCoordinates = coordinates * u_oneOverHeightMapSize;
}
```

Listing 13.11. The vertex shader used to update normals in a clipmap.

location. Another difference is that the texture coordinates are normalized to the range $[0, 1]$. This is necessary because the height-map texture is configured to use sampler2D instead of sampler2DRect.

The corresponding fragment shader is shown in Listing 13.12.

```
in vec2 fsTextureCoordinates;
out vec3 normalOutput;

uniform float u_heightExaggeration;
uniform float u_postDelta;
uniform vec2 u_oneOverHeightMapSize;
uniform sampler2D u_heightMap;

void main()
{
  float top = texture(
      u_heightMap,
      fsTextureCoordinates +
      vec2(0.0, u_oneOverHeightMapSize.y)).r;
  float bottom = texture(
      u_heightMap,
      fsTextureCoordinates +
      vec2(0.0, -u_oneOverHeightMapSize.y)).r;
  float left = texture(
      u_heightMap,
      fsTextureCoordinates +
      vec2(-u_oneOverHeightMapSize.x, 0.0)).r;
  float right = texture(
      u_heightMap,
      fsTextureCoordinates +
      vec2(u_oneOverHeightMapSize.x, 0.0)).r;

  vec2 xy = vec2(left - right, bottom - top) *
            u_heightExaggeration;

  normalOutput = vec3(xy, 2.0 * u_postDelta);
}
```

Listing 13.12. The fragment shader used to update heights in a clipmap.

The fragment shader computes the normal by sampling from the height map four times to obtain two partial derivatives by finite differencing and then taking their cross product. This is essentially the central-differencing technique used in Section 11.3.2. In order to compute an accurate normal, we need to account for the world-space distance between posts in the height map as well as the height exaggeration in use, so those quantities are passed to the fragment shader as `u_postDelta` and `u_heightExaggeration`, respectively.

Section 11.3.1 mentions a problem that can occur when computing normals by finite differencing, as we do in this fragment shader. At the topmost row of the height map, there is no texel available above the central one. Similarly, at the rightmost column there is no texel available to the right, at the bottom there is no texel available below, and at the left there is no texel to the left. With a sampling mode that clamps texture coordinates, the terrain at the edges will appear to be flat, creating a noticeable artifact.

Fortunately, we don't actually care very much about the accuracy of the normals at the edges of each clipmap level. Recall from Section 13.4 that the normals near the edges of each clipmap level are blended with the normals from the next-coarser clipmap level. In fact, the outer perimeter vertices all have an `alpha` value of zero, meaning that the normal is determined entirely from the coarser normal map. While some fragments just inside that outer perimeter will have an `alpha` value greater than zero, causing blending with the incorrect normal, the weighting will be so small as to be unnoticeable.

Care must be taken, however, when the viewer moves and the normal map is updated accordingly. At that time, harmless, incorrect normals at the perimeter effectively move toward the interior. Eventually, as the viewer continues to move, 100% weight will be given to the incorrect normal, leading to obvious artifacts.

The solution to this problem that we employ in OpenGlobe is to compute normals over a region that is slightly larger than the one that received updated heights. Adding a one-post border around the update region does the trick. Normals in bordering positions that were potentially computed without the benefit of correct height data are recomputed once the height data are available.

13.5.5   Multithreaded, Out-of-Core Update

Incremental clipmap updates are pretty fast. Assuming a height-map tile is already in GPU memory, copying a subset of it to the clipmap level's height texture and computing normals is fast enough that we can do it during the course of rendering.

But what if the tile is *not* in GPU memory already? Maybe it's in system memory and we need to stream it over the system bus to the GPU. That's going to take a little longer. Worse, maybe it's stored on disk and not even in system memory, yet. Initiating a read of the tile from disk in the render thread and then waiting for it to finish is definitely going to cause a noticeable stutter in the rendering. If you think it can't get any worse—what if that tile can only be found on a distant network server? It might take several seconds to download it, load it into memory, and send it to the GPU. In the meantime, the rendering thread better not be sitting there waiting for it.

For smooth rendering, we must, at a minimum, move the loading of new tiles into system memory off of the rendering thread and on to a worker thread.

In OpenGlobe, we take it one step further. The worker thread, in addition to loading the tile from disk or a network server, also creates a Texture2D and fills it with the tile's height data. This way, the tile's data are transferred from system memory to GPU memory under the control of the worker thread as well. Only when the tile is completely ready to go is it used by the render thread to update clipmap levels.

In effect, tiles cache height data in GPU memory so that it can be quickly used to update the clipmap. This is one part of a complete cache hierarchy for clipmap updates; additional tiles may be cached in system memory, on disk, or on a nearby network server.

> Can even more of the update process be moved to a worker thread? How would we update a level's height map in a worker thread while it is simultaneously being used to render?

◯◯◯◦ **Question**

Our multithreaded tile-loading strategy follows the outline presented in Section 10.4.3. A background thread with its own OpenGL context is responsible for loading the tiles and creating Texture2D instances with their contents.

The threads communicate using two message queues: a request queue and a done queue. The messages placed on both queues are instances of the TileLoadRequest class shown in Listing 13.13.

Initially, when a tile load is requested, the **Texture** field is *null*. The other two fields identify the tile that needs to be loaded and the clipmap level that needs it.

The **MessageReceived** handler for the request queue is straightforward. It creates a Texture2D instance with the dimensions of the tile and fills it

```
private class TileLoadRequest
{
  public ClipmapLevel Level;
  public RasterTerrainTile Tile;
  public Texture2D Texture;
}
```

Listing 13.13. The class used to send a tile-load request to the background tile-loading thread. The same class is used to return the tile once it has been loaded.

with the height data of the tile using a WritePixelBuffer, loading it from disk or a network connection if necessary. In OpenGlobe, we create the texture using TextureFormat.**Red32f**, so a 32-bit floating-point value is used to represent each height. If the terrain data were stored in a compressed form, they would be decompressed here as well.

Once the texture is created, we pass the TileLoadRequest instance, now with a nonnull **Texture** field, back to the render thread on the done queue. However, before we do that, we need to synchronize the two GL contexts, as explained in Section 10.4.3, by creating a Fence and waiting on it. Without this fence, the render thread could start rendering from the texture before it has been populated with height data.

Try This ○○○○

> OpenGlobe's ClipmapUpdater waits on the fence in the worker thread with a call to **ClientWait** with an infinite time out. While it's waiting, the worker thread does not do any useful work. Modify it to periodically poll the fence with a time out of zero and do other work, such as loading additional tiles, while waiting.

The render thread adds requests to the *request* queue in the course of updating the clipmap, as shown in Listing 13.14.

First, we check to see if the tile we need to update a region of the clipmap is already loaded. If it is, we render it to the level's height-map texture as explained in Section 13.5.3.

If the tile is not yet loaded, and it hasn't been added to the request queue yet either, we add it to the request queue. We quickly identify tiles that are on the request queue because those tiles exist in the _loadedTiles hashtable with a null Texture2D instance. For that reason, the first task of **RequestTileLoad** is to add the tile to _loadedTiles. Then, it creates a TileLoad Request and posts it to the request queue.

Next, whether or not the tile was already on the request queue, we update the tile's region in the clipmap with data upsampled from the next-

```
Dictionary<RasterTerrainTileIdentifier, Texture2D> _loadedTiles =
    // ...
Context context = // ...
ClipmapLevel level = // ...
RasterTerrainTileRegion region = // ...
Texture2D tileTexture;
bool loadingOrLoaded =
    _loadedTiles.TryGetValue(region.Tile.Identifier,
                             out tileTexture);
if (loadingOrLoaded && tileTexture != null)
{
  RenderTileToLevelHeightTexture(context, level,
                                 region, tileTexture);
}
else
{
  if (!loadingOrLoaded)
  {
    RequestTileLoad(level, region.Tile);
  }
  UpsampleTileData(context, level, region);
}
```

Listing 13.14. Updating a clipmap level's height texture. If a required tile is not loaded, it is added to the request queue, and the missing data are upsampled from the next coarser clipmap level.

coarser clipmap level. Upsampling gives us a reasonable approximation of the terrain detail without waiting for the tile to load. This upsampling process is described in detail in Section 13.5.6.

As mentioned previously, the worker thread posts tiles to the done queue when it finishes loading them. The done queue is processed synchronously in the render thread each render frame. In OpenGlobe, the ClipmapUpdater.`ApplyNewData` method is responsible for processing the done queue and updating the corresponding clipmap level with the new tile data. It is called after the `NextExtent` and `OriginInTextures` for all levels are computed, as described in Section 13.5.3.

When a new tile is available, the corresponding region in the clipmap level, which was previously populated with data upsampled from the next-coarser level, needs to be updated with the new data. Specifically, the region to update is the intersection of the extent of the new tile and the extent of the clipmap level in terrain coordinates, as shown in Figure 13.10. Of course, this region of intersection may not be the same as the region that motivated the loading of this tile in the first place because the viewer may have moved in the meantime. In a particularly unfortunate scenario, the tile won't contribute any posts to the new extent of the clipmap level that is in play when the tile is finally loaded.

If the region of intersection is not empty, it is updated as before. In addition, for nonempty intersection regions, this tile updating process is applied recursively to any of this tile's child tiles that are *not* loaded.

Figure 13.10. When a new tile is loaded, the intersection of the tile's extent (blue) with the clipmap extent (yellow) defines the region that needs to be updated (red).

This recursive application is necessary because, when the child tiles are not loaded, the data used to populate the clipmap in the regions covered by the child tiles are upsampled from this tile. Previously, this tile's data were upsampled from its parent tile. Now that this tile has better data, the children should be upsampled from the better data. Loaded tiles do not need their data reapplied, however, because either (a) the tile was loaded when the clipmap was last updated, so the clipmap already reflects the tile data, or (b) the tile is in the done queue and will be (or already has been) applied to the clipmap.

13.5.6 Upsampling

When a tile needed to update a clipmap level is not yet in GPU memory, we upsample that data from the next-coarser clipmap level. Upsampling is the process of predicting fine detail from a coarse representation. The upsampled heights will not match the real heights, of course, but rendering with imperfect heights is better than the alternatives, such as waiting for the tile to be loaded, rendering the tile as if all its heights were zero, or rendering nothing in the area of the tile.

In order to successfully upsample a subset of one clipmap level from the next-coarser clipmap level, it's important that the next-coarser clipmap level already be updated with correct height data. For that reason, we always update clipmap levels in coarse-to-fine order.

Consider the situation where we're updating the clipmap levels for the very first time, and only the tiles for clipmap level 0 are in GPU memory. Clipmap level 0 is updated as normal, as described in Section 13.5.3. Then, clipmap level 1 is to be updated. Its tiles are not yet resident in GPU memory, though, so instead we upsample the heights in clipmap level 0 and use them to fill clipmap level 1.

The tiles for clipmap level 2 are also not yet resident in GPU memory. So we upsample the heights in clipmap level 1. The heights in clipmap level 1, of course, were themselves upsampled from clipmap level 0. So effectively we upsample the level 0 tiles twice. This process will continue until all levels have been filled with our best approximation of their actual heights.

So how does upsampling actually work? Just as we've done previously with updating the height map and computing normals, we'll use the GPU to do the upsampling by rendering a quad to a framebuffer. In this case, the quad will be textured with the coarser height-map texture, and the framebuffer's color attachment will be the finer height-map texture. In other words, we're going to write heights into the finer height map by passing the coarser one through a fragment shader. The vertex shader for this scheme is shown in Listing 13.15.

This vertex shader is very similar to the one in Listing 13.11 that was used to compute normals. One key difference is that the position to read from in the source texture, the coarser height map, is not the same as the position to write to in the destination texture, the finer height map. Not only do the two textures have different origins but they also have different scales. Specifically, the coarser texture has double the real-world scale of the finer texture. The vertex shader takes all of these factors into account.

The fragment shader is more interesting because there are a number of reasonable things it could do. Perhaps the simplest approach is to just copy the nearest post in the coarse height map into the fine height map. A better approach is to linearly interpolate the fine heights from the coarse ones.

```
in vec2 position;
out vec2 fsTextureCoordinates;
uniform mat4 og_viewportOrthographicMatrix;
uniform vec2 u_sourceOrigin;
uniform vec2 u_updateSize;
uniform vec2 u_destinationOffset;
uniform vec2 u_oneOverHeightMapSize;
void main()
{
  vec2 scaledPosition = position * u_updateSize;
  gl_Position = og_viewportOrthographicMatrix *
                vec4(scaledPosition + u_destinationOffset,
                     0.0, 1.0);
  fsTextureCoordinates =
      (scaledPosition * 0.5 + u_sourceOrigin) *
      u_oneOverHeightMapSize;
}
```

Listing 13.15. The vertex shader for upsampling a clipmap level from the next coarser level.

```
in vec2 fsTextureCoordinates;
out float heightOutput;

uniform sampler2D og_coarseHeightMap;

void main()
{
  heightOutput = texture(og_coarseHeightMap,
                         fsTextureCoordinates).r;
}
```

Listing 13.16. A very simple fragment shader to upsample finer heights from coarser ones.

In fact, these two approaches can be implemented using the very same simple fragment shader, shown in Listing 13.16. The only difference between them is whether the sampler used to read the fine height map is `NearestClamp` or `LinearClamp`.

When this fragment shader is paired with a sampler that performs linear interpolation, the results are pretty good. Much more sophisticated fragment shaders are possible, however.

One possibility is to use cubic convolution. Cubic convolution over a height map involves computing five 1D cubic convolutions, using a total of 16 samples from the coarse height map [83].

Asirvatham and Hoppe recommend upsampling using the tensor product version of the well-known four-point subdivision curve interpolant and state that it has the desirable property of being C^1 smooth [9].

13.5.7   Replacement and Prefetching

Section 13.5.5 discussed how new height-map tiles are brought into GPU memory so that they can be used to update clipmap levels. But when are tiles removed from GPU memory?

Tiles need to be removed from GPU memory when the memory they occupy is needed for other, more important, purposes. For example, in an extreme case, a terrain tile halfway around the world is removed in order to make room for a terrain tile right beneath the viewer's feet. The policy for choosing which tiles to remove is known as a replacement policy, and several generalized replacement policies for terrain rendering are described in Section 12.3.3.

The arrangement of clipmap levels into concentric squares suggests a simple and effective replacement policy: the first tiles to be replaced are the ones that are farthest from the viewer.

Actually, we can do a bit better than this. Remember that the current height map for all clipmap levels is always stored in GPU memory, and we only load tiles into GPU memory in order to use them to update the

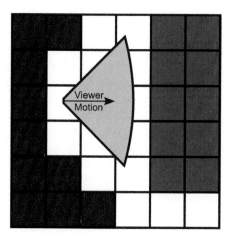

Figure 13.11. Viewer motion can be taken into account when determining replacement and prefetching of tiles. Tiles ahead of the viewer (red) are more likely to be useful in the near term than are tiles in the viewer's wake (blue).

clipmap levels. Furthermore, clipmap levels are always updated at their edges because interior heights already exist within the clipmap level and are moved via toroidal addressing, as described in Section 13.5.1. This means that the tiles we are most likely to need again soon are in fact the ones that are closest to the edges of the clipmap level, rather than the ones that are closest to the viewer. Of course, this distinction may be too small to worry about, especially if the clipmap size and tile size are similar.

Another possible improvement is to take into account the motion of the viewer, as shown in Figure 13.11. If the viewer has been heading east for the past few rendered frames, tiles to the west of the clipmap level are less likely to be needed again soon than are tiles to the east of the clipmap level.

These same principles can be applied to prefetching as well. The simple reactive tile-loading mechanism shown in Listing 13.14 is replaced with a more proactive process. Given a budget for tile data in GPU memory, we load all of the closest tiles up to that limit, optionally skewing for the motion of the viewer.

Prefetching offers the possibility of eliminating upsampling when viewer motion is relatively slow and is within a relatively confined area. In these cases, it significantly improves the user experience. Of course, in virtual globes in particular, prefetching is unlikely to completely eliminate all upsampling because camera motion is simply too fast and unpredictable.

13.5.8   Compression and Synthesis

Losasso and Hoppe, the inventors of the geometry-clipmapping technique, tested their implementation using a 40-gigabyte dataset covering the conterminous United States with approximately 30 m between posts [105]. While not huge by virtual globe standards, it is a bigger dataset than we would ordinarily expect to be able to fit into memory. Yet, they didn't implement any out-of-core data management techniques, and instead fit the dataset into a mere 355 megabytes. How did they do it?

The answer, in case you haven't already guessed, is that they used compression. Their compression scheme involved two parts, one of which relied on the unique structure of geometry clipmaps.

First, they used a particular, lossy image-compression algorithm called progressive transform coder (PTC) that is able to efficiently decode subsets of an image [108]. This allowed them to store the entire height-map image in system memory and decode portions of it as necessary to send to the GPU as textures. Since then, PTC has been standardized as part of JPEG XR.

Second, they noted that the unique organization of the clipmap levels allowed the height information to be stored more compactly.

Recall from Section 13.5.6 that we update clipmap levels in coarse-to-fine order and that this allows us to predict the heights of a fine clipmap level from the heights of a coarse one. As presented previously, we only do this prediction when the fine data are not available.

Instead, we can always predict the fine detail from the coarse while updating clipmap levels. This way, we don't need the actual heights in order to update a clipmap level; we only need the differences between the actual heights and the heights predicted from the coarser level. These differences, called *residuals*, are much more compressible than the complete height data. In fact, the data that Losasso and Hoppe compressed with PTC were residual data, not height data, and this is how they fit a 40-gigabyte uncompressed dataset into 355 megabytes.

For virtual globe-sized datasets, even these techniques will probably not allow us to completely avoid the need to store data on disk and on network servers. Furthermore, imagery data tend to be much less compressible than height data, so it's unlikely that we can fit an entire high-resolution color map of the United States in memory, even using these techniques. However, each byte that we don't need to use to represent the height data is one less byte that we need to read from slow hard disks or send over slow network connections, which means we can show more detail to our users sooner or preserve I/O bandwidth for other purposes.

The main cost of this approach is the additional time spent in the preprocessing step to compute the residuals and compress them. There is additional cost at runtime, too, because the residuals must be decompressed on the CPU. However, the additional CPU time necessary for decompression is typically more than made up for by the time saved reading the smaller dataset from disk or a network server.

Interestingly, this residual-based approach offers the possibility that the residuals for more detailed clipmap levels not be stored at all. Instead, the residuals are procedurally synthesized at runtime using fractal techniques. Losasso and Hoppe used uncorrelated Gaussian noise to synthesize fine detail when zoomed in close to the terrain surface. This produces a realistic-looking rough surface when the resolution of the terrain data is exceeded, even though it does not, strictly speaking, reflect the real world.

13.6 Shading

Terrain rendered by geometry clipmapping can be shaded in all of the ways described in Section 11.4. Perhaps the most common way to shade terrain in virtual globe applications, however, is to apply a color map derived from satellite imagery. In this case, the color map is probably even larger than the height map!

Recall from Section 12.2.1 that geometry clipmapping is derived from clipmapping, a similar technique used to manage enormous textures. Thus, it's natural to extend geometry clipmapping to also manage a color map.

If the color map and height map have identical extents, mipmap levels, and tile structures, things couldn't be easier. We simply extend the fragment shader in Listing 13.4 to sample from a color map in addition to the normal map. We use the `alpha` blend parameter from Listing 13.5 to blend between the finer and coarser color maps, just as we did with the normal map. Color maps are accessed toroidally and incrementally updated by rendering quads textured with tiled data. Color-map tiles are loaded into textures on the GPU in lockstep with the height-map tiles.

However, the color and height maps are unlikely to line up quite so nicely. They may come from different suppliers that use different schemes to tile their data. Or, one color-map sample per height might simply be too blurry. They could have significantly different resolutions. In these cases, we need to take steps to ensure that differences are accounted for in the shaders.

First, we select an appropriate size for the color clipmap. It must cover at least the world extent covered by the height clipmap, as shown in Figure 13.12. We compute the world location of the southwest post in the height clipmap and then transform it into texel indices in the color map.

If height-map posts don't line up precisely with color-map texels, the computed indices will not be integers, in which case the south and west

Figure 13.12. The color map used to shade a clipmap level must cover at least the extent covered by the height map. If it's slightly larger, offsets are computed and used in the vertex and fragment shaders to adjust texture coordinates. In this figure, color-map texels are shown as red dots and height-map texels are shown as blue dots.

coordinates are rounded toward the south and west. Similarly, the north and east coordinates are rounded toward the north and east.

The difference between the southwest corner of the color map and the southwest corner of the height map, both expressed in the coordinates of the color map, is passed to the shaders as a uniform. In addition, the ratio of distances between texels in the two maps, in world space, is passed to the shaders. Thus, to compute texture coordinates in the color map, multiply the texture cooordinates in the height map by the ratio between them and then add the offset.

13.7   Geometry Clipmapping on a Globe

Geometry clipmapping as presented thus far only renders terrain extruded from a flat, horizontal plane. The common myth notwithstanding, a flat Earth wasn't considered very accurate even before Christopher Columbus made his famous voyage to the Americas. How can we extend the algorithm to extrude terrain from a sphere or, better yet, an ellipsoid?

Many of the strengths of geometry clipmapping come from its regularity:

- Very little horizontal coordinate data are needed. A small set of vertices can express the horizontal coordinates for an entire terrain.

- The horizontal geometry does not inherently have a specific location in the world; only a translation added in the vertex shader specifies its absolute position. This translation can easily be expressed relative to the viewer rather than in world coordinates, meaning that the geometry itself is independent of the viewer's location. Thus, the geometry can be incrementally updated as the viewer moves because only the shader uniform changes, while avoiding the jittering that results from using absolute world coordinates in a world the size of Earth.

- The nested structure allows an efficient compression mechanism; only residuals need to be stored at each level.

- Terrain vertices are precisely aligned with height-map texels. If the height map is sampled at a different rate, aliasing artifacts will occur.

The challenge of extending geometry clipmapping to globes is in maintaining these strengths.

13.7.1 Mapping to the Ellipsoid in the Vertex Shader

Perhaps the most direct approach is to map vertices to the ellipsoid in the vertex shader. Consider the common case where the source terrain data are expressed with a WGS84 geographic projection, which means height samples form a regular grid in coordinates of longitude and latitude.

Using this regular grid, we can render the entire world using geometry clipmapping by interpreting the incoming (x, y) position in the vertex shader as longitude and latitude. Initially, this means that the entire world occupies a plane.

```glsl
vec3 GeodeticSurfaceNormal(vec3 geodetic)
{
    float cosLatitude = cos(geodetic.y);

    return vec3(
        cosLatitude * cos(geodetic.x),
        cosLatitude * sin(geodetic.x),
        sin(geodetic.y));
}

vec3 GeodeticToCartesian(vec3 globeRadiiSquared, vec3 geodetic)
{
    vec3 n = GeodeticSurfaceNormal(geodetic);
    vec3 k = globeRadiiSquared * n;
    vec3 g = k * n;
    float gamma = sqrt(g.x + g.y + g.z);

    vec3 rSurface = (k * n) / gamma;
    return rSurface + (geodetic.z * n);
}
```

Listing 13.17. GLSL function for transforming from geographic to Cartesian coordinates.

To map to an ellipsoid, we only need to add code to the vertex shader to transform the geographic vertex position (i.e., longitude, latitude, and height) into Cartesian coordinates. This can be done by porting Ellipsoid.`ToVector3D` from Section 2.3.2 from C# to GLSL, as shown in Listing 13.17.

There are two significant problems with this approach. The first is that it introduces precision problems. When using single-precision floating-point numbers, which is common on today's GPUs, `GeodeticToCartesian` produces a reasonable representation of the Cartesian positions of posts spaced at about 30 m.

For higher-resolution terrain, or for more accurate post placement, a different technique is required on 32-bit GPUs, such as emulating double precision as surveyed by Thall [167]. We've already explored similar techniques in Section 5.4 for emulating subtraction. With the growing availability of 64-bit, double-precision GPU hardware, however, this approach will become increasingly practical in the future.

The second significant problem with this approach, shown in Figure 13.13, is that the height samples, which are equally spaced in geodetic coordinates, are not equally spaced in Cartesian coordinates. In Cartesian space, the samples get closer together as we approach the poles. At the north and south poles, an entire row of height samples corresponds to a single point on the Earth.

This also means that the world extents of the clipmap regions become very small near the poles. In the extreme, imagine a viewer positioned

Figure 13.13. As we approach the north or south pole, clipmap vertices get closer together. At the poles, an entire row of height-map texels occupies the same location.

Figure 13.14. A viewer standing near the south pole and looking south can easily see the end of even the coarsest clipmap level.

just south of the north pole and looking north. The terrain will appear to come to a point at the north pole and drop off thereafter. This is shown in Figure 13.14.

In some sense, expressing the source data in a geographic projection implies that the poles are not particularly important, since geographic projections have oversampling problems at the poles. For that reason, it may be acceptable to use a rendering algorithm, such as the one described here, that has its own problems at the poles. If necessary, we can limit how close the clipmap can get to the poles by discarding fragments above or below a threshold latitude in the fragment shader, or by moving vertices outside of the latitude range to the center of the Earth in the vertex shader.

Unfortunately, this technique may be as good as it gets for preserving all of the advantages of geometry clipmapping for terrain on a globe. In a sense, we are sacrificing visual quality near the poles, and possibly elsewhere due to precision problems, in order to preserve all of the benefits of geometry clipmapping. This is a reasonable trade-off for many virtual globes. The remaining techniques to be described, while better than this one in certain ways, sacrifice one or more of these advantages.

13.7.2 Spherical Clipmapping

Spherical clipmapping covers the visible hemisphere of a spherical planet with a series of concentric rings centered around the viewer. Like in conventional, planar geometry clipmapping, the vertices that are displaced

Figure 13.15. In spherical clipmapping, a series of circular, concentric rings are centered around the viewer. These rings are static with respect to the viewer.

by reading from a height map are static with respect to the viewer. The arrangement of clipmap levels around the viewer is shown in Figure 13.15.

The details of implementing spherical clipmapping are described by Clasen and Hege [27]. Vertex data are created from a spherical parameterization. Like in the original geometry-clipmap algorithm, the vertex data do not need to change as the viewer moves. Unlike planar geometry clipmapping, however, spherical-clipmap vertices are not aligned with height-map texels. This leads to the major problem with spherical clipmaps: aliasing.

Consider a simple terrain peak, shown in the cross-section in Figure 13.16. The shape of the resulting terrain changes significantly depending

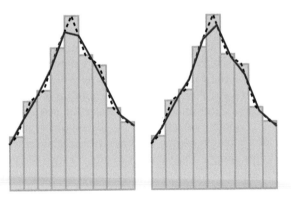

Figure 13.16. Because spherical clipmap vertices are not precisely aligned with height samples, the terrain can appear to change shape as the viewer moves. Two different views of the same peak are shown where different sampling creates a drastically different shape (red lines). The bars represent the height samples and the dashed black lines represent linear interpolation of the height samples.

on how the vertices land on height-map samples. As the viewer moves, the vertices slide across the height map and the rendered geometry changes as a result.

This effect becomes less noticeable as the resolution of the spherical clipmap is increased, and it will become nearly unnoticeable when triangles shrink to approximately the size of a pixel. Unfortunately, rendering such a spherical clipmap is very expensive.

Another problem with spherical clipmaps is that they are, as the name implies, designed to render terrain on a spherical globe. Extending the technique to an ellipsoid model of Earth, such as the WGS84 oblate spheroid, is challenging. For some applications, a spherical globe may be acceptable. Terrain and imagery rendered on a spherical versus WGS84 globe are subjectively identical. When combined with other data, though, the difference can be enormous. For example, a satellite or aircraft positioned with a precise geocentric, Cartesian coordinate system will appear to have an altitude as much as 21.3 km different from its actual altitude.

13.7.3   Coordinate Clipmapping

The most promising approach for mapping geometry clipmaps to a globe, which we call *coordinate clipmapping*, is described by Frühstück [53]. Instead of storing just heights in the vertex texture associated with each clipmap level, complete sets of Cartesian coordinates are stored in a three-channel floating-point texture.

The static clipmap geometry serves only as texture coordinates that are used to retrieve the full xyz-position of the vertex from the texture and to describe how the vertices form triangles. This approach enables an impressive degree of flexibility in how vertices in a clipmap level are arranged. They can represent a mapping of the height map onto a plane, sphere, ellipsoid, or any number of other shapes.

At first glance, this may not appear much different from the conventional rendering approach of storing a triangle mesh in a vertex buffer. In fact, by simply using a vertex buffer, we would avoid the need to do a vertex texture fetch in the vertex shader. There is an important difference, however. By storing our vertex positions in a texture, we can utilize toroidal texture addressing (see Section 13.5.1) to incrementally upload new positions to the GPU as the viewer moves around.

The biggest downside to this approach compared to conventional, planar geometry clipmapping is that the clipmap levels occupy at least three times the memory. In an environment where size is speed, this is a significant disadvantage.

Another concern is the precision with which the vertices can be represented. In order to support incremental updates, the vertices cannot be defined relative to the viewer. If they were, their positions would need to be updated every time the viewer moved, which would require rewriting the entire texture. If we define the vertices relative to the center of the Earth, however, we run into the same precision problems that we saw when mapping vertices to the ellipsoid in the vertex shader.

One solution is to use a floating origin. Vertices are defined relative to an origin that is near the viewer. As the viewer moves, the origin remains unchanged. At some point, when the viewer is too far from the origin, a new origin is selected and all of the vertices are updated to be relative to the new origin. Double buffering can be used to perform the update over the course of several frames rather than trying to update all clipmap levels in one frame, which would cause a noticeable stutter in the rendering. Microsoft Flight Simulator uses a similar technique [163] as described in Section 5.5.

Frühstück presents a variation of this idea in which the world is divided into regions, each of which has a fixed origin [53]. Depending on the position of the viewer, up to four regions need to be rendered, as shown in Figure 13.17.

Another solution is to store the vertices using simulated double precision and transform them to be relative to the viewer in the vertex shader, as described in Section 5.4. This solution is rather elegant, but it does double

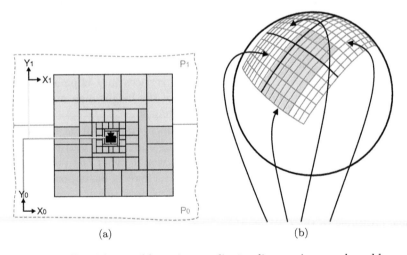

(a)   (b)

Figure 13.17. Precision problems in coordinate clipmapping can be addressed by dividing the world into multiple regions, each with their own origin. (a) Rendering clipmaps that overlap two terrain regions. (b) When the viewer is near the corner of a region, up to four regions need to be rendered. (Images courtesy of Anton Malischew.)

again the quantity of per-height data that are needed. Each height sample is represented by three components, x, y, and z, each of which is 8 bytes, for a total of 24 bytes per vertex, compared to 4 bytes per vertex for planar geometry clipmapping. An interesting variation is to use the precision LOD technique described in Section 5.4.2 to only render vertices with double precision when it is necessary. For example, the finest one or two levels are rendered with double precision and the rest are rendered with single precision.

In either approach, longitude, latitude, and height samples are transformed to Cartesian space in double precision on the CPU. When using the GPU RTE approach, the tiles sent to the GPU represent the vertex positions in full precision. When using a floating origin, the tile vertex positions on the GPU are relative to the floating origin as well, which means that they must be refreshed when the floating origin changes.

Coordinate clipmapping represents an effective solution to the problem of applying geometry clipmapping to a globe and offers a great deal of flexibility in exactly where the vertices are placed. In addition, it retains most of the advantages of planar geometry clipmapping. The primary weakness is that it requires that much more data be streamed to the GPU and stored in GPU memory.

13.8 Resources

Losasso and Hoppe's original paper on geometry clipmaps is a great resource [105], as is Asirvatham and Hoppe's article on extending it to make better use of the GPU [9]. The latter is freely available online, along with all of *GPU Gems 2*.

Clasen and Hege describe spherical clipmapping in detail [27]. In his master's thesis, Frühstück provides a detailed explanation of the technique that we've termed coordinate clipmapping [53].

○○○○○ **14**

Chunked LOD

Chunked LOD is a hierarchical LOD system that breaks an entire terrain into a quadtree of tiles, called *chunks* (see Figure 14.1). The root chunk is a low-detail representation of the entire world. The four child chunks of the root evenly divide the world into four equal-sized areas and provide a higher-detail representation. Each of those chunks then has four child chunks itself, which further divide the world.

Each node in the quadtree is generated in a preprocessing step by simplifying a subset of the overall terrain mesh to achieve a specific level of geometric error. At each lower-detail level of the quadtree, the geometric

Figure 14.1. A scene near the Sierra Nevada mountains rendered using chunked LOD. Chunked LOD renders massive terrains with precise control over the screen-space error of the terrain and makes good use of the GPU. (Image taken using STK.)

error is double what it is in the next-finer level. This is the generation portion of the LOD algorithm, as described in Section 12.1.

At runtime, chunks are selected for rendering by projecting their geometric error to screen space to compute the screen-space error, in pixels, of the chunk. If the error is too high, its children are visited instead, which is called *refinement*. This is the selection portion of the LOD algorithm. When two chunks of different LODs are adjacent to each other, their edge vertices will not necessarily coincide, leading to cracks between the chunks. A particular concern is how to fill these cracks so that the mesh appears seamless.

When zooming in toward a chunk, there comes a moment where the error estimate for the chunk gets too high, and at that moment, the chunk will subdivide into its children. Similarly when zooming out, four child chunks may be suddenly replaced with their parent chunk. This is likely to create an obvious and objectionable popping artifact. Chunked LOD addresses this problem by gradually morphing between the levels of detail. This is the switching portion of the LOD algorithm.

In many ways, chunked LOD is the direct application of the hierarchical LOD approach to terrain rendering (see Section 12.1.3). It is efficient on modern GPUs because it uses relatively large, static vertex buffers to render each chunk. Only a minimal amount of CPU time is required to determine which chunks to render in a given scene. It also does an excellent job of rendering a scene with guaranteed bounds on the amount of screen-space error introduced by LOD.

Ulrich introduced chunked LOD at SIGGRAPH in 2002 during a course on rendering massive terrains [170]. His approach pulled together various threads of research, plus a number of his own innovations, to create a very practical system for rendering terrain with out-of-core datasets. Even better, he backed it up with working, public domain source code.[1]

Since then, chunked LOD has seen significant success in virtual globes, games, and other applications that need to render massive terrains. Perhaps more surprisingly, considering the significant changes in GPUs since 2002, chunked LOD is still highly relevant today.

Kevin Says ○○○○

> Chunked LOD has been the basis for the terrain rendering in Insight3D and STK for a number of years now; prior to that, we used ROAM. I suspect that most commercial virtual globes available today use an algorithm similar to chunked LOD for their terrain rendering.

---
[1] http://tulrich.com/geekstuff/chunklod.html

14.1   Chunks

The process of generating chunks is largely orthogonal to the process of rendering them. In principal, chunks can be created from any input terrain dataset, including terrains with vertical or overhung features. However, the chunks, once created, must have the following characteristics:

- *Rectangular shape.* Each chunk must have a vertex in all four corners of the chunk's extent. In addition, there must be an edge between the chunk's adjacent corners. This guarantees that chunks form a watertight mesh, at least in the horizontal plane. We'll discuss the solution to cracks caused by differences in height between adjacent chunks in Section 14.2.

- *Monotonic geometric error.* We must know the maximum geometric error for each chunk. Geometric error is the maximum distance of any vertex in the full-detail model to the closest corresponding point in the reduced-detail model, and it must be monotonic relative to the chunk level. In other words, rendering the four children of a chunk must always result in *less* error than rendering the chunk itself. The geometric error is computed at the time the chunk is generated.

- *Known bounding volume.* Each chunk must be contained within a known bounding volume. The bounding volume, along with the geometric error, is used to compute the maximum screen-space error of the chunk. Furthermore, for best results, the bounding volume of child chunks should be fully contained within parent chunks. In practice, an AABB or bounding sphere for the chunk is computed while generating the chunks.

As shown in Figure 14.2, each chunk has a vertex buffer indicating the location of the vertices in the chunk and usually other per-vertex information, such as normals and texture coordinates. It also has an index buffer specifying how the vertices are connected to form the terrain surface. Finally, the maximum geometric error and bounding volume for each chunk are known.

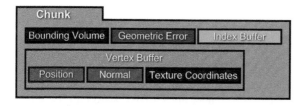

Figure 14.2. The data associated with each chunk.

In Section 14.5, we will show how a chunked LOD quadtree meeting these requirements is generated. First, we will assume that it already exists and show how to go about rendering it.

14.2   Selection

To render a chunked LOD quadtree, we first need to choose a value for our maximum tolerable screen-space error, tau. The value of tau can be tweaked for faster performance on slower hardware or for greater detail on fast hardware.

We render the chunk quadtree starting at the root, using an algorithm like the one shown in Listing 14.1. For each chunk, the screen-space error of the chunk is calculated from the chunk's maximum geometric error, as described in Section 12.1.4, and compared to tau. If the screen-space error of the chunk is under the allowable limit, the chunk is rendered. Otherwise, we refine by recursing on the chunk's four child chunks. This continues until the entire terrain is rendered at the lowest level of detail that meets our screen-space error requirements. Figure 14.3 shows the quadtree nodes rendered and refined in an example frame.

We prefer to render child nodes in front-to-back order to take advantage of GPU depth buffer optimizations. This is surprisingly efficient and easy to do in a quadtree structure like the one used in chunked LOD, as described in Section 12.4.5. It is also useful to perform view-frustum and horizon culling at this stage, as described in Section 12.4.2.

At this point, our entire terrain is rendered at appropriate levels of detail. This simple implementation exhibits two major artifacts, however: cracking and popping.

```
public void Render(ChunkNode node, SceneState sceneState)
{
  if (ScreenSpaceError(node, sceneState) <= tau)
  {
    RenderChunk(node, sceneState);
  }
  else
  {
    foreach (ChunkNode child in node.Children)
    {
      Render(child, sceneState);
    }
  }
}
```

Listing 14.1. Chunks to render are selected based on their screen-space error.

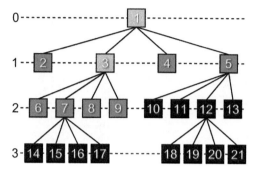

Figure 14.3. In this example, gray quadtree nodes did not meet screen-space error requirements and were refined. Green nodes did meet the error requirements and were rendered. Blue nodes were not visited because an ancestor was rendered.

14.3 Cracks between Chunks

When we render two chunks of different levels of detail next to each other, there is no guarantee that the heights along the adjacent edges are the same. In fact, the two chunks may not even have the same number of vertices along their edges!

These height differences lead to clearly visible cracks between the chunks. How can we fill these cracks? As discussed in Section 12.1.5, this is a common problem with hierarchical LOD algorithms, and there are several approaches to solving it. Ulrich fills the cracks by adding a "skirt" around the perimeter of each chunk [170].

A skirt is a strip of triangles that starts at the perimeter of the chunk and drapes down below it, as shown in Figures 14.4 and 14.5. To hide pinholes in the mesh caused by T-junctions between adjacent chunks, the

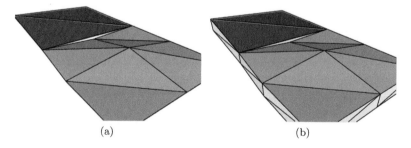

(a)   (b)

Figure 14.4. (a) Crack. (b) A skirt around the perimeter of each chunk fills the cracks that occur when adjacent chunks have different levels of detail.

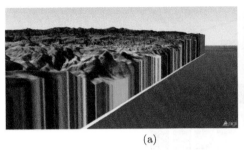
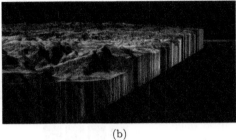

(a)   (b)

Figure 14.5. The skirt around tiles in a chunked LOD implementation. Notice that the skirts are shaded just like the chunk edges. (Images from STK.)

bottom edge of the skirt is angled slightly away from the center of the chunk. While this doesn't fill the pinholes, it hides them because the user sees the skirt instead of the background through the pinholes.

The length of the skirt is somewhat arbitrary, but it must be long enough to cover the height difference between the vertices in the adjacent chunks. In addition, we want it to be as short as possible to minimize the fill rate for triangles largely hidden under the chunks.

At chunk-generation time, we know the height at a given position at all levels of detail. So, we can precisely compute the length of the skirt and bake it into the chunk's static vertex buffer. Optionally, we can impose a limit on the maximum LOD difference between two adjacent chunks and take this limit into account as well when generating the skirt. At render time, the skirt is rendered along with the chunk, filling the chunks, with no additional draw calls.

14.4   Switching

As implemented so far, our chunked LOD algorithm can select an appropriate LOD based on the desired screen-space error and seamlessly join the chunks with skirts. Combined with an algorithm for bringing chunks into GPU memory as they're needed, as described in Section 14.7, this algorithm can be used to render enormous terrains with high visual fidelity. One problem remains, however.

As the viewer moves, the level of detail of the terrain will change as chunks are refined and merged. The changes will be small. In fact, the positions of individual vertices in screen space are guaranteed not to change by more than **tau** pixels, at least at the center of the screen. In most cases, the change resulting from an LOD switch will be much smaller than tau. If tau is a small value, the change might be nearly imperceptible.

The trouble, however, is that many vertices will change all at once as the region "pops" from one LOD to the other. While one vertex would likely be imperceptible, hundreds of vertices popping slightly all at once will be distracting.

In chunked LOD, the popping problem is actually easier to solve than it sounds. The solution is to subtly morph between LODs instead of switching all at once.

At chunk-generation time, we need to compute an extra value for each vertex in a given chunk: the morph delta for the vertex. The morph delta is the difference between the height of the vertex in the chunk and the height at the same (x, y) position in the parent chunk. Of course, the morph delta for the coarsest-detail chunk is zero, reflecting the fact that it has no coarser LOD to which to morph.

Then, at render time, we compute a morph parameter for the entire chunk according to Equation (14.1):

$$\text{morph} = \text{clamp}\left(\frac{2\rho}{\tau} - 1, 0, 1\right),\qquad(14.1)$$

where ρ is the maximum screen-space error of the chunk as computed in Equation (12.1), and τ is the maximum allowed screen-space error before the chunk is split into its four child chunks, or tau in previous code examples.

This equation produces $morph = 0$ at the distance that the chunk's parent is split to produce this chunk. It produces $morph = 1$ at the distance that the chunk itself is split into its four child chunks. As the viewer moves

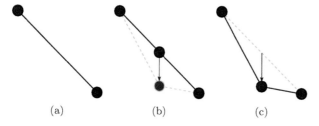

(a)   (b)   (c)

Figure 14.6. A side view of a small segment of terrain. Chunked LOD avoids vertex "pops" by smoothly morphing between LODs in the vertex shader. (a) A segment of a low-detail chunk. (b) When the chunk is first refined, a new vertex is added exactly where it appeared to be in the low-detail chunk, so no change is visible to the user. (c) As the viewer moves, the vertex smoothly morphs to its true position.

```
in vec3 position;
in float morphDelta;
// ...

uniform float u_morph;
// ...

void main()
{
    // ...

    float fineHeight = position.z;
    float coarseHeight = position.z + morphDelta;
    float height = mix(coarseHeight, fineHeight, u_morph);

    vec3 morphedPosition = vec3(position.xy, height);

    // ...
}
```

Listing 14.2. Chunks are smoothly morphed from one LOD to another with a simple modification to the vertex shader.

between those two distances, the value of *morph* smoothly moves between *morph* = 0 and *morph* = 1. This is shown in Figure 14.6.

This method of computing *morph* results in the morph occurring over the entire range of screen-space error for which the chunk is displayed, but other approaches are possible. It may be more desirable to quickly morph to full detail over a fraction of the chunk's screen-space error range. This will get more detail into the scene sooner, at the cost of a potentially more visible transition.

Another approach is to morph based on wall clock time instead of screen-space error. For example, *morph* = 0 when the chunk's parent is first refined and smoothly ramps to *morph* = 1 over the course of the next two seconds.

The morph itself is done in the vertex shader. The morph parameter, *morph*, is passed to the shader as a uniform. The morph delta for each vertex is passed to the shader as a vertex attribute. The vertex position is then computed in the shader as shown in Listing 14.2.

This morph based on the viewer position makes the LOD switch completely undetectable to the user.

14.5 Generation

While chunked LOD can be used to render terrain created from virtually any data source, Ulrich focused on generating chunks from height maps. Given that height maps are by far the most common terrain representation in virtual globe applications, we will follow that approach as well.

In addition to creating individual quadtree chunks with the spatial properties described previously, all the chunks at a given level are decimated to the same geometric error relative to the original height map. Specifically, given a level l, the geometric error, ϵ, at the next-coarser level, $l - 1$, is given by Equation (14.2) for l greater than zero:

$$\epsilon (l - 1) = 2\epsilon (l) . \qquad (14.2)$$

In other words, the geometric error in each level halves while going from coarser to finer levels. The geometric error for the finest level is chosen at the time the chunk tree is generated. A value of 0.5 means that the heights in the finest-detail chunks are allowed to deviate from the original mesh by up to 0.5 height units. Using a nonzero value for ϵ at the finest clipmap level allows the most detailed mesh to be simplified as well in flat or nearly flat areas, saving memory and bus bandwidth. This is an advantage of chunked LOD over geometry clipmapping, which requires uniform tessellation even where it does not provide any benefit.

In order to achieve this very specific level of error, Ulrich implemented an algorithm that is based on work by Lindstrom et al. and Duchaineau et al., and has similarities to the view-independent progressive mesh technique described by Bloom [18, 41, 102].

The algorithm consists of three parts: *updating*, *propagation*, and *meshing*.

14.5.1 Updating

The first part of the chunk-generation algorithm updates each vertex in the input height map with an *activation level*, which is the number of the coarsest level that must include the vertex in order to meet the geometric error requirements.

The input height map, which must have $2^n + 1$ heights on each side, is treated as an implicit binary triangle tree, or *bintree*. At the root, the bintree has two triangles formed by four posts from the height map, as shown in Figure 14.7. Each of these triangles has two child triangles formed by bisecting the triangle's longest edge. Subdivision continues recursively until all height-map posts are covered by triangle vertices. Due to the carefully chosen dimensions of the input height map, the bisecting vertex always falls on a post in the height map.

Updating proceeds recursively over the implicit triangle bintree, starting with the two triangles that form the quad that is the extent of the entire terrain region. This recursive update function is shown in Listing 14.3. The vertices of each triangle are labeled as shown in Figure 14.8.

Computing the activation level for the base vertex of a triangle is straightforward. First, we compute an estimate of the height at the base

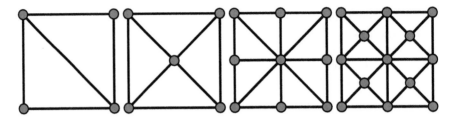

Figure 14.7. The input height map forms an implicit binary triangle tree (bintree). The root consists of two triangles formed by the posts at the four corners of the height map. Each triangle has two children formed by bisecting the triangle's longest edge. The first four levels of the implicit bintree are shown.

vertex by averaging the heights at the left and right vertices of the triangle. The difference between this estimated height and the actual height at the base vertex is the geometric error.

```
void Update(HeightMap heightMap,
            double baseMaxError,
            int baseLevel,
            int apexX, int apexY,
            int rightX, int rightY,
            int leftX, int leftY)
{
    // The base vertex is midway between left and right.
    int dx = leftX - rightX;
    int dy = leftY - rightY;
    if (Math.Abs(dx) <= 1 && Math.Abs(dy) <= 1)
    {
        return;
    }

    int baseX = rightX + (dx / 2);
    int baseY = rightY + (dy / 2);

    // Sample the heights of left, right, and base.
    // Also estimate the height at base by averaging the
    // left and right heights.
    short leftHeight = heightMap.GetHeight(leftX, leftY);
    short rightHeight = heightMap.GetHeight(rightX, rightY);
    short baseHeight = heightMap.GetHeight(baseX, baseY);
    double estimatedHeight = (leftHeight + rightHeight) / 2.0;

    // Compute the difference between the actual and estimated
    // heights at the base. This is the geometric error.
    double geometricError = Math.Abs(baseHeight - estimatedHeight);

    // If this error is larger than the error allowed at the
    // finest detail level, compute the coarsest detail level
    // that still must include this vertex and update the vertex
    // with this information.
    if (error >= baseMaxError)
    {
        int activationLevel =
```

```
        baseLevel - (int)Math.Round(Math.Log(geometricError /
                                         baseMaxError, 2.0));
    heightMap.Activate(baseX, baseY, activationLevel);
}

// Recurse to the triangle formed by base, apex, right.
Update(heightMap, baseMaxError, baseLevel, baseX, baseY,
       apexX, apexY, rightX, rightY);

// Recurse to the triangle formed by base, left, apex.
Update(heightMap, baseMaxError, baseLevel, baseX, baseY,
       leftX, leftY, apexX, apexY);
}
```

Listing 14.3. This recursive function assigns an activation level to each vertex in a height map. The activation level is the coarsest level that must include the vertex in order to meet geometric error requirements.

At first glance, it may seem strange to compute an error metric from one line segment. After all, we're choosing whether or not to include a vertex in a particular LOD, and removing a vertex actually affects all the triangles that share that vertex. Depending on the triangulation of the height map, a single vertex is included in up to eight triangles!

In reality, we're computing the error for a very specific situation. Given an existing mesh, we're computing how much error is introduced if we remove a single edge separating two triangles and fuse the triangles. Thus, the computed error is relative to the previous, more complicated mesh, not relative to the original, highest-detail mesh.

Lindstrom et al. found that this approach resulted in less than 5% of the displacements exceeding the target geometric error when compared against the original mesh [102]. Furthermore, the average geometric error was well below the target geometric error. If this is inadequate for a particular application, a more precise simplification algorithm can be used, such as

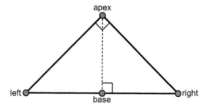

Figure 14.8. In the implicit bintree, triangle vertices are labeled *apex*, *left*, and *right*. The angle at the apex is always 90°, and the angles at *left* and *right* are equal. A fourth vertex, called *base*, bisects the line segment between the *left* and *right* vertices.

one of the algorithms mentioned in Section 12.2.3, at the cost of increased preprocessing time.

Using our rule that each successive level is allowed twice the geometric error of the prior level, we compute the number of levels past the finest level that need to also include this vertex. Finally, we subtract that from the level number of the finest level to find the actual number of the coarsest level that must include this vertex.

14.5.2 Propagation

The second part in the chunk-generation process, shown in Listing 14.4, propagates activation levels between vertices so that the simplified mesh at each level remains watertight.

```
void Propagate(HeightMap heightMap,
               int levels,
               int size)
{
  int increment = 2;
  for (int level = 0; level < levels; ++level)
  {
    for (int chunkCenterY = increment / 2;
         chunkCenterY < size;
         chunkCenterY += increment)
    {
      for (int chunkCenterX = increment / 2;
           chunkCenterX < size;
           chunkCenterX += increment)
      {
        PropagateInChunk(heightMap, level,
                         chunkCenterX, chunkCenterY);
      }
    }
    increment *= 2;
  }
}

void PropagateInChunk(HeightMap heightMap, int level,
                      int centerX, int centerY)
{
  int halfSize = 1 << level;
  int quarterSize = halfSize / 2;

  if (level > 0)
  {
    // Propagate child vertices to edge vertices
    int activationLevel;

    // Northeast child
    activationLevel =
        heightMap.GetActivationLevel(centerX + quarterSize,
                                     centerY - quarterSize);
    heightMap.Activate(centerX + halfSize,
                       centerY, activationLevel);
    heightMap.Activate(centerX,
                       centerY - halfSize, activationLevel);
```

```
    // Northwest child
    activationLevel =
        heightMap.GetActivationLevel(centerX - quarterSize,
                                     centerY - quarterSize);
    heightMap.Activate(centerX,
                       centerY - halfSize, activationLevel);
    heightMap.Activate(centerX - halfSize,
                       centerY, activationLevel);

    // Southwest child
    activationLevel =
        heightMap.GetActivationLevel(centerX - quarterSize,
                                     centerY + quarterSize);
    heightMap.Activate(centerX - halfSize,
                       centerY, activationLevel);
    heightMap.Activate(centerX,
                       centerY + halfSize, activationLevel);

    // Southeast child
    activationLevel =
        heightMap.GetActivationLevel(centerX + quarterSize,
                                     centerY + quarterSize);
    heightMap.Activate(centerX,
                       centerY + halfSize, activationLevel);
    heightMap.Activate(centerX + halfSize,
                       centerY, activationLevel);
}

// Propagate edge vertices to center.
heightMap.Activate(
    centerX, centerY,
    hf.GetActivationLevel(centerX + halfSize, centerY);
heightMap.Activate(
    centerX, centerY,
    hf.GetActivationLevel(centerX, centerY - halfSize);
heightMap.Activate(
    centerX, centerY,
    hf.GetActivationLevel(centerX, centerY + halfSize);
heightMap.Activate(
    centerX, centerY,
    hf.GetActivationLevel(centerX - halfSize, centerY);
}
```

Listing 14.4. This function propagates the activation levels computed by Update between vertices in order to maintain a watertight mesh.

To that end, propagation begins on the finest, most detailed level and proceeds up the hierarchy of levels from there. Propagation is completed for all chunks in a given level before proceeding to the next level.

Within a level, propagation proceeds chunk by chunk. The propagation process for a single chunk within a level is shown in Figure 14.9.

First, in all levels except the finest, the activation level of the vertex at the center of each of the four child quads is propagated to the adjacent edges. Then, in all levels including the finest, the activation levels of the four edge vertices are propagated to the center vertex.

When an activation level is propagated to a vertex, the vertex adopts the new activation level only if it is lower than its current activation level.

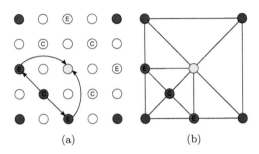

Figure 14.9. (a) Propagation for a 5×5 chunk. Child vertices are marked with C and edge vertices are marked with E. After the update step, five vertices (red) are activated. In order to create a watertight mesh at this level, the activation of the red child vertex is propagated to the blue edges. Then, the edge activation is propagated to the center (yellow). (b) The resulting tessellation.

In other words, propagation can cause a vertex to be included in coarser-detail levels but can never cause it to be removed from levels it already inhabits.

14.5.3   Meshing

In the final part of the chunk-generation process, we create a mesh from the activated vertices in each chunk.

Ulrich uses Lindstrom et al.'s algorithm to tessellate each chunk [102]. This algorithm produces one long triangle strip for the entire chunk but uses a large number of degenerate triangles to do so.

Another way to approach the problem is to tessellate the chunk using a straightforward indexed triangle list instead of a triangle strip. Then, the mesh can be handed off to an algorithm that optimizes it for the vertex cache. Forsyth and Sander et al. describe fast and effective ways to optimize arbitrary meshes for the vertex cache [50, 148].

Question ○○○○

This chunk-generation process is by far the most complicated part of a chunked LOD implementation. The preprocessing requirement also imposes limits on the user-supplied data that can be rendered. Is it necessary? Or are today's GPUs fast enough that we could simply use a tiled mipmap of the height map as a quadtree? What are the advantages and disadvantages of such an approach?

14.6   Shading

How can we apply textures to a chunked LOD terrain?

As a first cut, textures are chunked similar to geometry, and we simply associate a static texture with each chunk and map it to the chunk vertices based on their coordinates in the (x, y) plane. In fact, all of the shading techniques discussed in Section 11.4 are applicable to terrain rendered with chunked LOD. An obvious question, though, when applying many of these techniques is, what resolution texture should we use for any given chunk?

In much the same way that we determine which LOD to use for a chunk based on chunk geometric errors projected into screen space, we can also directly determine the texture resolution necessary to achieve a specific relationship between texels and pixels on the screen.

Recall that the screen-space geometric error of a chunk was computed according to Equation (12.1). Similarly, the size, t_s, of a texel projected onto the screen satisfies the inequality in Equation (14.3):

$$t_s \le \frac{t_g x}{2 d_{\min} \tan \frac{\theta}{2}},\tag{14.3}$$

where t_g is the geometric size of a texel.

By substituting τ for ρ in Equation (12.1) and rearranging it, we can determine the minimum distance, $d_{\min}$, at which the chunk will be displayed. This is Equation (14.4):

$$d_{\min} = \frac{\epsilon x}{2 \tau \tan \frac{\theta}{2}}.\tag{14.4}$$

If the distance from the viewer to the chunk is even slightly smaller than this value, the chunk's four children will be rendered instead.

Finally, we substitute $d_{\min}$ into Equation (14.3) and solve for T_g, yielding Equation (14.5):

$$t_g \le \frac{t_s \epsilon}{\tau}.\tag{14.5}$$

Using this equation, we directly compute the minimum texel size, t_g, necessary to achieve a given screen-space texel error, t_s, for a chunk with geometric error ϵ and maximum screen-space height error τ.

Conveniently, because ϵ doubles with each coarser quadtree level, the required texture resolution halves at each level.

This equation does not, however, take into account the effect of texture stretching on steeply sloped terrain features. The chunk-generation process could take this effect into account as well, if desired, at the cost of a more time-consuming preprocessing step. There is also the possibility that it could lead to an unreasonably large texture being selected for an entire chunk due to a small area of steep geometry. Texture stretching is discussed in more detail in Section 11.4.3.

Kevin Says ○○○○

It is, of course, quite reasonable for the maximum screen-space geometric error to be different from the maximum screen-space texel error. In STK and Insight3D, we use a default maximum screen-space geometric error of two pixels and a default maximum screen-space texel error of one pixel. We found that higher texel error resulted in more obvious imagery popping. However, both quantities can be configured by the user.

14.7   Out-of-Core Rendering

With a large terrain, it is not possible to fit all of the chunks into memory at once. Thus, an out-of-core algorithm is required in order to bring chunks into memory as they're needed.

We follow the outline given in Section 12.3, in which a loading thread is responsible for loading chunks that are requested by the rendering thread. This is shown in Listing 14.5.

As before, we render the first chunk that meets the screen-space error requirements, but now we also request residency for the child chunks of each rendered chunk. This acts as a form of prefetching because it increases the likelihood that the child chunks will already be resident when the viewer moves closer to the child chunks. If a chunk is already loading or resident, `RequestResidency` increases its loading priority or reduces its replacement priority, respectively (see Section 12.3.2).

In addition, we fall back on rendering the current chunk, regardless of screen-space error, anytime the current chunk's children are not loaded. Thus, on a cache miss, we render the best data that we have available for the region.

The ChunkNode instances in Listing 14.5 do not necessarily contain the chunk's vertex and index buffers. When `RequestResidency` is called by the rendering thread, the request is posted to the request queue. The loading thread selects chunks to load based on the load-ordering policy and populates ChunkNode instances with the loaded data.

A ChunkNode instance without vertex or index data is called a *skeleton node*. It contains information about the world-space extent and geometric error of the chunk, as well as the parent–child relationships between chunks, but does not contain the geometry for the chunk. The loading thread turns skeleton nodes into *loaded nodes*. It can also turn a loaded node back into a skeleton node if the node is selected for replacement according to the replacement policy (see Section 12.3.3).

```
public void Render(ChunkNode node, SceneState sceneState)
{
  RequestResidency(node);

  if (!node.AllChildChunksResident ||
      ScreenSpaceError(node, sceneState) <= tau)
  {
    RenderChunk(node, sceneState);
    foreach (ChunkNode child in node.Children)
    {
      RequestResidency(child);
    }
  }
  else
  {
    foreach (ChunkNode child in node.Children)
    {
      Render(child, sceneState);
    }
  }
}
```

Listing 14.5. The chunked LOD rendering algorithm is extended to support out-of-core rendering.

For a large terrain, there can easily be millions of nodes, so keeping all of them in memory, even just as skeletons, is prohibitive. Instead, nodes should only be kept in memory if they were recently rendered, if they are queued for loading, or if they are ancestors of loading or rendered nodes. In a garbage-collected language like C#, each node stores a WeakReference to each of its child nodes. A WeakReference does not prevent the object from being garbage collected and is automatically set to null when that happens.

Loaded and loading nodes are prevented from being garbage collected because the replacement and request queues hold normal (nonweak) references to them. In addition, nodes hold normal references to their parent nodes so that the parents are not garbage collected if the children are loading or loaded. Thus, skeleton nodes are garbage collected eventually if they don't have loaded or loading children. Since skeleton nodes, unlike loaded nodes, do not have any unmanaged resources, it is not necessary to explicitly dispose them.

In a language with explicit memory management, like C++, normal references are maintained by reference counting. To simulate the weak references from parent to child nodes, an uncounted pointer is used that is set to null by the child node's destructor.

Most chunked LOD implementations enforce the rule that if a chunk is loaded, all ancestor chunks all the way up to the root have their geometry resident as well. This is important when the viewer moves away from a chunk, causing its parent to be rendered instead of it and its three neighbors, and it is the reason for the call to `RequestResidency` at the top of Listing 14.5.

If the parent is not loaded, we have two options, neither of them very good. We can show the first ancestor that *is* loaded, resulting in changing the LOD by two or more levels at once and certainly creating objectionable popping. Or, we can continue to show the higher-detail child chunks until the parent loads. This second option leads to a potentially substantial increase in rendering load and can cause aliasing artifacts as well if the higher LOD has multiple triangles per pixel on the screen in the new view.

14.8   Chunked LOD on a Globe

Perhaps surprisingly, the chunked LOD algorithm presented so far can be used to render terrain extruded from an ellipsoid with few modifications. If the vertices in each chunk are already mapped to the ellipsoid, and the bounding volumes and geometric errors of each chunk reflect this, the core rendering algorithm is unchanged from what we've presented so far.

In fact, the only change we need to make is in morphing between levels of detail. When extruding from a plane, we simply maintain a morph delta per vertex, which is the difference in height between the coarser and the current LOD. When extruding from an ellipsoid, however, the up direction is not so easy to determine. Instead, we provide the corresponding position of the vertex in the next-coarser LOD and smoothly blend with that, as shown in Listing 14.6.

How do we create ellipsoid-mapped chunks during the chunk-generation process? The answer depends on the source terrain data, but in most cases, it's fairly straightforward.

```
in vec3 position;
in vec3 coarserPosition;
// ...

uniform float u_morph;
// ...

void main()
{
  // ...

  vec3 morphedPosition = mix(coarserPosition, position, u_morph);

  // ...
}
```

Listing 14.6. Chunks on an ellipsoid are smoothly morphed from one LOD to another given the corresponding position of the vertex in the coarser LOD mesh.

Figure 14.10. Two vertices have equal height but are far apart on the globe. Adding a vertex in the middle vastly improves the terrain's conformance to the globe, so this vertex should be active. If ellipsoid mapping is not considered when computing error, this vertex instead appears to have little impact on the error and is not activated.

For the common case where the source data is a height map in the geographic projection, the $(longitude, latitude, height)$ of each post is transformed to WGS84 Cartesian coordinates, as shown in Section 2.3. The transformed position is used for determining vertex activation during chunk generation so that large regions of equal height, such as oceans, maintain their curved appearance, as shown in Figure 14.10. In addition, the transformed coordinates are used to compute the bounding volume for the chunk.

Because a geographic projection is oversampled at the poles, chunks created in this way will be oversampled there as well. The mesh-simplification process during chunk generation goes a long way toward mitigating this, however. While a region of the input height map might have hundreds of coincident points at the poles, mesh simplification will eliminate most of these.

Care must be taken to ensure that sufficient numerical precision is preserved when mapping posts to the ellipsoid. Storing WGS84 Cartesian vertex positions as single-precision floating-point numbers will likely lead to jittering, as explained in Chapter 5. Any of the solutions explained there will address the precision problem for chunked LOD, but rendering relative to center is usually the best solution.

14.9 Chunked LOD Compared to Geometry Clipmapping

Chunked LOD and geometry clipmapping take two very different approaches to terrain LOD. The strengths and weaknesses of the two approaches

	Geometry Clipmapping	Chunked LOD
Preprocessing	Minimal. A mipmapped height map is all that is required.	Extensive. The terrain mesh is simplified to create chunks, and a bounding volume and error metric must be computed for each chunk.
Mesh Flexibility	None. The mesh must be a regular grid, and therefore, vertical and overhanging features are not possible.	Good. Chunks are irregular meshes and the topology is not important as long as a bounding volume and error metric can be determined.
Triangle Count	High. The regular grid structure burns a lot of triangles, including in areas where they provide no benefit. The triangles are pushed to the GPU very efficiently, however.	Lower. Mesh simplification eliminates unnecessary triangles.
Ellipsoid Mapping	Challenging. Solutions that preserve all of the advantages of the algorithm probably require 64-bit GPUs to be practical. Other solutions lose some of the advantages.	Straightforward. Vertices are mapped to the ellipsoid during chunk generation, and precision problems are managed using RTC or RTE techniques.
Error Control	Poor. The horizontal size of triangles can be controlled by changing the clipmap size, but extreme vertical features can make triangles arbitrarily large.	Excellent. Chunks are selected to achieve a maximum screen-space error, in pixels.
Frame-Rate Consistency	Excellent. Scenes have the same number of triangles regardless of the terrain features, so the frame rate is steady.	Poor. Because the screen-space error guides rendering, bumpy terrain will cause more high-detail chunks to be rendered and the frame rate will be lower than when rendering flatter terrain.
Mesh Continuity	Excellent. Blending between levels creates a smooth and watertight mesh.	Poor. Cracks and T-junctions are hidden rather than eliminated.
Terrain Data Size	Small. Only heights need to be stored, and the regular structure allows for impressive levels of compression. Some techniques for mapping the terrain to an ellipsoid will require that more data be stored, however.	Large. Full xyz-coordinates need to be stored, plus morph targets.
Legacy Hardware Support	Poor. GPU must be able to do efficient vertex texture fetch.	Good. Triangle throughput is important, but even the fixed-function pipeline can be used to render chunks.

Table 14.1. Comparing geometry clipmapping and chunked LOD.

are summarized in Table 14.1. In most cases, a virtual globe developer will not choose one of these algorithms and run with it, but will instead use these algorithms as starting points for developing their own, perhaps incorporating aspects of each.

14.10 Resources

Ulrich's original paper on chunked LOD is a good place to go for more information [170]. His website[2] also has prechunked test data and full source code to a public domain chunk-generation and rendering implementation.

---

[2]http://tulrich.com/geekstuff/chunklod.html

Implementing a Message Queue

Section 10.3.1 introduced **OpenGlobe.Core.**MessageQueue for communication between threads. We treated it as a black box that we posted messages to on one thread and usually processed in another thread. This appendix opens the black box and looks at the message queue's implementation. In particular, it looks at implementing the following public members:

```
public class MessageQueue : IDisposable
{
  public event EventHandler<MessageQueueEventArgs>
      MessageReceived;
  public void StartInAnotherThread();
  public void Post(object message);
  public void Post(Action<object> callback, object message);
  public void ProcessQueue();
  public void Terminate();
  // ...
}
```

These members serve the following purposes:

- **MessageReceived**. The event raised to process a message. It is raised in the thread that is processing the queue.

- **StartInAnotherThread**. Starts a dedicated thread to process messages in the queue.

- **Post**. Asynchronously adds a message. In one overload, just a message of type object is passed. In the other overload, a callback is also provided that is called instead of **MessageReceived** when the message is processed. This allows different messages to be processed in different ways. We will see its usefulness when implementing `Terminate`.

- **`ProcessQueue`.** Synchronously processes all messages in the queue in the calling thread.

- **`Terminate`.** Asynchronously signals the message queue to stop processing messages. The call returns immediately without waiting for `MessageQueue` processing to terminate. Any messages posted prior to calling this method will be processed prior to message queue termination.

The implementation of MessageQueue requires two private data members: a queue for messages and an indicator of the current state, as shown in Listing A.1.

The queue holds messages that have been posted but have not yet been processed. Conceptually, messages are added to the end and removed from the beginning. The message queue can be in one of three states:

- *Stopped.* The message queue is not processing messages. If new messages are posted, they will be added to the queue but not processed until later.

- *Running.* The message queue is running; posted messages are processed in the order they are received.

- *Stopping.* The message queue has been signaled to stop by a call to `Terminate`, but it has not yet stopped.

```
private struct MessageInfo
{
  public MessageInfo(Action<object> callback, object message)
  {
    Callback = callback;
    Message = message;
  }

  public Action<object> Callback;
  public object Message;
}

private enum State
{
  Stopped,
  Running,
  Stopping
}

private List<MessageInfo> _queue = new List<MessageInfo>();
private State _state;
```

Listing A.1. MessageQueue private data members

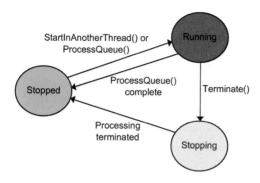

Figure A.1. Transitions between message queue states.

The transitions between these states are shown in Figure A.1. A message queue usually spends most of its time in the **Running** state, either waiting for messages or processing them.

Access to the two data members is protected by a mutex.[1] In C#, any reference type can be used as a mutex with the lock keyword, so we use _queue as the lock object for convenience. We are careful to hold the lock for the minimum amount of time necessary to achieve the necessary mutual exclusion.

The implementation for **Post** is shown in Listing A.2. The overload taking just a message is implemented in terms of the overload taking a callback and a message. The queue stores both the callback and the message using an instance of MessageInfo introduced in Listing A.1.

```
public void Post(object message)
{
    Post(null, message);
}

public void Post(Action<object> callback, object message)
{
    lock (_queue)
    {
        _queue.Add(new MessageInfo(callback, message));
        Monitor.Pulse(_queue);
    }
}
```

Listing A.2. MessageQueue.**Post** implementation.

[1] For best performance, a nonblocking message queue can be implemented using a singly linked list and atomic compare-and-swap operations. We use a mutex to keep our implementation simple; writing lock-free code can be error prone [162].

The most important part of the `Post` implementation is the thread synchronization. Since `Post` can be called from multiple threads at the same time, and the processing thread could also be accessing `_queue`, `_queue` is protected with a lock so only one thread at a time can access it. After obtaining the lock, `Post` adds an item to the queue and uses Monitor.`Pulse` to signal the potentially waiting processing thread that a message is now available for processing.[2] The object Monitor.`Pulse` only wakes up a thread that is already waiting on the same lock object. It will have no effect on a call to Monitor.`Wait` that happens after the pulse. For that reason, it is important that the call to Monitor.`Pulse` be inside the lock. Otherwise, the pulse could occur after the processing thread has decided to call Monitor.`Wait`, but before it actually does so.

Question ○○○○

> An alternative to the asynchronous `Post` is the synchronous `Send`, which blocks until the message is processed in the processing thread. How would you implement `Send`? When is deadlock possible? We've included an implementation of `Send` in MessageQueue. How does it compare to your implementation?

As described previously, messages are processed when `ProcessQueue` is called explicitly or continuously by a dedicated thread that is started with a call to `StartInAnotherThread`. The implementation of `ProcessQueue` is shown in Listing A.3.

We must first lock in order to avoid a race condition if `Post` is called in another thread, or if two different threads call `ProcessQueue` at the same time. Once we've established that no other thread is currently processing queued messages, we set the `_state` to `Running`, pull all of the currently queued messages into a local list, clear `_queue`, and release the lock. Notably, we release the lock before we actually process the messages. This maximizes parallelism by allowing other threads to `Post` messages even while previously posted messages are being processed.

Also note how we move the entire list of messages from `_queue` to `current` all at once, rather than removing one message and processing it before removing the next message. This has two advantages. First, it allows us to use List<MessageInfo>, a simple array-like data structure, rather

---

[2]In C++, the .NET lock keyword and Monitor class can be replaced with **boost::mutex** and **boost::condition_variable**, respectively. In Java, they can be replaced with the **synchronized** keyword and the **notify** method on **java.** lang.Object.

```
public void ProcessQueue ()
{
  List<MessageInfo> current = null;

  lock (_queue)
  {
    if (_state != State.Stopped)
    {
      throw new InvalidOperationException (
          "The MessageQueue is already running.");
    }

    if (_queue.Count > 0)
    {
      _state = State.Running;
      current = new List<MessageInfo>(_queue);
      _queue.Clear ();
    }
  }

  if (current != null)
  {
    ProcessCurrentQueue (current);
    lock (_queue)
    {
      _state = State.Stopped;
    }
  }
}
```

Listing A.3. MessageQueue.`ProcessQueue` implementation.

than a more complicated data structure that allows items to be efficiently removed from the beginning. Second, and more importantly, it minimizes locking overhead.

To avoid the lock overhead of **Post**, the rendering thread can add requests to a local queue and only lock once per frame when adding all requests. What are the trade-offs in doing so? How could this negatively affect the parallelism between the rendering thread and the worker thread? ○○○○ **Question**

Once the lock is released, the `current` queue is processed by Process CurrentQueue, shown in Listing A.4. Finally, the _state is transitioned back to Stopped.

We simply loop over the messages in the queue and process each one. A message is processed in one of two ways, depending on which overload of **Post** was used to post the message. If the message has a callback, the callback is invoked and passed the message data. Otherwise,

```
private void ProcessCurrentQueue(List<MessageInfo> currentQueue)
{
  for (int i = 0; i < currentQueue.Count; ++i)
  {
    if (_state == State.Stopping)
    {
      // Push the remainder of 'current' back into '_queue'.
      lock (_queue)
      {
        currentQueue.RemoveRange(0, i);
        _queue.InsertRange(0, currentQueue);
      }
      break;
    }

    MessageInfo message = currentQueue[i];
    if (message.Callback != null)
    {
      message.Callback(message.Message);
    }
    else
    {
      EventHandler<MessageQueueEventArgs> e = MessageReceived;
      if (e != null)
      {
        e(this, new MessageQueueEventArgs(message.Message));
      }
    }
  }
}
```

Listing A.4. MessageQueue.**ProcessCurrentQueue** implementation.

the **MessageReceived** event is raised with the message data in the event arguments.

Question ○○○○

> What happens if an exception is thrown while processing a message? How can the MessageQueue handle this possibility more robustly?

Processing the message can change the state to **Stopping**, in which case the unprocessed portion of the **current** queue is copied back to the _queue so that those messages will be available if and when processing is restarted and the for loop exits. Changing the state to **Stopping** is exactly how **Terminate** is implemented, as shown in Listing A.5.

It is not necessary to lock before updating _state in StopQueue because this method is posted to the queue as a message and executed by the thread that is processing messages. Since no other thread cares about this state transition, no thread synchronization is necessary.

```
public void Terminate()
{
  Post(StopQueue, null);
}

private void StopQueue(object userData)
{
  _state = State.Stopping;
}
```

Listing A.5. Implementing MessageQueue.**Terminate** with **Post** and a callback method.

> Given your knowledge of **Terminate** and **Send**, how would you implement the synchronous **TerminateAndWait**?

○○○○ Question

So far, we have a working MessageQueue that allows us to call **Process Queue** to synchronously process the messages posted to it. Listing A.6 shows **StartInAnotherThread**, which spins up a thread dedicated to processing messages posted to the queue.

Much like **ProcessQueue**, **StartInAnotherThread** takes the lock and then verifies that no other thread is already processing messages. It then transitions the _state to **Running** and starts a new thread. The thread is configured as a background thread, which means it will not stop the overall process in which it is running from terminating. The entry point for the new thread is a method called **Run**, shown in Listing A.7.

```
public void StartInAnotherThread()
{
  lock (_queue)
  {
    if (_state != State.Stopped)
    {
      throw new InvalidOperationException(
          "The MessageQueue is already running.");
    }

    _state = State.Running;
  }

  Thread thread = new Thread(Run);
  thread.IsBackground = true;
  thread.Start();
}

private void Run() { /* ... */ }
```

Listing A.6. MessageQueue.**StartInAnotherThread** implementation.

```
private void Run()
{
  List<MessageInfo> current = new List<MessageInfo>();

  do
  {
    lock (_queue)
    {
      if (_queue.Count > 0)
      {
        current.AddRange(_queue);
        _queue.Clear();
      }
      else
      {
        Monitor.Wait(_queue);

        current.AddRange(_queue);
        _queue.Clear();
      }
    }

    ProcessCurrentQueue(current);
    current.Clear();
  } while (_state == State.Running);

  lock (_queue)
  {
    _state = State.Stopped;
  }
}
```

Listing A.7. Processing messages in the queue with MessageQueue.**Run**.

The implementation of this method is in many ways similar to the implementation of **ProcessQueue** shown in Listing A.3. We still take the lock, copy the contents of the _queue into a local **current** queue, release the lock, and process the **current** queue. The difference, though, is that the whole process is now wrapped in a do...while loop that repeats the process until the queue is no longer in the **Running** state. That will happen when the message posted by **Terminate** is processed and transitions the state to **Stopping**. In other words, it keeps processing messages in a loop until terminated.

Another significant difference is the call to Monitor.Wait when the queue has no messages. Without that call, the MessageQueue would still work! It is very important to include it, though, because it keeps the do ...while loop from spinning at full speed, busy waiting, and using lots of CPU time without doing any useful work. The alert reader may have noticed that, until now, we have not justified the existence of the call to Monitor.**Pulse** in **Post**. The call to Monitor.**Wait** in this method is the reason that it exists. The object Monitor.**Wait** causes the thread to wait, *without using any CPU time*, until another thread calls Monitor.**Pulse**.

MessageQueue is unbounded; it can grow indefinitely if messages are posted faster than they are consumed. Add a `MaximumLength` property so `Post` blocks if adding a message to the queue will exceed this property's value.

○○○○ **Try This**

MessageQueue allows multiple threads to post to the same queue but only a single thread to process messages. To emulate multiple processing threads, we suggested multiple message queues, scheduled using round-robin in Section 10.3.2. Modify the message queue to support multiple processing threads. Start with round-robin scheduling, then modify the design to allow user-defined scheduling algorithms.

○○○○ **Try This**

Bibliography

[1] Kurt Akeley and Jonathan Su. "Minimum Triangle Separation for Correct z-Buffer Occlusion." In *Proceedings of the 21st ACM SIGGRAPH/EUROGRAPHICS Symposium on Graphics Hardware*. New York: ACM Press, 2006. Available at http://research.microsoft.com/pubs/79213/GH%202006%20p027%20Akeley%20Su.pdf.

[2] Kurt Akeley. "The Hidden Charms of the z-Buffer." In *IRIS Universe 11*, pp. 31–37, 1990.

[3] Tomas Akenine-Moller, Eric Haines, and Naty Hoffman. *Real-Time Rendering*, Third edition. Wellesley, MA: A K Peters, Ltd., 2008.

[4] Johan Andersson and Natalya Tatarchuk. "Frostbite: Rendering Architecture and Real-Time Procedural Shading & Texturing Techniques." In *Game Developers Conference*, 2007. Available at http://developer.amd.com/assets/Andersson-Tatarchuk-FrostbiteRenderingArchitecture%28GDC07_AMD_Session%29.pdf.

[5] Johan Andersson. "Terrain Rendering in Frostbite Using Procedural Shader Splatting." In *Proceedings of SIGGRAPH 07: ACM SIGGRAPH 2007 Courses*, pp. 38–58. New York: ACM Press, 2007. Available at http://developer.amd.com/media/gpu_assets/Andersson-TerrainRendering(Siggraph07).pdf.

[6] Johan Andersson. "Parallel Graphics in Frostbite—Current & Future." In *SIGGRAPH*, 2009. Available at http://s09.idav.ucdavis.edu/talks/04-JAndersson-ParallelFrostbite-Siggraph09.pdf.

[7] Edward Angel. *Interactive Computer Graphics: A Top-Down Approach Using OpenGL*, Fourth edition. Reading, MA: Addison Wesley, 2005.

[8] Apple. "Enabling Multi-Threaded Execution of the OpenGL Framework." Technical report, Apple, 2006. Available at http://developer.apple.com/library/mac/#technotes/tn2006/tn2085.html.

[9] Arul Asirvatham and Hugues Hoppe. "Terrain Rendering Using GPU-Based Geometry Clipmaps." In *GPU Gems 2*, edited by Matt Pharr, pp. 27–45. Reading, MA: Addison-Wesley, 2005. Available at http://http.developer.nvidia.com/GPUGems2/gpugems2_chapter02.html.

[10] Andreas Baerentzen, Steen Lund Nielsen, Mikkel Gjael, Bent D. Larsen, and Niels Jaergen Christensen. "Single-Pass Wireframe Rendering." In *Proceedings of SIGGRAPH 06: ACM SIGGRAPH 2006 Sketches*. New York: ACM Press, 2006. Available at http://www2.imm.dtu.dk/pubdb/views/publication_details.php?id=4884.

[11] Steve Baker. "Learning to Love Your z-Buffer." Available at http://www.sjbaker.org/steve/omniv/love_your_z_buffer.html, 1999.

[12] Avi Bar-Zeev. "How Google Earth Really Works." Available at http://www.realityprime.com/articles/how-google-earth-really-works, 2007.

[13] Sean Barrett. "Enumeration of Polygon Edges from Vertex Windings." Available at http://nothings.org/computer/edgeenum.html, 2008.

[14] Louis Bavoil. "Efficient Multifragment Effects on Graphics Processing Units." Master's thesis, University of Utah, 2007. Available at http://www.sci.utah.edu/~csilva/papers/thesis/louis-bavoil-ms-thesis.pdf.

[15] Bill Bilodeau and Mike Songy. "Real Time Shadows." In *Creativity '99, Creative Labs Inc. Sponsored Game Developer Conferences*, 1999.

[16] Ruzinoor bin Che Mat and Dr. Mahes Visvalingam. "Effectiveness of Silhouette Rendering Algorithms in Terrain Visualisation." In *Proceedings of the National Conference on Computer Graphics and Multimedia (CoGRAMM)*, 2002. Available at http://staf.uum.edu.my/ruzinoor/COGRAMM.pdf.

[17] Charles Bloom. "Terrain Texture Compositing by Blending in the Frame-Buffer." Available at http://www.cbloom.com/3d/techdocs/splatting.txt, 2000.

[18] Charles Bloom. "View Independent Progressive Meshes (VIPM)." Available at http://www.cbloom.com/3d/techdocs/vipm.txt, 2000.

[19] Jeff Bolz. "ARB_ES2_compatibility." Available at http://www.opengl.org/registry/specs/ARB/ES2_compatibility.txt, 2010.

[20] Jeff Bolz. "ARB_shader_subroutine." Available at http://www.opengl.org/registry/specs/ARB/shader_subroutine.txt, 2010.

[21] Flavien Brebion. "Tip of the Day: Logarithmic z-Buffer Artifacts Fix." Available at http://www.gamedev.net/blog/73/entry-2006307-tip-of-the-day-logarithmic-zbuffer-artifacts-fix/, 2009.

[22] Pat Brown. "EXT_texture_array." Available at http://www.opengl.org/registry/specs/EXT/texture_array.txt, 2008.

[23] Bruce M. Bush. "The Perils of Floating Point." Available at http://www.lahey.com/float.htm, 1996.

[24] Fay Chang, Jeffrey Dean, Sanjay Ghemawat, Wilson C. Hsieh, Deborah A. Wallach, Mike Burrows, Tushar Chandra, Andrew Fikes, and Robert E. Gruber. "Bigtable: A Distributed Storage System for Structured Data." In *Proceedings of the 7th Conference on USENIX Symposium on Operating Systems Design and Implementation*, 7, 7, pp. 205–218, 2006. Available at http://labs.google.com/papers/bigtable-osdi06.pdf.

[25] Bernard Chazelle. "Triangulating a Simple Polygon in Linear Time." *Discrete Comput. Geom.* 6:5 (1991), 485–524.

[26] M. Christen. "The Future of Virtual Globes—The Interactive Ray-Traced Digital Earth." In *ISPRS Congress Beijing 2008, Proceedings of Commission II, ThS 4*, 2008. Available at http://www.3dgi.ch/publications/chm/virtualglobesfuture.pdf.

[27] Malte Clasen and Hans-Christian Hege. "Terrain Rendering Using Spherical Clipmaps." In *Eurographics/IEEE-VGTC Symposium on Visualisation*, pp. 91–98, 2006. Available at http://www.zib.de/clasen/?page_id=6.

[28] Kate Compton, James Grieve, Ed Goldman, Ocean Quigley, Christian Stratton, Eric Todd, and Andrew Willmott. "Creating Spherical Worlds." In *Proceedings of SIGGRAPH 07: ACM SIGGRAPH 2007 Sketches*, p. 82. New York: ACM Press, 2007. Available at http://www.andrewwillmott.com/s2007.

[29] Wagner T. Correa, James T. Klosowski, and Claudio T. Silva. "Visibility-Based Prefetching for Interactive Out-of-Core Rendering." In *Proceedings of the 2003 IEEE Symposium on Parallel and Large-Data Visualization and Graphics*. Los Alamitos, CA: IEEE Computer Society, 2003. Available at http://www.evl.uic.edu/cavern/rg/20040525_renambot/Viz/parallel_volviz/prefetch_outofcore_viz_pvg03.pdf.

[30] Patrick Cozzi and Frank Stoner. "GPU Ray Casting of Virtual Globes." In *Proceedings of SIGGRAPH 10: ACM SIGGRAPH 2010 Posters*, p. 1. New York: ACM Press, 2010. Available at http://www.agi.com/gpuraycastingofvirtualglobes.

[31] Patrick Cozzi. "A Framework for GLSL Engine Uniforms." In *Game Engine Gems 2*, edited by Eric Lengyel. Natick, MA: A K Peters, Ltd., 2011. Available at http://www.gameenginegems.net/.

[32] Patrick Cozzi. "Delaying OpenGL Calls." In *Game Engine Gems 2*, edited by Eric Lengyel. Natick, MA: A K Peters, Ltd., 2011. Available at http://www.gameenginegems.net/.

[33] Matt Craighead, Mark Kilgard, and Pat Brown. "ARB_point_sprite." Available at http://www.opengl.org/registry/specs/ARB/point_sprite.txt, 2003.

[34] Matt Craighead. "NV_primitive_restart." Available at http://www.opengl.org/registry/specs/NV/primitive_restart.txt, 2002.

[35] Cyril Crassin, Fabrice Neyret, Sylvain Lefebvre, and Elmar Eisemann. "GigaVoxels: Ray-Guided Streaming for Efficient and Detailed Voxel Rendering." In *ACM SIGGRAPH Symposium on Interactive 3D Graphics and Games (I3D)*. New York: ACM Press, 2009. Available at http://artis.imag.fr/Publications/2009/CNLE09.

[36] Chenguang Dai, Yongsheng Zhang, and Jingyu Yang. "Rendering 3D Vector Data Using the Theory of Stencil Shadow Volumes." In *ISPRS Congress Beijing 2008, Proceedings of Commission II, WG II/5*, 2008. Available at http://www.isprs.org/proceedings/XXXVII/congress/2_pdf/5_WG-II-5/06.pdf.

[37] Christian Dick, Jens Krüger, and Rüdiger Westermann. "GPU Ray-Casting for Scalable Terrain Rendering." In *Proceedings of Eurographics 2009—Areas Papers*, pp. 43–50, 2009. Available at http://wwwcg.in.tum.de/Research/Publications/TerrainRayCasting.

[38] Christian Dick. "Interactive Methods in Scientific Visualization: Terrain Rendering." In *IEEE/VGTC Visualization Symposium*, 2009. Available at http://wwwcg.in.tum.de/Tutorials/PacificVis09/Terrain.pdf.

[39] Alan Neil Ditchfield. "Honeycomb Spherical Figure." Available at http://www.neubert.net/Download/global-grid.doc, 2001.

[40] David Douglas and Thomas Peucker. "Algorithms for the Reduction of the Number of Points Required to Represent a Digitized Line or its Caricature." *The Canadian Cartographer* 10:2 (1973), 112–122.

[41] Mark Duchaineau, Murray Wolinsky, David E. Sigeti, Mark C. Miller, Charles Aldrich, and Mark B. Mineev-Weinstein. "ROAMing Terrain: Real-Time Optimally Adapting Meshes." In *Proceedings of the 8th Conference on Visualization '97*, pp. 81–88. Los Alamitos, CA: IEEE Computer Society, 1997.

[42] Jonathan Dummer. "Cone Step Mapping: An Iterative Ray-Heightfield Intersection Algorithm." Available at http://www.lonesock.net/files/ConeStepMapping.pdf, 2006.

[43] David Eberly. *3D Game Engine Design: A Practical Approach to Real-Time Computer Graphics*, Second edition. San Francisco: Morgan Kaufmann, 2006.

[44] David Eberly. "Triangulation by Ear Clipping." Available at http://www.geometrictools.com/Documentation/TriangulationByEarClipping.pdf, 2008.

[45] David Eberly. "Wild Magic 5 Overview." Available at http://www.geometrictools.com/WildMagic5Overview.pdf, 2010.

[46] Christer Ericson. "Physics for Games Programmers: Numerical Robustness (for Geometric Calculations)." Available at http://realtimecollisiondetection.net/pubs/, 2007.

[47] Christer Ericson. "Order Your Graphics Draw Calls Around!" Available at http://realtimecollisiondetection.net/blog/?p=86, 2008.

[48] Carl Erikson, Dinesh Manocha, and William V. Baxter III. "HLODs for Faster Display of Large Static and Dynamic Environments." In *Proceedings of the 2001 Symposium on Interactive 3D Graphics*, pp. 111–120. New York: ACM Press, 2001.

[49] ESRI. "ESRI Shapefile Technical Description." Available at http://www.esri.com/library/whitepapers/pdfs/shapefile.pdf, 1998.

[50] Tom Forsyth. "Linear-Speed Vertex Cache Optimisation." Available at http://home.comcast.net/~tom_forsyth/papers/fast_vert_cache_opt.html, 2006.

[51] Tom Forsyth. "A Matter of Precision." Available at http://home.comcast.net/~tom_forsyth/blog.wiki.html#[[A%20matter%20of%20precision]], 2006.

[52] Tom Forsyth. "Renderstate Change Costs." Available at http://home.comcast.net/~tom_forsyth/blog.wiki.html#[[Renderstate%20change%20costs]], 2008.

[53] Anton Frühstück. "GPU Based Clipmaps." Master's thesis, Vienna University of Technology, 2008. Available at http://www.cg.tuwien.ac.at/research/publications/2008/fruehstueck-2008-gpu/.

[54] M. R. Garey and D. S. Johnson. *Computers and Intractability: A Guide to the Theory of NP-Completeness.* New York: W. H. Freeman, 1979.

[55] Michael Garland and Paul S. Heckbert. "Surface Simplification Using Quadric Error Metrics." In *Proceedings of SIGGRAPH 97, Computer Graphics Proceedings, Annual Conference Series*, edited by Turner Whitted, pp. 209–216. Reading, MA: Addison Wesley, 1997. Available at http://mgarland.org/files/papers/quadrics.pdf.

[56] Samuel Gateau. "Solid Wireframe." In *White Paper WP-03014-001 v01.* NVIDIA Corporation, 2007. Available at http://developer.download.nvidia.com/SDK/10.5/direct3d/Source/SolidWireframe/Doc/SolidWireframe.pdf.

[57] Nick Gebbie and Mike Bailey. "Fast Realistic Rendering of Global Worlds Using Programmable Graphics Hardware." *Journal of Game Development* 1:4 (2006), 5–28. Available at http://web.engr.oregonstate.edu/~mjb/WebMjb/Papers/globalworlds.pdf.

[58] Ryan Geiss and Michael Thompson. "NVIDIA Demo Team Secrets: Cascades." In *Game Developers Conference*, 2007. Available at http://www.slideshare.net/icastano/cascades-demo-secrets.

[59] Ryan Geiss. "Generating Complex Procedural Terrains Using the GPU." In *GPU Gems 3*, edited by Hubert Nguyen, pp. 7–37. Reading, MA: Addison Wesley, 2007. Available at http://http.developer.nvidia.com/GPUGems3/gpugems3_ch01.html.

[60] Philipp Gerasimov, Randima Fernando, and Simon Green. "Shader Model 3.0: Using Vertex Textures." Available at ftp://download.nvidia.com/developer/Papers/2004/Vertex_Textures/Vertex_Textures.pdf, 2004.

[61] Thomas Gerstner. "Multiresolution Visualization and Compression of Global Topographic Data." *GeoInformatica* 7:1 (2003), 7–32. Available at http://wissrech.iam.uni-bonn.de/research/pub/gerstner/globe.pdf.

[62] Benno Giesecke. "Space Vehicles in Virtual Globes: Recommendations for the Visualization Of Objects in Space." Available at http://www.agi.com/downloads/support/productSupport/literature/pdfs/whitePapers/2007-05-24_SpaceVehiclesinVirtualGlobes.pdf, 2007.

[63] Enrico Gobbetti, Dave Kasik, and Sung eui Yoon. "Technical Strategies for Massive Model Visualization." In *Proceedings of the 2008 ACM Symposium on Solid and Physical Modeling*, pp. 405–415. New York: ACM Press, 2008.

[64] Google. "KML Reference." Available at http://code.google.com/apis/kml/documentation/kmlreference.html, 2010.

[65] K. M. Gorski, E. Hivon, A. J. Banday, B. D. Wandelt, F. K. Hansen, M. Reinecke, and M. Bartelmann. "HEALPix: A Framework for High-Resolution Discretization and Fast Analysis of Data Distributed on the Sphere." *The Astrophysical Journal* 622:2 (2005), 759–771. Available at http://stacks.iop.org/0004-637X/622/759.

[66] Evan Hart. "OpenGL SDK Guide." Available at http://developer.download.nvidia.com/SDK/10.5/opengl/OpenGL_SDK_Guide.pdf, 2008.

[67] Chris Hecker. "Let's Get to the (Floating) Point." *Game Developer Magazine* February (1996), 19–24. Available at http://chrishecker.com/Miscellaneous_Technical_Articles#Floating_Point.

[68] Tim Heidmann. "Real Shadows, Real Time." In *IRIS Universe*, 18, 18, pp. 23–31. Fremont, CA: Silicon Graphics Inc., 1991.

[69] Martin Held. "FIST: Fast Industrial-Strength Triangulation of Polygons." Technical report, Algorithmica, 2000. Available at http://cgm.cs.mcgill.ca/~godfried/publications/triangulation.held.ps.gz.

[70] Martin Held. "FIST: Fast Industrial-Strength Triangulation of Polygons." Available at http://www.cosy.sbg.ac.at/~held/projects/triang/triang.html, 2008.

[71] Martin Held. "Algorithmische Geometrie: Triangulation." Available at http://www.cosy.sbg.ac.at/~held/teaching/compgeo/slides/triang_slides.pdf, 2010.

[72] John L. Hennessy and David A. Patterson. *Computer Architecture: A Quantitative Approach*, Fourth edition. San Francisco: Morgan Kaufmann, 2006.

[73] John Hershberger and Jack Snoeyink. "Speeding Up the Douglas-Peucker Line-Simplification Algorithm." In *Proc. 5th Intl. Symp. on Spatial Data Handling*, pp. 134–143, 1992. Available at http://www.bowdoin.edu/~ltoma/teaching/cs350/spring06/Lecture-Handouts/hershberger92speeding.pdf.

[74] Hugues Hoppe. "Smooth View-Dependent Level-of-Detail Control and Its Application to Terrain Rendering." In *Proceedings of the Conference on Visualization '98*, pp. 35–42. Los Alamitos, CA: IEEE Computer Society, 1998.

[75] Takeo Igarashi and Dennis Cosgrove. "Adaptive Unwrapping for Interactive Texture Painting." In *ACM Symposium on Interactive 3D Graphics*, pp. 209–216. New York: ACM Press, 2001. Available at http://www-ui.is.s.u-tokyo.ac.jp/~takeo/papers/i3dg2001.pdf.

[76] Ivan-Assen Ivanov. "Practical Texture Atlases." In *Gamasutra*, 2006. Available at http://www.gamasutra.com/features/20060126/ivanov_01.shtml.

[77] Tim Jenks. "Terrain Texture Blending on a Programmable GPU." Available at http://www.jenkz.org/articles/terraintexture.htm, 2005.

[78] Jukka Jylänki. "A Thousand Ways to Pack the Bin—A Practical Approach to Two-Dimensional Rectangle Bin Packing." Available at http://clb.demon.fi/files/RectangleBinPack.pdf, 2010.

[79] Jukka Jylänki. "Even More Rectangle Bin Packing." Available at http://clb.demon.fi/projects/even-more-rectangle-bin-packing, 2010.

[80] Anu Kalra and J.M.P. van Waveren. "Threading Game Engines: QUAKE 4 & Enemy Territory QUAKE Wars." In *Game Developer Conference*, 2008. Available at http://mrelusive.com/publications/presentations/2008_gdc/GDC%2008%20Threading%20QUAKE%204%20and%20ETQW%20Final.pdf.

[81] Brano Kemen. "Floating Point Depth Buffer." Available at http://outerra.blogspot.com/2009/12/floating-point-depth-buffer.html, 2009.

[82] Brano Kemen. "Logarithmic Depth Buffer." Available at http://outerra.blogspot.com/2009/08/logarithmic-z-buffer.html, 2009.

[83] R. Keys. "Cubic Convolution Interpolation for Digital Image Processing." *IEEE Transactions on Acoustics, Speech and Signal Processing* ASSP-29 (1981), 1153–1160.

[84] Mark Kilgard and Daniel Koch. "ARB_fragment_coord_conventions." Available at http://www.opengl.org/registry/specs/ARB/fragment_coord_conventions.txt, 2009.

[85] Mark Kilgard and Daniel Koch. "ARB_provoking_vertex." Available at http://www.opengl.org/registry/specs/ARB/provoking_vertex.txt, 2009.

[86] Mark Kilgard and Daniel Koch. "ARB_vertex_array_bgra." Available at http://www.opengl.org/registry/specs/ARB/vertex_array_bgra.txt, 2009.

[87] Mark Kilgard, Greg Roth, and Pat Brown. "GL_ARB_separate_shader_objects." Available at http://www.opengl.org/registry/specs/ARB/separate_shader_objects.txt, 2010.

[88] Mark Kilgard. "Avoiding 16 Common OpenGL Pitfalls." Available at http://www.opengl.org/resources/features/KilgardTechniques/oglpitfall/, 2000.

[89] Mark Kilgard. "More Advanced Hardware Rendering Techniques." In *Game Developers Conference*, 2001. Available at http://developer.download.nvidia.com/assets/gamedev/docs/GDC01_md2shader_PDF.zip.

[90] Mark Kilgard. "OpenGL 3: Revolution through Evolution." In *Proceedings of SIGGRAPH Asia*, 2008. Available at http://www.khronos.org/developers/library/2008_siggraph_asia/OpenGL%20Overview%20SIGGRAPH%20Asia%20Dec08%20.pdf.

[91] Mark Kilgard. "EXT_direct_state_access." Available at http://www.opengl.org/registry/specs/EXT/direct_state_access.txt, 2010.

[92] Donald E. Knuth. *Seminumerical Algorithms*, Second edition. The Art of Computer Programming, vol. 2, Reading, MA: Addison-Wesley, 1997.

[93] Jaakko Konttinen. "ARB_debug_output." Available at http://www.opengl.org/registry/specs/ARB/debug_output.txt, 2010.

[94] Eugene Lapidous and Guofang Jiao. "Optimal Depth Buffer for Low-Cost Graphics Hardware." In *Proceedings of the ACM SIGGRAPH/EUROGRAPHICS Workshop on Graphics Hardware*, pp. 67–73, 1999. Available at http://www.graphicshardware.org/previous/www_1999/presentations/d-buffer/.

[95] Cedric Laugerotte. "Tessellation of Sphere by a Recursive Method." Available at http://student.ulb.ac.be/~claugero/sphere/, 2001.

[96] Jon Leech. "ARB_sync." Available at http://www.opengl.org/registry/specs/ARB/sync.txt, 2009.

[97] Aaron Lefohn and Mike Houston. "Beyond Programmable Shading." In *ACM SIGGRAPH 2008 Courses*, 2008. Available at http://s08.idav.ucdavis.edu/.

[98] Aaron Lefohn and Mike Houston. "Beyond Programmable Shading." In *ACM SIGGRAPH 2009 Courses*, 2009. Available at http://s09.idav.ucdavis.edu/.

[99] Aaron Lefohn and Mike Houston. "Beyond Programmable Shading." In *ACM SIGGRAPH 2010 Courses*, 2010. Available at http://bps10.idav.ucdavis.edu/.

[100] Ian Lewis. "Getting More From Multicore." In *Game Developer Conference*, 2008. Available at http://www.microsoft.com/downloads/en/details.aspx?FamilyId=A36FE736-5FE7-4E08-84CF-ACCF801538EB&displaylang=en.

[101] Peter Lindstrom and Jonathan D. Cohen. "On-the-Fly Decompression and Rendering of Multiresolution Terrain." In *Proceedings of the 2010 ACM SIGGRAPH Symposium on Interactive 3D Graphics and Games*, pp. 65–73. New York: ACM Press, 2010. Available at https://e-reports-ext.llnl.gov/pdf/371781.pdf.

[102] Peter Lindstrom, David Koller, William Ribarsky, Larry F. Hodges, Nick Faust, and Gregory A. Turner. "Real-Time, Continuous Level of Detail Rendering of Height Fields." In *Proceedings of SIGGRAPH 96, Computer Graphics Proceedings, Annual Conference Series*, edited by Holly Rushmeier, pp. 109–118. Reading, MA: Addison Wesley, 1996.

[103] Benj Lipchak, Greg Roth, and Piers Daniell. "ARB_get_program_binary." Available at http://www.opengl.org/registry/specs/ARB/get_program_binary.txt, 2010.

[104] William E. Lorensen and Harvey E. Cline. "Marching Cubes: A High Resolution 3D Surface Construction Algorithm." *SIGGRAPH Computer Graphics* 21:4 (1987), 163–169.

[105] Frank Losasso and Hugues Hoppe. "Geometry Clipmaps: Terrain Rendering Using Nested Regular Grids." In *Proceedings of SIGGRAPH 2004 Papers*, pp. 769–776. New York: ACM Press, 2004. Available at http://research.microsoft.com/en-us/um/people/hoppe/proj/geomclipmap/.

[106] Kok-Lim Low and Tiow-Seng Tan. "Model Simplification Using Vertex-Clustering." In *I3D 97: Proceedings of the 1997 Symposium on Interactive 3D Graphics*, pp. 75–ff. New York: ACM Press, 1997.

[107] David Luebke, Martin Reddy, Jonathan Cohen, Amitabh Varshney, Benjamin Watson, and Robert Huebner. *Level of Detail for 3D Graphics*. San Francisco: Morgan Kaufmann, 2002. Available at http://lodbook.com/.

[108] Henrique S. Malvar. "Fast Progressive Image Coding without Wavelets." In *Proceedings of the Conference on Data Compression*. Washington, DC: IEEE Computer Society, 2000. Available at http://research.microsoft.com/apps/pubs/?id=101991.

[109] CCP Mannapi. "Awesome Looking Planets." Available at http://www.eveonline.com/devblog.asp?a=blog&bid=724, 2010.

[110] Paul Martz. "OpenGL FAQ: The Depth Buffer." Available at http://www.opengl.org/resources/faq/technical/depthbuffer.htm, 2000.

[111] Grégory Massal. "Depth Buffer—The Gritty Details." Available at http://www.codermind.com/articles/Depth-buffer-tutorial.html, 2006.

[112] Morgan McGuire and Kyle Whitson. "Indirection Mapping for Quasi-Conformal Relief Mapping." In *ACM SIGGRAPH Symposium on Interactive 3D Graphics and Games (I3D '08)*, 2008. Available at http://graphics.cs.williams.edu/papers/IndirectionI3D08/.

[113] Tom McReynolds and David Blythe. *Advanced Graphics Programming Using OpenGL, The Morgan Kaufmann Series in Computer Graphics*. San Francisco: Morgan Kaufmann, 2005.

[114] "Direct3D 11 MultiThreading." Available at http://msdn.microsoft.com/en-us/library/ff476884.aspx, 2010.

[115] Microsoft. "DirectX Software Development Kit—RaycastTerrain Sample." Available at http://msdn.microsoft.com/en-us/library/ee416425(v=VS.85).aspx, 2008.

[116] James R. Miller and Tom Gaskins. "Computations on an Ellipsoid for GIS." *Computer-Aided Design and Applications* 6:4 (2009), 575–583. Available at http://people.eecs.ku.edu/~miller/Papers/CAD_6_4__575-583.pdf.

[117] NASA Solar System Exploration. "Mars: Moons: Phobos." Available at http://solarsystem.nasa.gov/planets/profile.cfm?Object=Mar_Phobos, 2003.

[118] National Imagery and Mapping Agency. *Department of Defense World Geodetic System 1984: Its Definition and Relationships with Local Geodetic Systems*, Third edition. National Imagery and Mapping Agency, 2000. Available at http://earth-info.nga.mil/GandG/publications/tr8350.2/wgs84fin.pdf.

[119] Kris Nicholson. "GPU Based Algorithms for Terrain Texturing." Master's thesis, University of Canterbury, 2008. Available at http://www.cosc.canterbury.ac.nz/research/reports/HonsReps/2008/hons_0801.pdf.

[120] NVIDIA. "Using Vertex Buffer Objects." Available at http://www.nvidia.com/object/using_VBOs.html, 2003.

[121] NVIDIA. "Improve Batching Using Texture Atlases." Available at http://developer.download.nvidia.com/SDK/9.5/Samples/DEMOS/Direct3D9/src/BatchingViaTextureAtlases/AtlasCreationTool/Docs/Batching_Via_Texture_Atlases.pdf, 2004.

[122] NVIDIA. "NVIDIA CUDA Compute Unified Device Architecture Programming Guide." Available at http://developer.download.nvidia.com/compute/cuda/2_0/docs/NVIDIA_CUDA_Programming_Guide_2.0.pdf, 2008.

[123] NVIDIA. "NVIDIA GPU Programming Guide." Available at http://www.nvidia.com/object/gpu_programming_guide.html, 2008.

[124] NVIDIA. "Bindless Graphics Tutorial." Available at http://www.nvidia.com/object/bindless_graphics.html, 2009.

[125] Deron Ohlarik. "Horizon Culling." Available at http://blogs.agi.com/insight3d/index.php/2008/04/18/horizon-culling/, 2008.

[126] Deron Ohlarik. "Precisions, Precisions." Available at http://blogs.agi.com/insight3d/index.php/2008/09/03/precisions-precisions/, 2008.

[127] Deron Ohlarik. "Triangulation Rhymes with Strangulation." Available at http://blogs.agi.com/insight3d/index.php/2008/03/20/triangulation-rhymes-with-strangulation/, 2008.

[128] Deron Ohlarik. "Horizon Culling 2." Available at http://blogs.agi.com/insight3d/index.php/2009/03/25/horizon-culling-2/, 2009.

[129] Jon Olick. "Next Generation Parallelism in Games." In *Proceedings of ACM SIGGRAPH 2008 Courses.* New York: ACM Press, 2008. Available at http://s08.idav.ucdavis.edu/olick-current-and-next-generation-parallelism-in-games.pdf.

[130] Sean O'Neil. "A Real-Time Procedural Universe, Part Three: Matters of Scale." Available at http://www.gamasutra.com/view/feature/2984/a_realtime_procedural_universe_.php, 2002.

[131] Open Geospatial Consortium Inc. "OSG KML." Available at http://portal.opengeospatial.org/files/?artifact_id=27810, 2008.

[132] Charles B. Owen. "CSE 872 Advanced Computer Graphics—Tutorial 4: Texture Mapping." Available at http://www.cse.msu.edu/~cse872/tutorial4.html, 2008.

[133] David Pangerl. "Practical Thread Rendering for DirectX 9." In *GPU Pro,* edited by Wolfgang Engel. Natick, MA: A K Peters, Ltd., 2010. Available at http://www.akpeters.com/gpupro/.

[134] Steven G. Parker, James Bigler, Andreas Dietrich, Heiko Friedrich, Jared Hoberock, David Luebke, David McAllister, Morgan McGuire, Keith Morley, Austin Robison, and Martin Stich. "OptiX: A General Purpose Ray Tracing Engine." *ACM Transactions on Graphics.* Available at http://graphics.cs.williams.edu/papers/OptiXSIGGRAPH10/.

[135] Emil Persson. "ATI Radeon HD 2000 Programming Guide." Available at http://developer.amd.com/media/gpu_assets/ATI_Radeon_HD_2000_programming_guide.pdf, 2007.

[136] Emil Persson. "Depth In-Depth." Available at http://developer.amd.com/media/gpu_assets/Depth_in-depth.pdf, 2007.

[137] Emil Persson. "A Couple of Notes about z." Available at http://www.humus.name/index.php?page=News&ID=255, 2009.

[138] Emil Persson. "New Particle Trimming Tool." Available at http://www.humus.name/index.php?page=News&ID=266, 2009.

[139] Emil Persson. "Triangulation." Available at http://www.humus.name/index.php?page=News&ID=228, 2009.

[140] Emil Persson. "Making it Large, Beautiful, Fast, and Consistent: Lessons Learned Developing Just Cause 2." In *GPU Pro*, edited by Wolfgang Engel, pp. 571–596. Natick, MA: A K Peters, Ltd., 2010. Available at http://www.akpeters.com/gpupro/.

[141] Matt Pettineo. "Attack of the Depth Buffer." Available at http://mynameismjp.wordpress.com/2010/03/22/attack-of-the-depth-buffer/, 2010.

[142] Frank Puig Placeres. "Fast Per-Pixel Lighting with Many Lights." In *Game Programming Gems 6*, edited by Michael Dickheiser, pp. 489–499. Hingham, MA: Charles River Media, 2006.

[143] John Ratcliff. "Texture Packing: A Code Snippet to Compute a Texture Atlas." Available at http://codesuppository.blogspot.com/2009/04/texture-packing-code-snippet-to-compute.html, 2009.

[144] Ashu Rege. "Shader Model 3.0." Available at ftp://download.nvidia.com/developer/presentations/2004/GPU_Jackpot/Shader_Model_3.pdf, 2004.

[145] Marek Rosa. "Destructible Volumetric Terrain." In *GPU Pro*, edited by Wolfgang Engel, pp. 597–608. Natick, MA: A K Peters, Ltd., 2010. Available at http://www.akpeters.com/gpupro/.

[146] Jarek Rossignac and Paul Borrel. "Multi-Resolution 3D Approximations for Rendering Complex Scenes." In *Modeling in Computer Graphics: Methods and Applications*, edited by B. Falcidieno and T. Kunii, pp. 455–465. Berlin: Springer-Verlag, 1993. Available at http://www.cs.uu.nl/docs/vakken/ddm/slides/papers/rossignac.pdf.

[147] Randi J. Rost, Bill Licea-Kane, Dan Ginsburg, John M. Kessenich, Barthold Lichtenbelt, Hugh Malan, and Mike Weiblen. *OpenGL Shading Language*, Third edition. Reading, MA: Addison-Wesley, 2009.

[148] Pedro V. Sander, Diego Nehab, and Joshua Barczak. "Fast Triangle Reordering for Vertex Locality and Reduced Overdraw." *Proc. SIGGRAPH 07, Transactions on Graphics* 26:3 (2007), 1–9.

[149] Martin Schneider and Reinhard Klein. "Efficient and Accurate Rendering of Vector Data on Virtual Landscapes." *Journal of WSCG* 15:1–3 (2007), 59–64. Available at http://cg.cs.uni-bonn.de/de/publikationen/paper-details/schneider-2007-efficient/.

[150] William J. Schroeder, Jonathan A. Zarge, and William E. Lorensen. "Decimation of Triangle Meshes." *Proc. SIGGRAPH 92, Computer Graphics* 26:2 (1992), 65–70.

[151] Jim Scott. "Packing Lightmaps." Available at http://www.blackpawn.com/texts/lightmaps/, 2001.

[152] Mark Segal and Kurt Akeley. "The OpenGL Graphics System: A Specification (Version 3.3 Core Profile)." Available at http://www.opengl.org/registry/doc/glspec33.core.20100311.pdf, 2010.

[153] Dean Sekulic. "Efficient Occlusion Culling." In *GPU Gems*, edited by Randima Fernando. Reading, MA: Addison-Wesley, 2004. Available at http://http.developer.nvidia.com/GPUGems/gpugems_ch29.html.

[154] Jason Shankel. "Fast Heightfield Normal Calculation." In *Game Programming Gems 3*, edited by Dante Treglia. Hingham, MA: Charles River Media, 2002.

[155] Dave Shreiner and The Khronos OpenGL ARB Working Group. *OpenGL Programming Guide: The Official Guide to Learning OpenGL, Versions 3.0 and 3.1.*, Seventh edition. Reading, MA: Addison-Wesley, 2009.

[156] Irvin Sobel. "An Isotropic 3×3 Image Gradient Operator." In *Machine Vision for Three-Dimensional Scenes*, pp. 376–379. Orlando, FL: Academic Press, 1990.

[157] Wojciech Sterna. "Porting Code between Direct3D 9 and OpenGL 2.0." In *GPU Pro*, edited by Wolfgang Engel, pp. 529–540. Natick, MA: A K Peters, Ltd., 2010. Available at http://www.akpeters.com/gpupro/.

[158] David Stevenson. "A Report on the Proposed IEEE Floating Point Standard (IEEE Task p754)." *ACM SIGARCH Computer Architecture News* 8:5.

[159] Benjamin Supnik. "The Hacks of Life." Available at http://hacksoflife.blogspot.com/.

[160] Benjamin Supnik. "OpenGL and Threads: What's Wrong." Available at http://hacksoflife.blogspot.com/2008/01/opengl-and-threads-whats-wrong.html, 2008.

[161] Herb Sutter. "The Free Lunch Is Over: A Fundamental Turn toward Concurrency in Software." *Dr. Dobb's Journal* 30:3. Available at http://www.gotw.ca/publications/concurrency-ddj.htm.

[162] Herb Sutter. "Lock-Free Code: A False Sense of Security." *Dr. Dobb's Journal* 33:9. Available at http://www.drdobbs.com/cpp/210600279.

[163] Adam Szofran. "Global Terrain Technology for Flight Simulation." In *Game Developers Conference*, 2006. Available at http://www.microsoft.com/Products/Games/FSInsider/developers/Pages/GlobalTerrain.aspx.

[164] Christopher C. Tanner, Christopher J. Migdal, and Michael T. Jones. "The Clipmap: a Virtual Mipmap." In *Proceedings of SIGGRAPH 98, Computer Graphics Proceedings, Annual Conference Series*, edited by Michael Cohen, pp. 151–158. Reading, MA: Addison Wesley, 1998.

[165] Terathon Software. "C4 Engine Wiki: Editing Terrain." Available at http://www.terathon.com/wiki/index.php/Editing_Terrain, 2010.

[166] Art Tevs, Ivo Ihrke, and Hans-Peter Seidel. "Maximum Mipmaps for Fast, Accurate, and Scalable Dynamic Height Field Rendering." In *Symposium on Interactive 3D Graphics and Games (i3D'08)*, pp. 183–190, 2008. Available at http://www.tevs.eu/project_i3d08.html.

[167] Andrew Thall. "Extended-Precision Floating-Point Numbers for GPU Computation." Technical report, Alma College, 2007. Available at http://andrewthall.org/papers/.

[168] Nick Thibieroz. "Clever Shader Tricks." *Game Developers Conference.* Available at http://developer.amd.com/media/gpu_assets/04%20Clever%20Shader%20Tricks.pdf.

[169] Chris Thorne. "Origin-Centric Techniques for Optimising Scalability and the Fidelity of Motion, Interaction and Rendering." Ph.D. thesis, University of Western Australia, 2007. Available at http://www.floatingorigin.com/.

[170] Thatcher Ulrich. "Rendering Massive Terrains Using Chunked Level of Detail Control." In *SIGGRAPH 2002 Super-Size It! Scaling Up to Massive Virtual Worlds Course Notes*. New York: ACM Press, 2002. Available at http://tulrich.com/geekstuff/sig-notes.pdf.

[171] David A. Vallado and Wayne D. McClain. *Fundamentals of Astrodynamics and Applications*, Third edition. New York: Springer-Verlag, 2007. Available at http://celestrak.com/software/vallado-sw.asp.

[172] J.M.P. van Waveren. "Real-Time Texture Streaming & Decompression." Available at http://software.intel.com/en-us/articles/real-time-texture-streaming-decompression/, 2007.

[173] J.M.P. van Waveren. "Geospatial Texture Streaming from Slow Storage Devices." Available at http://software.intel.com/en-us/articles/geospatial-texture-streaming-from-slow-storage-devices/, 2008.

[174] J.M.P. van Waveren. "id Tech 5 Challenges: From Texture Virtualization to Massive Parallelization." In *SIGGRAPH*, 2009. Available at http://s09.idav.ucdavis.edu/talks/05-JP_id_Tech_5_Challenges.pdf.

[175] Gokul Varadhan and Dinesh Manocha. "Out-of-Core Rendering of Massive Geometric Environments." In *Proceedings of the Conference on Visualization '02*, pp. 69–76. Los Alamitos, CA: IEEE Computer Society, 2002.

[176] James M. Van Verth and Lars M. Bishop. *Essential Mathematics for Games and Interactive Applications*, Second edition. San Francisco: Morgan Kaufmann, 2008.

[177] Harald Vistnes. "GPU Terrain Rendering." In *Game Programming Gems 6*, edited by Michael Dickheiser, pp. 461–471. Hingham, MA: Charles River Media, 2006.

[178] Ingo Wald. "Realtime Ray Tracing and Interactive Global Illumination." Ph.D. thesis, Saarland University, 2004. Available at http://www.sci.utah.edu/~wald/PhD/.

[179] Bill Whitacre. "Spheres Through Triangle Tessellation." Available at http://musingsofninjarat.wordpress.com/spheres-through-triangle-tessellation/, 2008.

[180] David R. Williams. "Moon Fact Sheet." Available at http://nssdc.gsfc.nasa.gov/planetary/factsheet/moonfact.html, 2006.

[181] Michael Wimmer and Jiří Bittner. "Hardware Occlusion Queries Made Useful." In *GPU Gems 2: Programming Techniques for High-Performance Graphics and General-Purpose Computation*, edited by Matt Pharr and Randima Fernando. Reading, MA: Addison-Wesley, 2005. Available at http://www.cg.tuwien.ac.at/research/publications/2005/Wimmer-2005-HOQ/.

[182] Steven Wittens. "Making Worlds: 1—Of Spheres and Cubes." Available at http://acko.net/blog/making-worlds-part-1-of-spheres-and-cubes, 2009.

[183] Matthias Wloka. "Batch, Batch, Batch: What Does It Really Mean?" *Game Developers Conference.* Available at http://developer.nvidia.com/docs/IO/8230/BatchBatchBatch.pdf.

Index

About the Authors

Patrick Cozzi is a senior software developer on the 3D team at Analytical Graphics, Inc. (AGI) and a part-time lecturer in computer graphics at the University of Pennsylvania. He is a contributor to SIGGRAPH and the *Game Engine Gems* series. Before joining AGI in 2004, he worked on storage systems in IBM's Extreme Blue internship program at the Almaden Research Lab, interned with IBM's z/VM operating system team, and interned with the chipset validation group at Intel. Patrick has a master's degree in computer and information science from the University of Pennsylvania and a bachelor's degree in computer science from Penn State. His email address is pjcozzi@siggraph.org.

Kevin Ring first picked up a book on computer programming around the same time he learned to read, and he hasn't looked back since. In his software development career, he has worked on a wide range of software systems, from class libraries to web applications to 3D game engines to interplanetary spacecraft trajectory design systems, and most things in between. Kevin has a bachelor's degree in computer science from Rensselaer Polytechnic Institute and is currently the lead architect of AGI Components at Analytical Graphics, Inc. His email address is kevin@kotachrome.com.

Printed and bound by CPI Group (UK) Ltd, Croydon, CR0 4YY

23/10/2024

01777690-0001